REMOTE SENSING OF IMPERVIOUS SURFACES

Taylor & Francis Series in Remote Sensing Applications

Series Editor

Qihao Weng

Indiana State University
Terre Haute, Indiana, U.S.A.

Taylor & Francis Series in Remote Sensing Applications

Qihao Weng, Series Editor

REMOTE SENSING OF IMPERVIOUS SURFACES

Edited by

Qihao Weng

CRC Press
Taylor & Francis Group
Boca Raton London New York

CRC Press is an imprint of the
Taylor & Francis Group, an **informa** business

CRC Press
2385 NW Executive Center Drive, Suite 320, Boca Raton FL 33431

and by CRC Press
4 Park Square, Milton Park, Abingdon, Oxon, OX14 4RN

CRC Press is an imprint of Taylor & Francis Group, LLC

© 2008 Taylor & Francis Group, LLC

First issued in paperback 2020

No claim to original U.S. Government works

Library of Congress Cataloging-in-Publication Data

Remote sensing of impervious surfaces / edited by Qihao Weng.
    p. cm.
  Includes bibliographical references and index.
  ISBN-13: 978-1-4200-4374-7
  ISBN-10: 1-4200-4374-9
    1. Pavements--Location--Remote sensing. 2. Buildings--Location--Remote sensing. 3. Land use, Urban--Remote sensing. 4. Urban hydrology--Remote sensing. I. Weng, Qihao. II. Title.

TE153.W46 2008
624--dc22   2007015254

ISBN-13: 978-0-367-57766-7 (pbk)
ISBN-13: 978-1-4200-4374-7 (hbk)

Visit the Taylor & Francis Web site at
http://www.taylorandfrancis.com

and the CRC Press Web site at
http://www.crcpress.com

Contents

Part III Transport-Related Impervious Surfaces

Part IV Roof-Related Impervious Surfaces

Part V Impervious Surface Data Applications

Editor

 **Qihao Weng** was born in Fuzhou, China in 1964. He received an AS in geography from Minjiang University in 1984, an MS in physical geography from South China Normal University in 1990, an MA in geography from the University of Arizona in 1996, and a PhD in geography from the University of Georgia in 1999. He is currently an associate professor of geography and director of the Center for Urban and Environmental Change at Indiana State University, United States. He is also a guest/adjunct professor at Wuhan University and Beijing Normal University, China. His research focuses on remote sensing and geographic information system analysis of urban ecological and environmental systems, land-use and land-cover change, urbanization impacts, and human–environment interactions. Dr Weng is the author of more than 75 peer-reviewed journal articles and other publications, and is series editor for *Taylor & Francis Series in Remote Sensing Applications*. He has published the books of *Urban Remote Sensing* (October 2006) and *Remote Sensing of Impervious Surfaces* (October 2007) by CRC Press, an imprint of Taylor & Francis. In September 2006, he coedited, with Dale A. Quattrocchi, a special issue of Thermal Remote Sensing of Urban Areas in *Remote Sensing of Environment*. Dr Weng has been the recipient of the Robert E. Altenhofen Memorial Scholarship Award by the American Society for Photogrammetry and Remote Sensing (1999) and the Best Student-Authored Paper Award from the International Geographic Information Foundation (1998). In 2006, he received the Theodore Dreiser Distinguished Research Award by Indiana State University, the university's highest research honor bestowed to faculty. He has worked extensively with optical and thermal remote sensing data, primarily for urban heat island study, land-cover and impervious surface mapping, urban growth detection, subpixel image analysis, and the integration with socioeconomic characteristics, with financial support from agencies that include: NSF, NASA, USGS, USAID, National Geographic Society, and Indiana Department of Natural Resources. Dr Weng is a national director of American Society for Photogrammetry and Remote Sensing (2007–2010).

Contributors

Marvin E. Bauer Department of Forest Resources, University of Minnesota, St. Paul, Minnesota

Toby N. Carlson Department of Meteorology, Pennsylvania State University, Pennsylvania

Fabio Dell'Acqua Department of Electronics, University of Pavia, Pavia, Italy

Giorgio Franceschetti University of Naples, "Federico II", Naples, Italy, and UCLA, Los Angeles, CA

Paolo Gamba Department of Electronics, University of Pavia, Pavia, Italy

James Gerjevic Department of Geography, University of Northern Iowa, Cedar Falls, Iowa

Robert R. Gillies Utah State University, Utah Climate Centre, Logan, Utah

Armin Gruen Swiss Federal Institute of Technology, Zurich, Switzerland

Karin Hedman Institute of Astronomical and Physical Geodesy, Munich University of Technology, Munich, Germany

Martin Herold Department of Geography, Friedrich Schiller University Jena, Jena, Germany

Stefan Hinz Institute of Photogrammetry and Cartography, Munich University of Technology, Munich, Germany

Xuefei Hu Department of Geography, Indiana State University, Terre Haute, Indiana

Antonio Iodice Department of Electronic and Telecommunication Engineering, University of Naples, "Federico II", Naples, Italy

Bingqing Liang Department of Geography, Indiana State University, Terre Haute, Indiana

Brian C. Loffelholz Department of Forest Resource, University of Minnesota, St. Paul, Minnesota

Dengsheng Lu School of Forestry and Wildlife Sciences, Auburn University, Auburn, Alabama

Travis Maxwell Department of Geodesy and Geomatics Engineering, University of New Brunswick, Fredericton, New Brunswick, Canada

Assefa M. Melesse Department of Environmental Studies, Florida International University, Miami, Florida

Rama Prasada Mohapatra Department of Geography, University of Wisconsin–Milwaukee, Milwaukee, Wisconsin

Renaud Péteri University of La Rochelle, Mathematics, Image and Applications Laboratory (MIA), La Rochelle, France

Lindi J. Quackenbush State University of New York, College of Environmental Science and Forestry, Syracuse, New York

Thierry Ranchin Ecole des Mines de Paris, Center for Energy and Processes, Sophia Antipolis, France

Daniele Riccio Department of Electronics and Telecommunications Engineering, University of Naples, "Federico II", Naples, Italy

Uwe Stilla Institute of Photogrammetry and Cartography, Munich University of Technology, Munich, Germany

Ramanathan Sugumaran Department of Geography, University of Northern Iowa, Cedar Falls, Iowa

Matthew Voss Department of Geography, University of Northern Iowa, Cedar Falls, Iowa

Xixi Wang Energy & Environmental Research Center, University of North Dakota, Grand Forks, North Dakota

Yeqiao Wang Department of Natural Resources Science, University of Rhode Island, Kingston, Rhode Island

Birgit Wessel German Aerospace Centre, Oberpfaffenhofen, Germany

Bruce Wilson   Minnesota Pollution Control Agency, St. Paul, Minnesota

Changshan Wu   Department of Geography, University of Wisconsin–Milwaukee, Milwaukee, Wisconsin

George Xian   Science Applications International Corporation/U.S. Geological Survey (USGS) Center for Earth Resources Observation and Science, Sioux Falls, South Dakota

Xinsheng Zhang   EarthData International, Frederick, Maryland

Yun Zhang   Department of Geodesy and Geomatics Engineering, University of New Brunswick, Fredericton, Canada

Guoqing Zhou   Department of Civil Engineering and Technology, Old Dominion University, Norfolk, Virginia

Yuyu Zhou   Department of Natural Resources Science, University of Rhode Island, Kingston, Rhode Island

Remote Sensing of Impervious Surfaces: An Overview

Qihao Weng

1 Introduction

Impervious surfaces are anthropogenic features through which water cannot infiltrate into the soil, such as roads, driveways, sidewalks, parking lots, rooftops, and so on. In recent years, impervious surface has emerged not only as an indicator of the degree of urbanization, but also as a major indicator of environmental quality (Arnold and Gibbons, 1996). Impervious surface is a unifying theme for all participants at all watershed scales, including planners, engineers, landscape architects, scientists, social scientists, local officials, and others (Schueler, 1994). The magnitude, location, geometry, and spatial pattern of impervious surfaces, and the pervious–impervious ratio in a watershed have hydrological impacts. Although land-use zoning emphasizes roof-related impervious surfaces, transport-related impervious surfaces could have a greater impact. The increase in impervious cover would lead to the increase in the volume, duration, and intensity of urban runoff (Weng, 2001). Watersheds with large amounts of impervious cover may experience an overall decrease in groundwater recharge and baseflow and an increase in stormflow and flood frequency (Brun and Band, 2000). Furthermore, imperviousness is related to the water quality of a drainage basin and its receiving streams, lakes, and ponds. Increase in impervious cover and runoff directly impacts the transport of nonpoint source pollutants including pathogens, nutrients, toxic contaminants, and sediment (Hurd and Civco, 2004). Increases in runoff volume and discharge rates, in conjunction with nonpoint source pollution, will inevitably alter in-stream and riparian habitats, and result in the loss of some critical aquatic habits (Gillies et al., 2003). In addition, the areal extent and spatial occurrence of impervious surfaces may significantly influence urban climate by altering sensible and latent heat fluxes within the urban canopy and boundary layers (Yang et al., 2003). Impervious surface is found inversely related to vegetation cover in urban areas.

In other words, as impervious cover increases within a watershed/ administrative unit, vegetation cover would decrease. The percentage of land covered by impervious surfaces varies significantly with land-use categories and subcategories (Soil Conservation Service, 1975). Therefore, estimating and mapping (detecting, monitoring, and analyzing) impervious surface is valuable not only for environmental management, for example, water-quality assessment and stormwater taxation, but also for urban planning, for example, building infrastructure and sustainable urban growth.

In spite of its significance, the methods for estimating and mapping impervious surfaces and applications of impervious surface data have not been sufficiently explored. Many techniques have been applied to characterize and quantify impervious surfaces using either ground measurements or remotely sensed data. Field survey with global positioning system (GPS), although expensive and time-consuming, can provide reliable information on impervious surfaces. Manual digitizing from hardcopy maps or remote sensing imagery (especially aerial photographs) has also been used for mapping imperviousness. This technique has become more heavily involved with automation methods such as scanning and the use of feature extraction algorithms in recent years. During the 1970s and 1980s, remotely sensed data started to gain popularity in natural resources and environmental studies and were used in interpretive applications, spectral applications, and modeling applications of impervious surfaces (Slonecker et al., 2001). In reviewing the methods of impervious surface mapping, Brabec et al. (2002) identified four different approaches: (1) using a planimeter to measure impervious surface on aerial photography, (2) counting the number of intersections on the overlain grid on aerial photography, (3) conducting image classification, and (4) estimating impervious surface coverage through the percentage of urbanization in a region. These reviews concluded that in the 1970s and 1980s, aerial photography was the main source of remote sensing data for estimating and mapping impervious surfaces (Slonecker et al., 2001; Brabec et al., 2002).

With the advent of high-resolution imagery and more capable techniques recently, remote sensing of impervious surfaces is rapidly gaining interest in the remote sensing community and beyond. Driven by societal needs and technological advances, many municipal government agencies have started to collect and map impervious surface data for civic and environmental uses. Given its importance but lack of books in the market that systematically examine the contents of the field, it is urgent to publish such a book. Through review of basic concepts and methodologies, analysis of case studies, and examination of methods for applying up-to-date techniques to impervious surface estimation and mapping, this book may serve undergraduate and graduate students as a textbook, or be used as a reference book for professionals, researchers, and alike in academics, government, industries, and beyond.

2 Digital Remote Sensing Methods

Various digital remote sensing approaches have been developed to measure impervious surfaces, including mainly: (1) image classification, (2) multiple regression, (3) subpixel classification, (4) artificial neural network, and (5) classification and regression tree (CART) algorithm. The image classification approach utilizes image classifiers such as maximum likelihood classifier, spectral clustering, or other supervised/unsupervised classifiers to categorize and extract impervious surfaces as land-cover or land-use type(s) (Fankhauser, 1999; Hodgson et al., 2003; Dougherty et al., 2004). The multiple regression approach relates percent impervious surface to remote sensing and geographic information system (GIS) variables (Bauer et al., 2004; Chabaeva et al., 2004). The subpixel classification decomposes an image pixel into fractional components, assuming that the spectrum measured by a remote sensor is a linear combination of the spectra of all components within the pixel (Ji and Jensen, 1999; Wu and Murray, 2003; Lu and Weng, 2004). The artificial neural network approach applies advanced machine learning algorithms to derive impervious surface coverage. Flanagan and Civco (2001) developed an artificial neural network (ANN)-based impervious surface prediction model, which consisted of a two-tier neural network series, with the final output to be per-pixel impervious predictions and the training data from Landsat TM spectral reflectance values. The CART approach produces a rule-based model for prediction of continuous variables based on training data, and yields the spatial estimates of subpixel percent imperviousness (Yang et al., 2003).

Image classification is one of the most widely used methods in the extraction of impervious surfaces (Fankhauser, 1999; Slonecker et al., 2001; Brabec et al., 2002; Yang et al., 2003), but results are often not satisfactory because of the limitation of spatial resolution in remotely sensed imagery and the heterogeneity of urban landscapes. Various impervious surfaces may be mixed with other land-cover types, such as trees, grasses, and soils. Moreover, the difficulty in selecting training areas could also lower the accuracy of image classification. As fine spatial resolution data (mostly better than 5 m in spatial resolution), such as IKONOS and QuickBird, become available, they are increasingly employed for different applications including impervious surface mapping. A major advantage of these images is that such data greatly reduce the mixed pixel problem, providing a greater potential to extract more detailed information on land covers. However, new problems associated with these image data need to be considered, notably the shades caused by topography, tall buildings, or trees (Dare, 2005), and the high spectral variation within the same land-cover class.

Because of the inverse correlation between impervious surface and vegetation cover in urban areas, one potential approach for impervious surface extraction is through information on vegetation distribution (Gillies et al., 2003; Bauer et al., 2004). The Normalized Difference

Vegetation Index (NDVI) or greenness from tasseled cap transformation or principal component analysis may be used to represent vegetation distribution. Impervious surface can then be estimated based on regression models with vegetation indices. This approach, however, has a major drawback. Different seasons of satellite images could result in large variations in impervious surface estimation. In the leaf-on season, vegetation may be considerably overestimated, whereas in the leaf-off season, vegetation tends to be underestimated, leading to the overestimation of impervious surface coverage.

3 Use of Medium-Resolution Satellite Imagery

Most previous researches for extraction of impervious surfaces in urban areas used the medium spatial resolution (10–100 m) images, such as Landsat TM/ETM+ and Terra's ASTER images (Wu and Murray, 2003; Yang et al., 2003; Lu and Weng, 2006a,b). However, their spatial resolutions are regarded as too coarse, due to the heterogeneity of urban landscapes and the complexity of impervious surface materials. Urban landscapes are typically composed of features that are smaller than the spatial resolution of such sensors and are a complex combination of buildings, roads, grass, trees, soil, water, and so on. Strahler et al. (1986) described H- and L-resolution scene models based on the relationships between the size of the scene elements and the resolution cell of the sensor. The scene elements in the H-resolution model are larger than the resolution cell and can therefore be directly detected. In contrast, the elements in the L-resolution model are smaller than the resolution cells and are not detectable. When the objects in the scene become increasingly smaller relative to the resolution cell size, they may be no longer regarded as objects individually. Hence, the reflectance measured by the sensor can be treated as a sum of interactions among various classes of scene elements as weighted by their relative proportions (Strahler et al., 1986). The medium-resolution satellite images are attributed to L-resolution model. As the spatial resolution interacts with the fabric of urban landscapes, a special problem of mixed pixels is created, where several land-use and land-cover types are contained in one pixel. Such a mixture becomes especially prevalent in residential areas, where buildings, trees, lawns, concrete, and asphalt can all occur within a pixel. The mixed pixel has been recognized as a major problem affecting the effective use of remotely sensed data in thematic information extraction (Fisher, 1997; Cracknell, 1998).

Because of its effectiveness in handling the mixed pixel problem, spectral mixture analysis (SMA), as a subpixel classifier, is gaining great interest in the remote sensing community in recent years. As a physically based image analysis procedure, it supports repeatable and accurate extraction of quantitative subpixel information (Roberts et al., 1998). The SMA approach may be linear or nonlinear. However, for most remote sensing applications, a linear SMA approach is employed. The linear approach assumes that the spectrum measured by a sensor is a linear combination of the spectra of all components within the pixel (Adams et al., 1995). Different methods of impervious surface extraction based on the linear SMA model have been developed. For example, impervious surface may be extracted as one of the endmembers in the standard SMA model (Phinn et al., 2002). Impervious surface estimation can also be done by the addition of high-albedo and low-albedo fraction images, with both as the SMA endmembers (Wu and Murray, 2003). Moreover, a multiple endmember SMA (MESMA) method has been developed (Rashed et al., 2003), in which several impervious surface endmembers can be extracted and combined. However, these SMA-based methods have a common problem, that is, impervious surface tends to be overestimated in the areas with small amounts of impervious surface, but is underestimated in the areas with large amounts of impervious surface. The similarity in spectral properties among non-photosynthetic vegetation, soil, and different kinds of impervious surface materials makes it difficult to distinguish impervious from nonimpervious materials. In addition, shadows caused by tall buildings and large tree crowns in the urban areas may also lead to an underestimation of impervious surface area.

In particular, the addition of low-albedo and high-albedo fraction images has been proven effective in estimating and mapping impervious surface to a certain degree. However, the impervious surface may be overestimated in the areas where a low amount of imperviousness is detected. This is because the low-albedo fraction image may relate to different kinds of materials/covers, including water, canopy shadows, building shadows, moisture in grass or crops, and dark impervious surface materials. On the other hand, although the high-albedo image is largely associated with impervious surfaces, it may be confused with dry soils. Additional data or improved techniques are therefore necessary to separate impervious surfaces from others. Slonecker et al. (2001) noted that the use of thermal infrared and radar imagery may aid greatly in impervious surface estimation, possibly through data fusion. Lu and Weng (2006a) have successfully utilized Landsat ETM+ thermal infrared data to enhance their estimation of impervious surfaces based on the differences in land surface temperature between impervious and nonimpervious surfaces. The land surface temperature image was used as a threshold to remove dry soils from the high-albedo fraction image and to remove water and shadows from the low-albedo fraction image. In addition, radar data have inherent advantages in the identification and estimation of impervious surfaces because of the high dielectric properties of most construction materials and the unique geometry of man-made features (Slonecker et al., 2001).

4   The Structure of the Book

This book consists of five parts. Part I introduces various methods of remote sensing digital image processing for impervious surface estimation and mapping. Part II exemplifies most recent technological advances in the field of remote sensing of impervious surfaces. Part III presents techniques for extracting and mapping transport-related impervious surfaces using different types of remotely sensed data. The techniques and case studies for estimating and mapping roof-related impervious surfaces are contained in Part IV. The final part, Part V, examines some major application areas of impervious surface data, including impact analysis of water quality, hydrological modeling of water flow, examination of the effect on aquatic fauna, and population estimation.

The chapters in Part I are concerned with four different approaches to impervious surface estimation and mapping. These methods, including regression, CART, SMA, and artificial neural network, have received wide recognition in the remote sensing community. Chapter 1 describes the method and some results of estimation and mapping of impervious surface area for the state of Minnesota in 1990 and 2000. The method uses a regression modeling to estimate the percent of imperviousness for each pixel based on its inverse relation with the greenness component of the tasseled cap transformation of Landsat TM/ETM+ data. Although this project employed satellite images in spring, summer, and fall seasons, only summer images were used for the modeling to have the greatest contrast between imperviousness and vegetation responses. In Chapter 2, a method based on ANN is established to estimate the subpixel imperviousness from IKONOS imagery. The case study conducted in Grafton, Wisconsin, which had diverse land-use types, shows that the ANN model produced reasonable high accuracy with a mean error of 7.78. The model performed even better in the urban areas, where the satellite data were highly nonlinear. Chapter 3 applies the CART algorithm developed by the United States Geological Survey (Yang et al., 2003; Xian, 2006) to two fast-growing regions, Seattle-Tacoma, Washington and Las Vegas, Nevada. This method first classified high-resolution imagery. Pixels classified as urban were totaled to calculate impervious surface as a percentage and rescaled to match the pixel size of the medium-resolution satellite imagery. The percent imperviousness datasets were then used as dependent variables in the regression tree models, while the medium-resolution image data and derived variables (such as NDVI) together with other geospatial data (such as slope) were used as the independent variables. This chapter further examines how urban growth, as indicated by impervious surface data, related to housing density in Seattle-Tacoma between 1986 and 2002 and in Las Vegas between 1984 and 2002. Chapter 4 applies SMA to Landsat ETM+ imagery to derive fraction images, which are further used to compute impervious surfaces. This approach has demonstrated its effectiveness with reasonably high accuracy (Wu and Murray,

2003; Lu and Weng, 2006a). A major drawback with this approach lies in its difficulty in extracting dark impervious surface areas, which are confused with water and shadows. Therefore, the authors of this chapter further employed IKONOS data to extract impervious surface data by using a hybrid method of decision tree classifier and unsupervised ISODATA classifier for the purpose of validation.

Part II presents new developments in remote sensing of impervious surfaces, especially in the areas of ANN-based model, object-oriented detection, fractal analysis, image fusion, and use of hyperspectral imagery. In Chapter 5, two new models are introduced for extraction of impervious surfaces using remote sensing data of multiple sources. The ANN-based subpixel proportional land-cover information transformation (SPLIT) model establishes spectral relationship between Landsat TM pixel values and the corresponding high spatial resolution, airborne, digital multispectral videography data, so that proportions of impervious surface within the Landsat TM imagery can be extracted. The object-oriented multiple agent segmentation (MASC) model extracts impervious surfaces from true-color digital orthophotos by imbedded segmentation, shadow-effect, MANOVA-based classification, and postclassification submodels. Chapter 6 examines the benefits of hyperspectral imagery for extracting impervious surfaces with a case study in Indianapolis, United States. SMA was applied to EO-1 Hyperion imagery to calculate the fraction images of green vegetation, soil, high albedo, and low albedo. The fraction images of high albedo and low albedo are then used to estimate impervious surfaces. In comparison with ALI (multispectral) imagery, the Hyperion image was found to be more effective in discerning low-albedo surface materials, which have been a major obstacle for impervious surface estimation with medium-resolution multispectral images (refer to Chapter 4). Chapter 7 applies fractal analysis (triangular prism method) for separating types of impervious land cover with a primary focus on the separation of roofs, roads, and driveways in a suburban area of New York. It demonstrates that there were statistical differences between fractal dimensions calculated for different classes of impervious land cover. Roofs and roads were found generally separable on a pairwise basis, while driveways were more frequently confused. Recognizing the strengths and limitations of synthetic aperture radar (SAR) and optical remote sensing data, Chapter 8 demonstrates how the techniques of data fusion can be applied for better feature extraction in urban areas. The methodology was based on texture analysis of both SAR and optical images for detection and exploitation of spatial patterns, followed by a joint classification of the extracted spatial features together with the original spectral features. The Markov random field classifier used for this research was found to provide better accuracy than the neuro-fuzzy classifier and to have a similar capability to cope with multiple inputs.

The next two parts of the book focus each on one functional type of impervious surface, transport-related impervious surfaces in Part III and roof-related impervious surfaces in Part IV. Specifically, Part III explores the spectral characteristics of roads under different conditions and introduces unique techniques for extracting roads by using various remotely sensed data. Chapter 9 applies hyperspectral imagery (airborne visible/infrared imaging spectrometer—AVIRIS) for road extraction and assesses the effectiveness of four advanced image classifiers. Minimum noise fraction (refer to Chapters 4 and 6) and CART (refer to Chapter 3) were used to reduce the number of spectral dimensions to be analyzed. The four classifiers examined are the spectral angle mapper (SAM), object-oriented nearest neighbor, mixture-tuned matched filtering (MTMF), and the combination method of mixture-tuned matched filtering and CART (MTMF-CART). This study found that the object-oriented nearest neighbor classification method produced the best overall result compared with MTMF, SAM, and MTMF-CART classifiers. The overall accuracies for the four classifications were 93.2%, 81.89%, 88.92%, and 84.32%, respectively. Chapter 10 first describes the strengths and limitations (arising from the side-looking illumination of the sensor) of SAR imagery for road extraction. Then, it applies the "TUM-LOREX" method of road extraction (Steger et al., 1997; Wiedemann and Hinz, 1999; Wiedemann and Ebner, 2000) to SAR imagery. In order to compensate for possible gaps caused by adjacent high buildings and narrow streets, dominant scattering caused by building structures, traffic signs, and metallic objects in cities, the authors suggest that additional information is needed for better extraction through data integration. SAR imagery may be combined for use with context information, road class-specific modeling, and use of multiview imagery. As high spatial resolution satellite imagery (better than 5 m in the panchromatic channel), such as SPOT 5, IKONOS, QuickBird, OrbView, or EROS, becomes available for civilian applications, there is a shift in road extraction algorithms from linear to surface models. Chapter 11 proposes a new method of road extraction from high-resolution imagery, which integrates a linear representation of the road (graph module) with a surface representation (reconstruction module). In order to overcome local artifacts, the method makes use of advanced image-processing algorithms such as active contours and the wavelet transform. Its application and evaluation on a QuickBird image over an urban area in France achieved an acceptable accuracy. The last chapter in Part III, Chapter 12, discusses common spectral characteristics of asphalt roads and the impact of different road conditions and distresses, based on the Santa Barbara asphalt road spectra library (Herold and Roberts, 2005). It further evaluates the potential of hyperspectral remote sensing to study transportation infrastructure and road surfaces (also refer to Chapter 9).

Part IV is concerned with various methods and techniques for estimation and mapping of roof-related impervious surfaces. Chapter 13 presents a method for generation of an urban 3D model, especially a digital building model via integrating image knowledge and LiDAR data. The main contribution lies in the development of an object-oriented building extraction model, which defines roof types, roof boundary coordinates, planar equation, and other parameters. These parameters were extracted from the combined processing of LiDAR and orthoimage data. In Chapter 14, state-of-the-art building extraction and city modeling techniques are presented. Aerial images, with semiautomated photogrammetric techniques, are currently regarded to be most important for city modeling. This chapter evaluates the potential of aerial laser scan data for modeling cities. Chapter 15 discusses main characteristics of the electromagnetic models with respect to SAR data. The rationale and results for a sound electromagnetic SAR modeling of the dihedral canonical elements are provided. The models discussed are appropriate for airborne and the upcoming generation of high-resolution spaceborne SAR sensors. To achieve accurate roof-mapping results, it is often necessary to fuse the spectral information of the multi-spectral image and the spatial information of the panchromatic image of a given sensor or sensors into one image. This method is called pan-sharpening the multispectral image using the pan image. In Chapter 16, two examples of pan-sharpening for roof mapping are illustrated: pixel-based postclassification for small-scale roof mapping using Landsat TM and SPOT Pan fused images and object-oriented classification for medium-scale roof mapping using pan-sharpened QuickBird images. Both case studies demonstrate that an effective image fusion can significantly contribute to an improvement in the accuracy of roof mapping, although the final results may still contain errors.

Part V presents some examples of applications of remotely sensed impervious surface data. Impervious surface coverage affects both water quality and water abundance through its influence on surface runoff. Chapter 17 discusses this relationship in detail and illustrates it through a case study in Pennsylvania. The project aimed at the development of a software that would estimate the potential impact of urbanization on water quality within the watersheds in the state. By creating an ISA map for the year 2002 and by comparing with the existing 1985 and 1995 maps, development trends though a longer time period may also be determined. Chapter 18 is closely related to Chapter 17, but focuses on the storm runoff effect of impervious surface dynamics. It provides an overview of the Simple Method and Soil Conservation Service-based hydrologic models, which were widely used to predict the effects of urbanization on precipitation-runoff processes. In addition, this chapter introduces a remote sensing-based technique for determining the extents of impervious surface in the watershed using its inverse relationship with fraction vegetation cover (Carlson and Ripley, 1997). Case studies were conducted to demonstrate runoff responses to the increased impervious areas under different climate conditions, one in the Red River of the North Basin along the state border between North Dakota and Minnesota and the other in Simms Creek watershed in Florida. Chapter 19 reviews literatures regarding the effect of growth of impervious surface coverage on the biodiversity of terrestrial and aquatic fauna. The case study illustrated in this chapter resulted from the author's articles published in 2003 (Gillies et al., 2003). By computing a time line of impervious surface area and combining with historical freshwater mussel data, it examines the effect of impervious surface area on aquatic fauna in the Flint River tributaries over a period of extensive urban growth (1979–1997) associated with the Peachtree City of Atlanta metropolitan area. The last chapter of this book, Chapter 20, presents a method of population estimation by using impervious surface data derived from a Landsat ETM+ image. The research conducted population modeling in Marion County, Indiana, United States, at all census levels (census block, block group, and census tract). The performance of models was evaluated by several criteria (i.e., relative error, mean relative error, median relative error, and the error of total in percentage). Better models were found to have higher analytical scales, and the performance reached the best at the census tract level.

Acknowledgments

I wish to extend my thanks to all the contributors of this book for making this endeavor possible. Moreover, I offer my deepest appreciation to all the reviewers, who have taken precious time from their busy schedules to review the chapters submitted for this book. Finally, I am indebted to my family for their enduring love and support. It is my hope that the publication of this book will provide stimulation to students and researchers to perform more in-depth work and analysis on the applications of remote sensing to impervious surfaces. The reviewers of the chapters for this book are listed here in alphabetical order:

Toby Carlson, Lee Decola, Fabio Dell'Acqua, Charles Emerson, Paolo Gamba, Martin Herold, Junchang Ju, Jonathan Li, Weiguo Liu, Desheng Liu, Dengsheng Lu, Assefa M. Melesse, Soe Wint Myint, Maik Netzband, Renaud Peteri, Rebecca Powell, Ruiliang Pu, Fang Qiu, Umamaheshwaren Rajasekar, Sebastian Schiefer, Jie Shan, Guofang Shao, Conghe Song, Uwe Sorgel, Uwe Stilla, Ramanathan Sugumaran, Le Wang, Yeqiao Wang, Changshan Wu, George Xian, Honglin Xiao, Jingfeng Xiao, Jiansheng Yang, Stephen Yool, Fei Yuan, Yun Zhang, and Guoqing Zhou.

My work on remote sensing of impervious surfaces has been supported by various funding sources. These include a fund from National Science Foundation (BCS-0521734) for a project entitled "Role of Urban Canopy Composition and Structure in Determining Heat Islands: A Synthesis of Remote Sensing and Landscape Ecology Approach," and funds from the USGS IndianaView program and from the NASA's Indiana Space Grant program for a project entitled "Indiana Impervious Surface Mapping Initiative." Dr Dengsheng Lu, Auburn University, was an excellent collaborator in all these projects.

References

Adams, J.B., Sabol, D.E., Kapos, V., Filho, R.A., Roberts, D.A., Smith, M.O., and Gillespie, A.R., 1995, Classification of multispectral images based on fractions of endmembers: application to land cover change in the Brazilian Amazon. *Remote Sensing of Environment*, 52, 137–154.

Arnold, C.L. Jr., and Gibbons, C.J., 1996, Impervious surface coverage: the emergence of a key environmental indicator. *Journal of the American Planning Association*, 62, 243–258.

Bauer, M.E., Heiner, N.J., Doyle, J.K., and Yuan, F., 2004, Impervious surface mapping and change monitoring using Landsat remote sensing. *ASPRS Annual Conference Proceedings*, Denver, Colorado, May 2004 (Unpaginated CD ROM).

Brabec, E., Schulte, S., and Richards, P.L., 2002, Impervious surface and water quality: a review of current literature and its implications for watershed planning. *Journal of Planning Literature*, 16, 499–514.

Brun, S.E. and Band, L.E., 2000, Simulating runoff behavior in an urbanizing watershed. *Computers, Environment and Urban Systems*, 24, 5–22.

Carlson, T.N. and Ripley, A.J., 1997, On the relationship between fractional vegetation cover, leaf area index and NDVI. *Remote Sensing of Environment*, 62, 241–252.

Chabaeva, A.A., Civco, D.L., and Prisloe, S., 2004, Development of a population density and land used based regression model to calculate the amount of imperviousness. *ASPRS Annual Conference Proceedings*, Denver, Colorado, May 2004 (Unpaginated CD ROM).

Cracknell, A.P., 1998, Synergy in remote sensing—what's in a pixel? *International Journal of Remote Sensing*, 19, 2025–2047.

Dare, P.M., 2005, Shadow analysis in high-resolution satellite imagery of urban areas. *Photogrammetric Engineering and Remote Sensing*, 71, 169–177.

Dougherty, M., Dymond, R.L., Goetz, S.J., Jantz, C.A., and Goulet, N., 2004, Evaluation of impervious surface estimates in a rapidly urbanizing watershed. *Photogrammetric Engineering and Remote Sensing*, 70, 1275–1284.

Fankhauser, R., 1999, Automatic determination of imperviousness in urban areas from digital orthophotos. *Water Science and Technology*, 39, 81–86.

Fisher, P., 1997, The pixel: a snare and a delusion. *International Journal of Remote Sensing*, 18, 679–685.

Flanagan, M. and Civco, D.L., 2001, Subpixel impervious surface mapping. *ASPRS Annual Conference Proceedings*, St. Louis, Missouri, April 2001 (Unpaginated CD ROM).

Gillies, R.R., Box, J.B., Symanzik, J., and Rodemaker, E.J., 2003, Effects of urbanization on the aquatic fauna of the Line Creek watershed, Atlanta—a satellite perspective. *Remote Sensing of Environment*, 86, 411–422.

Herold, M. and Roberts, D.A., 2005, Spectral characteristics of asphalt road aging and deterioration: implications for remote sensing applications. *Applied Optics*, 44(20), 4327–4334.

Hodgson, M.E., Jensen, J.R., Tullis, J.A., Riordan, K.D., and Archer, C.M., 2003, Synergistic use of Lidar and color aerial photography for mapping urban parcel imperviousness. *Photogrammetric Engineering and Remote Sensing*, 69, 973–980.

Hurd, J.D. and Civco, D.L., 2004, Temporal characterization of impervious surfaces for the State of Connecticut. *ASPRS Annual Conference Proceedings*, Denver, Colorado, May 2004 (Unpaginated CD ROM).

Ji, M. and Jensen, J.R., 1999, Effectiveness of subpixel analysis in detecting and quantifying urban imperviousness from Landsat Thematic Mapper imagery. *Geocarto International*, 14, 33–41.

Lu, D. and Weng, Q., 2004, Spectral mixure analysis of the urban landscape in Indianapolis with Landsat ETM+ imagery. *Photogrammetric Engineering and Remote Sensing*, 70(9), 1053–1062.

Lu, D. and Weng, Q., 2006a, Use of impervious surface in urban land use classification. *Remote Sensing of Environment*, 102(1–2), 146–160.

Lu, D. and Weng, Q., 2006b, Spectral mixture analysis of ASTER imagery for examining the relationship between thermal features and biophysical descriptors in Indianapolis, Indiana. *Remote Sensing of Environment*, 104(2), 157–167.

Phinn, S., Stanford, M., Scarth, P., Murray, A.T., and Shyy, P.T., 2002, Monitoring the composition of urban environments based on the vegetation-impervious surface-soil (VIS) model by subpixel analysis techniques. *International Journal of Remote Sensing*, 23, 4131–4153.

Rashed, T., Weeks, J.R., Roberts, D., Rogan, J., and Powell, R., 2003, Measuring the physical composition of urban morphology using multiple endmember spectral mixture models. *Photogrammetric Engineering and Remote Sensing*, 69, 1011–1020.

Roberts, D.A., Batista, G.T., Pereira, J.L.G., Waller, E.K., and Nelson, B.W., 1998, Change identification using multitemporal spectral mixture analysis: applications in eastern Amazônia. In: *Remote Sensing Change Detection: Environmental Monitoring Methods and Applications* (R.S. Lunetta and C.D. Elvidge, editors), Ann Arbor Press, Ann Arbor, MI, pp. 137–161.

Schueler, T.R., 1994, The importance of imperviousness. *Watershed Protection Techniques*, 1, 100–111.

Slonecker, E.T., Jennings, D., and Garofalo, D., 2001, Remote sensing of impervious surface: a review. *Remote Sensing Reviews*, 20, 227–255.

Soil Conservation Service, 1975, Urban Hydrology for Small Watersheds, USDA Soil Conservation Service Technical Release No. 55. Washington, D.C.

Steger, C., Mayer, H., and Radig, B., 1997, The role of grouping for road extraction. In: *Automatic Extraction of Man-Made Objects from Aerial and Space Images*, Birkhauser Verlag, Basel, Switzerland, pp. 245–256.

Strahler, A.H., Woodcock, C.E., and Smith, J.A., 1986, On the nature of models in remote sensing. *Remote Sensing of Environment*, 70, 121–139.

Weng, Q., 2001, Modeling urban growth effect on surface runoff with the integration of remote sensing and GIS. *Environmental Management*, 28, 737–748.

Wiedemann, C. and Ebner, H., 2000, Automatic completion and evaluation of road networks. *International Archives of Photogrammetry and Remote Sensing*, 33(B3/2), 979–986.

Wiedemann, C. and Hinz, S., 1999, Automatic extraction and evaluation of road networks from satellite imagery. *International Archives of Photogrammetry and Remote Sensing*, 32(3-2W5), 95–100.

Wu, C. and Murray, A.T., 2003, Estimating impervious surface distribution by spectral mixture analysis. *Remote Sensing of Environment*, 84, 493–505.

Xian, G., 2006, Assessing urban growth with sub-pixel impervious surface coverage. In: *Urban Remote Sensing* (Weng and Quattrochi, editors), CRC Press/Taylor & Francis Group, Boca Raton, FL, pp. 179–199.

Yang, L., Huang, C., Homer, C.G., Wylie, B.K., and Coan, M.J., 2003, An approach for mapping large-scale impervious surfaces: synergistic use of Landsat-7 ETM+ and high spatial resolution imagery. *Canadian Journal of Remote Sensing*, 29, 230–240.

Part I

Digital Remote Sensing Methods

1

Estimating and Mapping Impervious Surface Area by Regression Analysis of Landsat Imagery

Marvin E. Bauer, Brian C. Loffelholz, and Bruce Wilson

CONTENTS

1.1   Introduction

Impervious surfaces are defined as any surface that water cannot infiltrate. These surfaces are primarily associated with transportation (streets, highways, parking lots, sidewalks) and buildings. Expansion of impervious surfaces increases water runoff and is a primary determinant of stormwater runoff volumes, water quality of lakes and streams, and stream habitat quality in urbanized areas. Increases in impervious surfaces, and accompanying phosphorous, sediment, and thermal loads, can have profound negative impacts on lakes and streams and habitat for fisheries. Percent impervious surface area has emerged as a key factor to explain and generally predict the degree of impact severity on streams and watersheds. It has been generally found that most stream health indicators decline when the impervious area of a watershed exceeds 10% (Schueler, 1994). Arnold and Gibbons (1996) suggest that impervious surface area provides a measure of land use that is closely correlated with these impacts and more generally that the amount of impervious surface in a landscape is an important indicator of environmental and habitat quality in urban areas. In the area of urban climate, Yuan and Bauer (2006) have recently documented a strong relationship between the amount of impervious surface area and land surface temperatures or the urban heat island effect. It follows that impervious surface information is fundamental for watershed planning and management and for urban planning and policy.

Continued urban growth, expected to occur over the next three decades, should be accompanied by carefully designed and maintained stormwater runoff controls as required by new federal and state stormwater permits and total maximum daily load (TMDL) allocations for municipal stormwater sources. In Minnesota, there are more than 200 Municipal Separate Storm Sewer System (MS4) communities that are required by the Stormwater Program to begin stormwater pollution prevention planning and implementing urban best management practices. The MS4 cities must identify best management practices and measurable goals associated with each minimum control measure. Quantifying impervious cover should be one of the first steps for these areas. Given the number and size of the areas of interest, an economical and consistent method for mapping impervious surface area is needed.

Since the formulation by Ridd (1995) of a conceptual model of urban landscapes as a spectral mixture of vegetation, impervious surfaces, and soil, a growing number of researchers have used Landsat data to map impervious surface area. A variety of approaches, including spectral mixture analysis (Wu and Murray, 2003; Wu, 2004; Lu and Weng, 2006), regression tree modeling (Yang et al., 2003a,b; Xian and Crane, 2005), decision tree classification (Dougherty et al., 2004; Jantz et al., 2005), subpixel classification (Civco et al., 2002), neural network classification (Civco and Hurd, 1997), and multiple regression (Bauer et al., 2004, 2005) have shown that Landsat remote sensing has the potential for mapping and monitoring impervious surface area. Landsat Thematic Mapper (TM) and Enhanced Thematic Mapper Plus (ETM+) data have several advantages for this application: synoptic view of multicounty areas, digital, GIS compatible data, availability of data since 1984, and economical costs.

In an urban area where most pixels of Landsat data are mixed pixels with mixtures of vegetation (particularly grass and trees), water, and impervious surfaces, we believe the best approach is to consider impervious as a continuous variable. By treating impervious as a continuous variable, the errors associated with assigning a mixed pixel to a single nominal class with a range of impervious amounts or in assigning an average impervious value to each land cover/use class are avoided. Our approach has been to use a regression model to estimate the percentage of impervious for each pixel. The theoretical basis for the approach is illustrated in Figure 1.1.

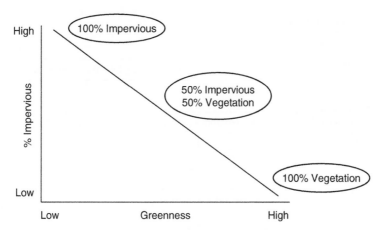

FIGURE 1.1
Conceptual model for estimating percent impervious surface area at the pixel level.

The greenness component of the tasseled cap transformation of Landsat TM/ETM+ data is sensitive to the amount of green vegetation and inversely related to the amount of impervious surface area. The resulting classification provides a continuous range of impervious area from 0% to 100%.

This chapter describes the methods and results for estimation and mapping of impervious surface area, using multiple regression modeling, for the state of Minnesota for two time periods, 1990 and 2000. Minnesota has a wide variety of rural and urban landscapes, making it a near-ideal setting to implement and evaluate the use of Landsat remote sensing for land cover and impervious surface mapping. The rural areas include agricultural cropland, forests, and wetlands cover types, interspersed with towns. The urban areas range from low to high intensity development and from small towns in rural areas to regional center cities to the Twin Cities metropolitan area. Although the primary impetus for our work has been to quantify and map impervious surface area in support of watershed management and planning, imperviousness is also important in relation to aesthetics, habitat, and urban climate.

1.2 Methods

Landsat TM/ETM+ digital imagery were acquired and analyzed for two time periods, 1990 and 2000. The key steps in the procedures were image acquisition; rectification, land cover classification, development, and application of a regression model relating percent impervious to Landsat TM tasseled cap greenness, and accuracy assessment (Figure 1.2). Image processing was performed in ERDAS Imagine, GIS operations in ArcGIS, and statistical analyses in SAS.

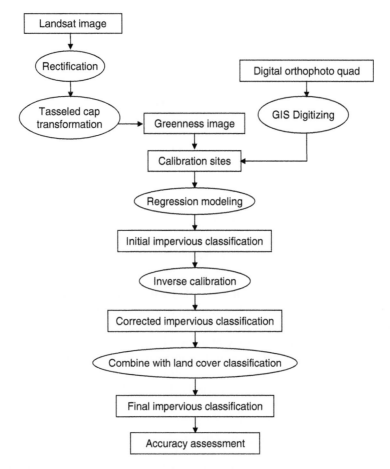

FIGURE 1.2
Flowchart of image processing and classification procedures for mapping impervious surface area. (From Bauer, M., Loeffelholz, B., and Wilson, B., *Proceedings, Pecora 16 Conference, American Society of Photogrammetry and Remote Sensing*, October 23–27, 2005, Sioux Falls, South Dakota, 2005. With permission.)

1.2.1 Landsat Image Acquisition, Rectification, and Land Cover Classification

Nineteen images of Landsat are required to cover the state of Minnesota. Selection of clear, cloud- and haze-free imagery was a high priority and the selected images had only a few areas with clouds. In those areas, the clouds and cloud shadows were manually digitized to create a cloud mask, which was overlaid on the impervious classification and all pixels within it were assigned a value of zero. It should be noted, however, that there were very few areas where clouds and urban overlapped.

The 19 images were rectified to the Universal Transverse Mercator (UTM) coordinate and projection system using ~35 ground control points per image and nearest neighbor resampling to a 30 m pixel size with an root-mean-square (RMS) error of $^1_4$ pixel (7.5 m) or less. The coordinates of the final images were adjusted to values evenly divisible by 30. Following rectification, the imagery was transformed to unsigned 8-bit Landsat TM/ETM+ tasseled cap values (Crist and Cicone, 1984; Huang et al., 2002).

Our approach for mapping impervious surfaces applies an impervious estimation model to developed and urban areas, thereby requiring a concurrent land cover map to separate rural areas from developed/urban areas. We used a multitemporal, multispectral image classification with a combination of spring, summer, and fall Landsat TM images acquired in ~2000 to classify land cover. The images were stratified into spectrally consistent classification units (SCCU) based on the Landsat image acquisition dates and paths, ecoregions, and vegetation phenology (Figure 1.3). The tasseled cap features of greenness, brightness, and wetness for the three-date multitemporal images were used with a k-nearest (kNN) classifier to generate a land cover classification of the state with seven classes: agriculture, grassland, forest, wetland, water, extraction, and urban. The kNN classifier assigns each unknown pixel of the satellite image the attributes of the most similar reference pixels for which field data exist. The similarity is defined by the Mahalanobis distances between classes. The kNN method has proven to be an accurate and cost-efficient method for extending field inventory data to landscape scales (McRoberts et al., 2002). The average statewide overall accuracy for the level 1 cover type classifications was 84.5% with a kappa statistic of 0.81. The average producer and user accuracies for the urban class were 91.7% and 95.4%, respectively.

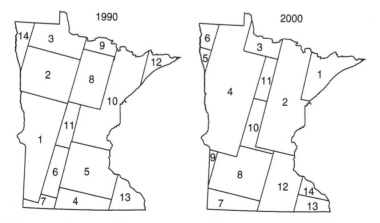

FIGURE 1.3
Strata, based on Landsat image acquisition dates, ecoregions, and vegetation phenology, used for land cover and impervious classifications. The Landsat image paths and rows and acquisition dates are listed in Table 1.1.

1.2.2 Development of Impervious Surface Regression Models

Model calibration sites were selected separately for each Landsat image with ~50 sites for each Landsat image. The selection of sites was stratified by the range of amounts and types of impervious cover (e.g., parks, residential housing with varying densities, commercial, and industrial land uses) as well as by variations in amounts and kinds of vegetative cover (e.g., grass, forest, shrub). Stratification by vegetation cover type was done to account for seasonal variability in greenness between vegetative cover types.

The calibration sites were typically 40–100 Landsat pixels or ~2.5–10 ha in size. Further, the boundaries of the calibration sites were "snapped" to the 30-m Landsat grid to ensure that the calibration sites in the high-resolution images matched the Landsat images. The impervious surface area of each site digitized from 1991 to 1992 1 m panchromatic digital orthophoto quadrangles (DOQs) for 1990 and from 2003 1 m color DOQs for 2000 to determine the percent impervious surface area within each site. Sites where the land cover or impervious area might have changed between acquisition of the aerial and Landsat imagery were not included.

The measurements of impervious surface area from the calibration sites were used to develop a least squares regression model relating percent impervious to the Landsat tasseled cap greenness responses for each SCCU image. Greenness is sensitive to the amount of green vegetation and therefore is inversely related to the amount of impervious surface area. The summer images provide the greatest contrast between impervious and vegetation responses. The images used for the impervious classifications are listed in Table 1.1.

TABLE 1.1

Landsat Image Acquisition Dates and Paths and Rows

Strata	~1990 Date	~1990 Path/Row	~2000 Date	~2000 Path/Row
1	30 August 1990	29/28–29	07 August 2001	26/27
2	30 August 1990	29/27	12 September 2000	27/26–28
3	30 August 1990	29/26	26 August 2000	28/26
4	04 September 1991	27/30	28 August 2001	29/26–28
5	04 September 1991	27/29	24 August 2000	30/27
6	26 August 1991	28/29	24 August 2000	30/26
7	07 August 1990	28/30	10 August 2000	28/30
8	07 August 1990	28/27	10 August 2000	28/29
9	10 August 1991	28/26	28 August 2001	29/29
10	04 September 1991	27/26–28	10 August 2000	28/28
11	07 August 1990	28/28	26 August 2000	28/27
12	09 August 1990	26/27	12 September 2000	27/29–30
13	25 August 1990	26/29–30	11 September 1999	26/30
14	23 July 1991	30/26–27	11 September 1999	26/29

Note: Strata refer to the maps in Figure 1.3.

1.2.3 Impervious Surface Classification

Classification of impervious surface was performed using an ERDAS Imagine Spatial Model with the Landsat tasseled cap greenness values for the calibration sites used as the input values for the impervious estimation models. Values generated represented the percent of impervious surface within the area of each pixel.

To remove estimation bias, an inverse calibration was computed from the linear fit of measured vs. Landsat-estimated plots and applied to the impervious surface classification (Walsh and Burk, 1993). Following the inverse calibration, the accuracy of the Landsat-derived impervious surface estimates was reassessed. The inverse calibration process did not significantly affect the R^2 or standard error values, but decreased the intercept and increased the slope of the regression equations, reducing the overall bias of the models and improving the final classification accuracy.

The land cover classification was used to mask and reclassify the non-urban areas to 0% impervious surface values. The 2000 land cover classification map was used as the primary identifier of urban to have consistent comparisons of the urban areas between the two years. We assumed that areas identified as urban in 2000, but not developed in 1990, would have a high greenness value (due to vegetative cover) in the 1990 imagery and would be modeled as having low to no impervious surface in the 1990 images. However, areas of bare soil in agricultural fields in 1990 that changed to urban by 2000 would have low greenness values on each date, causing errors in the modeling of impervious surface for 1990. A land cover map for the early 1990s, the Minnesota GAP land cover classification (Lillesand et al., 1998; Minnesota Department of Natural Resources (DNR), 2002), was used to remove the cropland and grassland areas from the areas considered as urban for 1990 to minimize this error.

Mines (gravel and sand quarries and iron ore open pit mines), considered as developed or urban in the 2000 land cover classifications, were identified for further processing in the impervious surface classification. Bare soil is classified in the impervious models as having a high degree of impervious surface due to its low greenness value. Much of the area of mines is bare soil, gravel, and related materials, making separation of the impervious surface from bare soil difficult. Data identifying the location and extent of all mines in the state do not exist; however, there were data produced by the Minnesota DNR–Division of Lands and Minerals that identified the locations of active mining areas in the Mesabi Iron Range where the majority of open pit mines are located. This dataset was used to force the pixel values that fell within the iron mines data to an impervious value of zero.

The last processing procedure established a minimum and maximum for the modeled impervious values. Although the regression modeled estimate of percent impervious for a pixel might be less than 0% or more than 100%, this is not physically possible. Therefore, pixels with estimated impervious surface values greater than 100% were reclassified to 100% impervious and those with less than 0% were reclassified as 0% impervious.

1.2.4 Accuracy Assessment

An independent random sample of ~25 accuracy assessment sites was selected from each of the Landsat images. The impervious surface values for these sites were determined in the same manner as the calibration sites described earlier. These sites were used for performing inverse calibration to remove estimation bias and to measure the accuracy of the final Landsat-derived impervious surface estimates. Accuracy was evaluated by regression analyses of measured vs. predicted amounts of impervious area.

1.3 Results and Discussion

We have found a strong relationship between Landsat tasseled cap greenness and percent impervious surface area. An example of the relationship of greenness to percent impervious surface is shown in Figure 1.4. The second-order regression model has an R^2 of 0.91 and standard error of 10.7. By considering greenness and percent impervious area as continuous variables, we can use a regression model to estimate the percent impervious area of each Landsat pixel. The resulting classification provides a continuous range of impervious surface area from 0% to 100%.

Figure 1.5 evaluates the agreement between the measured and Landsat-estimated percent impervious area for the same image as in Figure 1.4 following the inverse calibration. Similar results were obtained for the other 1990 and 2000 images. Figure 1.6 compares part of a DOQ image and the Landsat classification of percent impervious at the pixel level for an urban area. Although the Landsat classification is at a coarse resolution compared with the DOQ, the correspondence of features, particularly the pattern of streets and other urban features such as parks, residential areas, and commercial and industrial areas, is readily apparent in the two images.

The statistics for all of the images for both the 1990 and 2000 classifications were consistent with R^2 values ranging from 0.80 to 0.94 and standard errors of 7.7–15.9 (Table 1.2). Figure 1.7 combines the data from all classifications for 1990 and 2000 to assess the overall accuracy of the Landsat estimates. The overall agreement between measured and Landsat estimates of percent impervious was high for both time periods with R^2 values of 0.86 and standard errors of 11.8 and 11.7.

The statistics, as well as the image comparisons, of Landsat estimates and DOQ measurements of impervious area indicate strong agreement; however, there are several known sources of error. These include: (1) land cover classification errors in urban/developed vs. rural/nonurban areas. Our

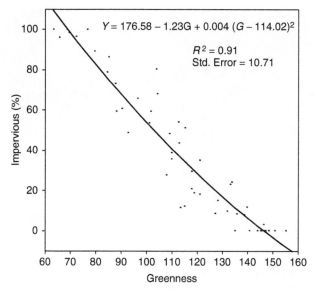

FIGURE 1.4
Example of relationship of Landsat greenness to percent impervious surface area (ETM+ data, path 28/row 28, August 10, 2000). (From Bauer, M., Loeffelholz, B., and Wilson, B., *Proceedings, Pecora 16 Conference, American Society of Photogrammetry and Remote Sensing*, October 23–27, 2005, Sioux Falls, South Dakota, 2005. With permission.)

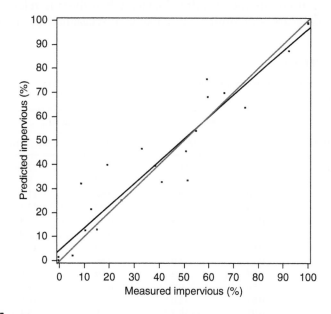

FIGURE 1.5
Comparison of measured to Landsat estimated impervious surface area (ETM+ data, path 28/row 28, August 10, 2000). (From Bauer, M., Loeffelholz, B., and Wilson, B., *Proceedings, Pecora 16 Conference, American Society of Photogrammetry and Remote Sensing*, October 23–27, 2005, Sioux Falls, South Dakota, 2005. With permission.)

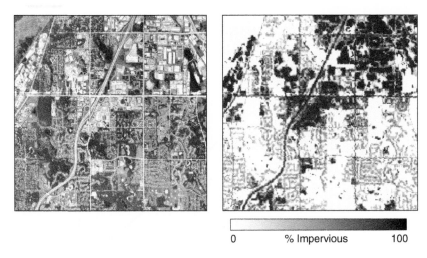

0   % Impervious   100

FIGURE 1.6
Comparison of a high-resolution DOQ of a local area in Eagan (*left*) to the Landsat-derived classification of intensity of impervious surface area.

approach estimates impervious for only the urban class so errors in classification of urban vs. nonurban lead to errors in the location and amount of impervious area. (2) Bare soil is spectrally similar to impervious surfaces. Although we used summer Landsat images when there is relatively little bare soil, some are still present and likely misclassified as impervious.

TABLE 1.2

Accuracy of Impervious Surface Classifications by Strata and Year

	1990		2000	
Strata	R^2	Std. Error	R^2	Std. Error
1	0.86	11.2	0.94	7.7
2	0.89	11.2	0.87	8.9
3	0.90	10.3	0.87	11.7
4	0.83	12.8	0.82	13.4
5	0.82	12.9	0.94	7.8
6	0.89	10.2	0.92	8.9
7	0.94	7.8	0.90	10.1
8	0.85	12.9	0.85	12.8
9	0.87	12.8	0.89	9.6
10	0.86	9.4	0.91	9.4
11	0.84	12.0	0.89	10.8
12	0.81	15.2	0.81	13.1
13	0.82	14.7	0.80	13.9
14	0.90	9.7	0.80	15.9

Note: The locations of strata are shown in Figure 1.2.

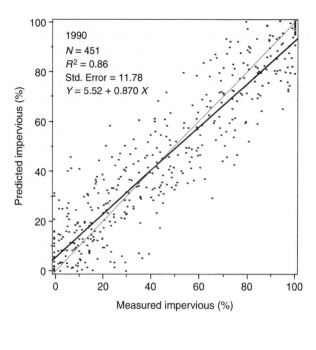

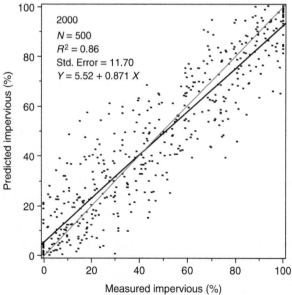

FIGURE 1.7
Evaluation of accuracy assessment statistics for 1990 (*top*) and 2000 (*bottom*) for the entire state. (From Bauer, M., Loeffelholz, B., and Wilson, B., *Proceedings, Pecora 16 Conference, American Society of Photogrammetry and Remote Sensing*, October 23–27, 2005, Sioux Falls, South Dakota, 2005. With permission.)

(3) Tree cover that obscures impervious areas. Although tree-covered areas are included in the calibration models, they are still a likely source of error. However, the error was <6% in a preliminary evaluation of this effect in an urban area with varying amounts of tree cover. (4) Differences in image acquisition dates and vegetation condition and phenology within images. We used mostly August images, but several are early September, which likely had somewhat less greenness for grass that was not irrigated than irrigated lawns. Similarly, early senescing trees would have less greenness. Both these conditions may cause an overestimate of imperviousness.

Examples of the impervious classifications and change maps for two areas, St. Cloud and Rochester, in east central and southeast Minnesota, which are experiencing significant growth, are shown in Figure 1.8. The growth of the urban area and accompanying increases in the amount of impervious surface area are readily apparent with a 40% increase for St. Cloud and 28% for Rochester. The large area covered by the classifications makes it impossible to show here their relevant spatial detail, especially at county to state scales. However, maps of the entire state with capability to roam and zoom can be viewed at: http://land.umn.edu/, along with statistics on amounts and changes in impervious area. The maps and statistics can be viewed, printed, and downloaded for county, city/township, ecoregion, watershed, and lakeshed units.

Table 1.3 lists the amount of impervious surface area for several representative cities, ecoregions, watersheds, and the state. Between 1990 and 2000 the amount of impervious area for the entire state increased 118,464 ha from 1.31% to 1.88% of the total land area, a 44% increase. However, it is the increases at the local, city, and watershed scales that are most critical to the water quality and other environmental effects. At the major watershed level, 20 of 81 watersheds had increases in total impervious area of more than 100% between 1990 and 2000, with 23 experiencing increases of 50%–99% and 22 with 10%–49%. Only 16 had increases of <10% or a small decrease. At the city scale, many cities, especially in the suburbs surrounding Minneapolis–St. Paul, as well as in regional center cities, had increases of 50% or more.

However, increases are not restricted to the larger urban centers. The area and degree of imperviousness also increased in and around many of the smaller towns. Of particular concern is the lake-rich areas of northern Minnesota, where, for example, in the Northern Lakes and Forest Ecoregion, the impervious area increased more than 13,000 ha, a 32.5% increase. Impervious cover increased ∼56% in 25 lake watersheds in north central Minnesota with ∼1%–4% of the watersheds remaining impervious. In the Crow River Watershed (the Crow River is an impaired water body for one to three parameters) west of the Twin Cities, 23 cities and towns and 7 associated townships had impervious area increases of 48% for the municipal areas and 129% for the townships. In 71 non-Twin Cities metro area cities and associated townships, the amount of imperviousness increased 69%. These examples illustrate that relatively large percentage increases in impervious cover have been occurring over the past decade and

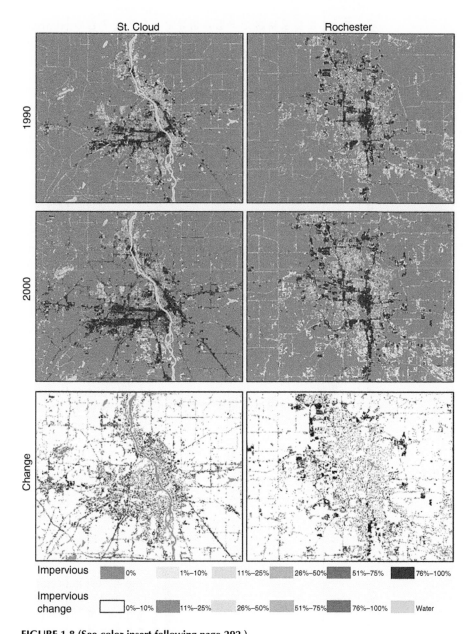

FIGURE 1.8 (See color insert following page 292.)
Impervious classifications of St. Cloud and Rochester for 1990 and 2000 and change maps. (From Bauer, M., Loeffelholz, B., and Wilson, B., *Proceedings, Pecora 16 Conference, American Society of Photogrammetry and Remote Sensing*, October 23–27, 2005, Sioux Falls, South Dakota, 2005. With permission.)

TABLE 1.3

Impervious Surface Area (ISA) Statistics for Selected Cities, Counties, Ecoregions, Watersheds, and State of Minnesota for 1990 and 2000

Area	Total Area (ha)	1990 ISA (ha)	2000 ISA (ha)	Change (ha)	1990 % ISA	2000 % ISA	Percent Change
St. Cloud	10,405	2,045	2,862	817	20.25	28.09	39.95
Rochester	11,932	2,285	2,921	636	20.16	24.74	27.83
Alexandria	2,562	377	667	290	15.58	27.41	76.92
Bemidji	3,448	607	676	69	19.21	21.27	11.37
Brainerd	2,936	514	568	54	18.47	20.24	10.51
Fergus Falls	3,878	399	625	226	11.14	17.39	56.64
Elk River	11,343	793	1,278	485	7.25	11.55	61.16
Sauk Rapids	1,409	316	483	167	23.67	36.06	52.85
Duluth	22,600	3,047	3,032	-15	17.33	17.23	-0.49
Mankato	4,290	929	1,389	460	21.81	32.69	49.52
Owatonna	3,436	746	973	227	21.83	28.44	30.43
Northern Lakes and Forest Ecoregion	6,823,378	41,308	54,738	13,430	0.67	0.88	32.51
N. Central Hardwoods Ecoregion	4,332,968	109,779	158,808	49,029	2.70	3.89	44.66
Western Corn Belt Ecoregion	4,152,960	50,195	88,895	38,700	1.22	2.16	77.10
Mississippi River—St. Cloud Watershed	290,477	7,330	15,671	8,341	2.64	5.58	113.79
St. Croix River—Stillwater Watershed	238,997	4,774	9,107	4,333	2.12	4.03	90.76
Canon River Watershed	380,867	5,530	9,821	4,291	1.49	2.64	77.59
Crow Wing River Watershed	503,935	3,925	6,596	2,671	0.84	1.40	68.05
State	21,852,928	269,649	388,700	119,051	1.31	1.88	44.15

that watershed management efforts may need more rapid updating of land cover information than on 10 or 20 year cycles.

1.4 Conclusions

A strong relationship between impervious surface area and greenness enables percent impervious area on a pixel basis to be mapped with Landsat TM/ETM+ data. Classification of the Landsat data provides a means to map and quantify the degree of impervious surface area, an indicator of environmental quality, over large geographic areas and over time at modest cost. This chapter has described work concentrated on mapping imperviousness over large areas using Landsat data; however, we have previously reported (Sawaya et al., 2003) that the same methods can be successfully applied to high-resolution IKONOS satellite imagery of local areas.

Although we are at an early stage in the analysis of spatial and temporal patterns of urban growth and imperviousness, the Minnesota Pollution Control Agency is incorporating the impervious cover data, obtained from Landsat satellite remote sensing, into watershed management efforts and stormwater best management practice planning and monitoring efforts. An increasing number of future community stormwater management efforts are expected to have phosphorus- and sediment-loading rates determined by formal TMDL allocation processes to restore and/or protect receiving water quality and habitat—based on impervious cover and associated stormwater management practices. The consistent impervious surface data provided by the Landsat classifications for over 200 MS4 communities, covered by the phase II stormwater regulations, are a new foundational data layer needed for refining watershed management strategies for protection as well as for rehabilitation.

Increasing population, new development in lake and river recreation areas, and growing cities and towns all translate into increasing impervious surface areas across Minnesota. The Landsat classifications provide critically important, consistent, and multidate impervious surface area maps and statistics for any area of Minnesota. It is envisioned that these data and updates will be an important foundation of Minnesota's stormwater management efforts. As urban stormwater runoff from impervious areas can have profound negative impacts to receiving waters, it is a critical new component of statewide stormwater education and management efforts.

Acknowledgments

The support of the University of Minnesota Agricultural Experiment Station, project MN-42-037, and the Minnesota Pollution Control Agency,

Stormwater Program, is gratefully acknowledged. The contributions of Trent Erickson and Perry Nacionales to the development of databases and web-based mapping are gratefully acknowledged. An earlier version of this chapter was published in the *Proceedings of the Pecora 16 Symposium*, October 23–27, 2005, Sioux Falls, South Dakota, by the American Society of Photogrammetry and Remote Sensing, Bethesda, Maryland.

References

Arnold, C.L. and Gibbons, C.J., Impervious surface coverage: the emergence of a key environmental indicator, *Journal of the American Planning Association*, 62(2):243–258, 1996.

Bauer, M.E., Heinert, N.J., Doyle, J.K., and Yuan, F., Impervious surface mapping and change monitoring using satellite remote sensing, *Proceedings, American Society of Photogrammetry and Remote Sensing Annual Conference*, May 24–28, Denver, Colorado. Unpaginated CD ROM, 10 pp, 2004.

Bauer, M., Loeffelholz, B., and Wilson, B., Estimation, mapping and change analysis of impervious surface area by Landsat remote sensing, *Proceedings, Pecora 16 Conference, American Society of Photogrammetry and Remote Sensing*, October 23–27, 2005, Sioux Falls, South Dakota. Unpaginated CD ROM, 9 pp, 2005.

Civco, D.L. and Hurd, J.D., Impervious surface mapping for the state of Connecticut, *Proceedings, American Society for Photogrammetry and Remote Sensing Annual Conference*, April 7–10, Seattle, Washington, Missouri, 3:124–135, 1997.

Civco, D.L., Hurd, J.D., Wilson, E.H., Arnold, C.L., and Prisloe, M.P. Jr., Quantifying and describing landscapes in the northeast United States, *Photogrammetric Engineering and Remote Sensing*, 68(10):1083–1090, 2002.

Crist, E.P. and Cicone, R.C., A physically-based transformation of Thematic Mapper date-the TM tasselled cap, *IEEE Transactions on Geoscience and Remote Sensing*, GE-22(3):256–263, 1984.

Dougherty, M., Dymond, R.L., Goetz, S.J., Jantz, C.A., and Goulet, N., Evaluation of impervious surface estimates in a rapidly urbanizing watershed, *Photogrammetric Engineering and Remote Sensing*, 70(11):1275–1284, 2004.

Huang, C., Wylie, B., Yang, L., Homer, C., and Zylstra G., Derivation of a tasselled cap transformation based on Landsat 7 at-satellite reflectance, *International Journal of Remote Sensing*, 23(8):1741–1748, 2002.

Jantz, P., Goetz S., and Jantz, C., Urbanization and the loss of resource lands in the Chesapeake Bay Watershed, *Environmental Management*, 36(3):1–19, 2005.

Lillesand, T.M., Chipman, J.W., Nagel, D.E., Reese, H.M., Bobo, M.R., and Goldmann, R.A., *Upper Midwest Gap Analysis Program Image Processing Protocol*. U.S. Geological Survey, Environmental Management Technical Center, Onalaska, Wisconsin, EMTC 98-G001, 25 pp.+Appendices, 1998.

Lu, D. and Weng, Q.H., Use of impervious surface in urban land-use classification, *Remote Sensing of Environment*, 102(1–2):146–160, 2006.

McRoberts, R.E., Nelson, M.D., and Wendt, D.G., Stratified estimation of forest area using satellite imagery, inventory data, and the k-nearest neighbors technique, *Remote Sensing of Environment*, 82(2–3):457–468, 2002.

Minnesota Department of Natural Resources, GAP (1993) land cover—tiled raster, accessible at: http://deli.dnr.state.mn.us/metadata.html?id = L390002710606, 2002.

Ridd, M.K., Exploring a V-I-S (vegetation-impervious surface-soil) model for urban ecosystem analysis through remote sensing: comparative anatomy for cities, *International Journal of Remote Sensing*, 16(12):2165–2185, 1995.

Sawaya, K., Olmanson, L., Heinert, N., Brezonik, P., and Bauer, M., Extending satellite remote sensing to local scales: land and water resource monitoring using high-resolution imagery, *Remote Sensing of Environment*, 88:144–156, 2003.

Schueler, T., The importance of imperviousness, *Watershed Protection Techniques*, 1(3):100–111, 1994.

Walsh, T.A. and Burk, T.E., Calibration of satellite classifications of land area, *Remote Sensing of Environment*, 46(3):281–290, 1993.

Wu, C. Normalized spectral mixture analysis for monitoring urban composition using ETM+ imagery, *Remote Sensing of Environment*, 93(4):480–492, 2004.

Wu, C. and Murray, A.T., Estimating impervious surface distribution by spectral mixture analysis, *Remote Sensing of Environment*, 84(4):493–505, 2003.

Xian, G. and Crain, M., Assessments of urban growth in Tampa Bay watershed using remote sensing data, *Remote Sensing of Environment*, 97(2):203–215, 2005.

Yang, L., Huang, C., Homer, C.G., Wylie, B.K., and Coan, M.J., An approach for mapping large-area impervious surfaces: synergistic use of Landsat-7 ETM+ and high spatial resolution imagery, *Canadian Journal of Remote Sensing*, 29(2):230–240, 2003a.

Yang, L., Xian, G., Klaver, J.M., and Deal, B., Urban land cover-change detection through sub-pixel imperviousness mapping using remotely sensed data, *Photogrammetric Engineering and Remote Sensing*, 69(9):1003–1010, 2003b.

Yuan, F. and Bauer, M.E., Comparison of impervious surface area and normalized difference vegetation index as indicators of surface urban heat island effects in Landsat imagery, *Remote Sensing of Environment*, 106(3):375–386, 2006.

2

Subpixel Imperviousness Estimation with IKONOS Imagery: An Artificial Neural Network Approach

Rama Prasada Mohapatra and Changshan Wu

CONTENTS

2.1 Introduction

Over the last century, especially since World War II, the process of urbanization has increased and intensified all over the world. In addition to the internal growth and restructuring of cities, emigration and demand for more living space (Clarke and Gaydos, 1998) play a vital role in the spread of urbanized areas. It is likely that this trend will continue in the twenty-first century, eventually leading to the formation of Gigalopolises or supercities containing hundreds of millions of people (Clarke and Gaydos, 1998). According to a United Nations Population Division report, by the year 2015, the number of cities that have an urban population >5 million will go up to 58 from 39 in 2000 (UN, 2001). One of the problems associated with rapid urbanization is the concurrent increase (Xian and Crane, 2005) in impervious surfaces, which include roads, sidewalks, parking lots, and various rooftops, through which water cannot infiltrate into soil (Arnold and Gibbons, 1996). A study (Elvidge et al., 2004) partly funded by NASA's land-cover land-use change program found that the aggregate impervious surface area of continental United States was slightly less than the total area of the state of Ohio. Although the percentage of impervious surface to the total landmass of the United States is just over 1%, it is a matter of great concern as impervious surface limits soil infiltration, threatens water quality through increased flow of polluted runoff, contributes to flooding, and creates heat islands (USEPA, 2003). Therefore, accurate estimation of impervious surface is crucial for sustainable urban development and planning (Arnold and Gibbons, 1996; Flanagan and Civco, 2001; Goetz et al., 2003; Wu and Murray, 2003; Dougherty et al., 2004; Xian and Crane, 2005; Yang, 2006).

Motivated by the importance of impervious surfaces, many researchers from various disciplines have attempted to estimate the amount and distribution of urban impervious surfaces. In these studies, satellite imagery has been very helpful as it could successfully characterize different land-cover and land-use types (Ward et al., 2000; Yang, 2006). The coarse and medium resolution of satellite imagery, however, has always been a hurdle to urban planners and led to underutilization of satellite imagery in urban applications (Mesev, 1997; Carlson, 2003). Recently, greater accessibility to higher-resolution satellite imagery and advanced computational techniques has opened new frontiers for estimating and monitoring imperviousness around urban growth centers (Yuan and Bauer, 2006). Utilization of high-resolution satellite imagery in extracting impervious surfaces is likely to be cost and time efficient when compared with the processes of manual digitization over aerial photographs. Research findings indicate that high-resolution IKONOS multispectral imagery allows measuring the relative contributions of different materials in urban/ex-urban areas; furthermore, their abundance could also be mapped in heterogeneous land-cover areas (Small, 2003). IKONOS imagery has proved its worthiness in estimating urban imperviousness (Goetz et al., 2003), although its high spatial resolution increases the spectral complexity and variability in urban areas (Herold et al., 2004).

For the purpose of urban imperviousness estimation, the vegetation-impervious surface-soil (VIS) model proposed by Ridd (1995) seems to be accepted by many scholars. The VIS model considers the combination of impervious surface material, green vegetation, and exposed soil as the most fundamental components of urban ecosystems if water surfaces are ignored. However, there exists the problem of mixed pixels as they relate to urban environments (Ridd, 1995; Wu, 2004; Song, 2005; Xian and Crane, 2005). Mixed pixels occur when the acquired signal in a pixel results from various land-cover types on the ground. Mixed pixels, associated with the difficulty in distinguishing various surface materials in urban areas and sometimes unavailability of adequate training data for some classes, have always hindered scholars in achieving satisfactory results. Earlier attempts to solve these problems include better endmember selection, normalization of the original image, or employing various computational techniques (e.g., regression tree analysis). In most of these attempts, spectral mixing among endmembers is predominantly considered to be linear although there are issues related to nonlinear mixing as evident in both medium (Wu and Murray, 2003) and high-resolution (Small, 2003) satellite data. In this context, powerful computational techniques like artificial neural networks (ANNs) may yield better result in mapping nonlinear relations in complex urban environments. ANNs provide a better alternative to statistical classification techniques as they could map complex relationships among variables without making any assumptions about the data (Ji, 2000; Linderman et al., 2004; Ingram et al., 2005) and require fewer training samples (Pal and Mather, 2003).

ANN classifiers have proven to be superior for classifying per-pixel Landsat TM and ETM+ data to traditional classifiers such as maximum likelihood and others (Ji, 2000; Dwivedi et al., 2004). ANNs are also capable of estimating subpixel imperviousness from medium-resolution satellite imagery such as Landsat TM and ETM+ (Civco and Hurd, 1997; Flanagan and Civco, 2001; Lee and Lathrop, 2006). Moreover, a comparison of spectral unmixing models states that neural network models outperform the traditional linear unmixing models (Liu and Wu, 2005). Although ANN has the ability to generate detailed land-use and land-cover information, its potential in extracting subpixel impervious surface information from high-resolution remote sensing imagery is still in question. Traditionally, ANN has been utilized for per-pixel classification (hard classification), in which spectral values represented as digital numbers or reflectance values are grouped together to form certain classes. The ANN, however, could also be used to perform spectral unmixing through subpixel classification (soft classification). Subpixel classification employs pure land-cover classes as inputs and yields a group of images representing the degree of membership of a pixel to each possible class (Mertens et al., 2004; Lee and Lathrop, 2006).

The main objective of this chapter is to establish a neural network model to estimate subpixel urban imperviousness from high spatial resolution remotely sensed imagery. A three-layered feed-forward back-propagation model, the most widely used neural network model for soft classification (Liu and Wu, 2005; Lee and Lathrop, 2006), was employed in the present study. Activation-level maps (Eastman, 2003), which explain the degree of membership of each pixel to each class (e.g., impervious surface), were generated with the ANN model. The soft ANN classification method was utilized in the present study, in which small numbers of pure vegetation, impervious surface, and soil samples were used as training data to estimate the fraction of these land-cover types. Three different methods were explored to convert the degree of membership of each pixel into percentage imperviousness. The estimated imperviousness was compared with the true imperviousness for accuracy assessment. The rest of the chapter is organized into the following sections: Section 2 describes the study area and data. The methodology including a general description of neural network and activation-level maps is described in Section 3. Section 4 reports the results and verification of the results. Finally, in Section 5, the chapter ends with conclusions and future scope.

2.2   Study Area and Data

The study area (Figure 2.1) covers Grafton village and township in Ozaukee County, Wisconsin, United States. This area was chosen because of the availability of the IKONOS image and its diversified urban and rural land uses. The total land area of Grafton is roughly 24 mi^2 (Census, 2000). Grafton is one of the rapidly growing (SEWRPC, 2004) suburban areas around the city of Milwaukee, Wisconsin, and exhibits a mixed land-use pattern comprising agriculture, forestry, residential, commercial, transportation, and so on. An IKONOS satellite image (Figure 2.2) of September 3, 2002, and a color aerial photograph for the same year were collected from the American Geographical Society Library (AGSL) at the University of Wisconsin–Milwaukee. The multispectral IKONOS image comprises

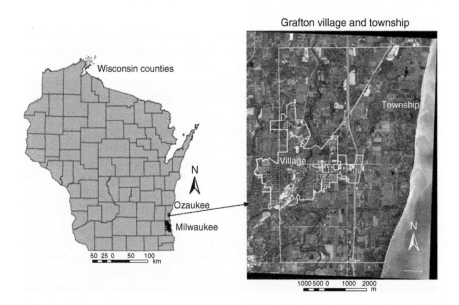

FIGURE 2.1
Grafton village and township in Ozaukee county of Wisconsin (map on the *left* shows location of Grafton in Wisconsin; map on the *right* shows the boundary of Grafton village and township over the color aerial photograph).

FIGURE 2.2
IKONOS imagery (reflectance values) for the study area obtained on September 3, 2002.

four bands (blue, green, red, and near-infrared [NIR]) and the color aerial photograph, with a resolution of 2 ft, has three bands: blue, green, and red. Because of these three channels, the color aerial photograph is very helpful in identifying various impervious surfaces for choosing training and testing data sets. Both the IKONOS image and aerial photograph are in the State Plane projection. For this study, they were reprojected to UTM (WGS84, Zone 16) to match the land-use and land-cover data that were used for reference. Visual inspection of IKONOS over aerial photograph revealed a slight misregistration. To avoid issues related to misregistration and ensure a higher level of accuracy, the IKONOS image was georeferenced using the aerial photograph as the reference image. According to Song et al. (2001), atmospheric correction will not affect the results using image endmembers, thus it is not applied to the image. The radiance values of the IKONOS image were converted to exoatmospheric reflectance values following the standard procedure provided in the Space Imaging documents (Space Imaging, 2005). Water features were masked out from the original image to avoid confusion during classification and impervious surface percentage calculation.

2.3 Methodology

2.3.1 Artificial Neural Networks

ANNs are composed of simple processing units called nodes (or artificial neurons) and operating links. Two or more nodes are combined to form a layer; a neural network may contain two or more layers. The nodes organized in layers and the links between successive layers provide ANNs a learning ability similar to humans; they are taught by sample data in a similar fashion as the human brain learns. ANNs mostly try to mimic the functioning of the human brain while exploring geographical data. During training, initial weights assigned to interconnection links are modified repeatedly until the ANN can produce an acceptable result that matches the testing samples. The feed-forward ANN models are well trained through back-propagation algorithms known as the delta rule (Eastman, 2003). A feed-forward back-propagation neural network calculates a pixel's activation values for all land-cover classes under consideration. In a hard ANN classification, the highest activation value is selected and the pixel is assigned to the corresponding class. However, the same activation values can be used to represent land-cover class memberships (Lee and Lathrop, 2006) and this is known as soft ANN classification approach. In this study, a soft ANN classification approach was followed to estimate the proportion of impervious surface for each individual pixel of the IKONOS image.

Several researchers have explored the effects of the number of layers and nodes on analyzing complex nonlinear data (Kavzoglu and Mather, 2003; Linderman et al., 2004). Typically, three-layer networks are widely used in image processing and subpixel information extraction from satellite imagery (Civco and Hurd, 1997; Flanagan and Civco, 2001; Kavzoglu and Mather, 2003; Liu and Wu, 2005), where the three types of layers are input layer, hidden layer, and output layer (Figure 2.3). The function of the input layer is to provide data (e.g., satellite imagery) into the model. The hidden layer, an invisible layer between input and output layers, is used to process the data and pass the results to the output layer. The output layer then constructs the output pattern (e.g., land-cover classes) based on the combination of information received from the hidden layer. The number of nodes in the input layer equals the number of bands of satellite image

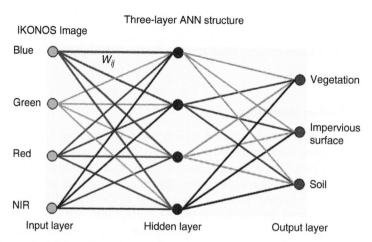

FIGURE 2.3 (See color insert following page 292.)
Artificial neural network structure.

and the number of nodes in the output layer is equivalent to the number of desired land-cover classes. The number of hidden layers and nodes required to obtain an accurate estimation could not be unanimously decided yet and there is no optimal structure that could be used in every situation (Li and Yeh, 2002; Kavzoglu and Mather, 2003). Too many hidden layer nodes amplify the amount of information leading to overfitting, whereas fewer nodes in the hidden layer reduce the amount of data that are necessary to identify the internal structure and lead to underfitting (Kavzoglu and Mather, 2003).

In this chapter, NEURALNET module available in IDRISI Kilimanjaro software made available by the Clark Labs was utilized to obtain the subpixel imperviousness. The NEURALNET (neural network) module uses the popular back propagation learning algorithm for classifying remotely sensed imagery (Eastman, 2003). The ANN classifier used the following algorithm to quantify the information at a single node j in the output layer:

$$\text{net}_j = \sum_i w_{ij} I_i,$$

where
  net_j is the information that an output node j receives
  w_{ij} represents the weights between node i and node j
  I_i refers to the information from node i of a sender layer (e.g., input or hidden layer)

The receiver node (artificial neuron in the output layer) creates a certain activation level in response to the incoming information net_j. The output at given node $j(O_j)$ is computed from

$$O_j = f(\text{net}_j),$$

where f is a nonlinear sigmoidal function popularly known as activation function.

2.3.2 Activation-Level Maps

The activation function is applied to the weighted sum of inputs before passing the information to the next layer. When an input pixel is presented to ANN, each output node is assigned a value that is compared with the expected value. With few exceptions, the assigned value would always differ from the expected value at the output node; the difference is the error propagated backward for relevant corrections via the delta rule in NEURALNET (Eastman, 2003). When the output result achieves a predefined accuracy level, the input pixel is assigned an activation value ranging between 0 and 1 where larger values represent a higher degree of membership belonging to the corresponding node (land-cover class). Output from NEURALNET can be binary (hard classification) or continuous (soft classification) value. The result from hard classification gives a thematic map where the pixels are grouped into desired classes based on the association of highest activation levels of the nodes. In the soft classification, the pixels are assigned the highest activation value of each node and produce as many images (activation-level maps) as the total number of nodes in the output layer. These activation-level maps express to what degree a pixel belongs to each node (land-cover class) in the output layer. For example, a pixel with a value of 0.82 at a particular node expresses the degree of membership of that pixel belonging to the corresponding land-cover class as 0.82. Although the output from ANNs in the form of activation-level maps ranges between 0 and 1, it is not necessary that the sum of activation values of all the land-cover classes for any given pixel will be equal to one as the outputs are obtained by fuzzying the input data into values in the range 0–1. As the study is focused toward subpixel information extraction, only the soft classification results were obtained from NEURALNET.

2.3.3 Neural Network Structure

To achieve higher accuracies while extracting subpixel fraction of various land-cover classes from remote sensing imagery through ANN classifiers, it is necessary to design an optimal network structure and set proper learning parameters. The structure of the neural network used in this study is described in Figure 2.3. In the input layer, four nodes represent four spectral bands of the IKONOS image. In NEURALNET, by default, the number of input images determines the number of input layer nodes; in this case, it is four: blue, green, red, and NIR. The number of output layer nodes is dependent on the number of training categories defined in the training site file; in this case, it is three corresponding to the three land-cover features vegetation, impervious surface, and soil. There is a need to determine the number of nodes for the hidden layer. For this study, it is estimated to be between 3 and 4 but the larger number was used as it helps in differentiating complex land covers (Kavzoglu and Mather, 2003). The number of nodes for the hidden layer was estimated by the following equation

$$N_h = \text{INT}\sqrt{N_i \times N_o},$$

where
   N_h is the number of hidden layer nodes
   N_i is the number of input layer nodes
   N_o is the number of output layer nodes

Besides ANN structure, several other parameters, such as sigmoid function constant, learning rate, and momentum factor, influence the performance of the classifier. The sigmoid function constant determines the shape of the sigmoidal curve and the gradient. The learning rate parameter controls the connecting weights. The momentum factor helps in avoiding oscillation problems during the search for minimum value on the error surface. According to Kavzoglu and Mather (2003), a value in the range of 0.1–0.2 is appropriate for the learning rate and a value of 0.5–0.6 for the momentum factor might yield a better result. As obtaining the proper combination of these parameters is just a trial and error process, various combinations of these parameters were tried to find the best result.

2.3.4   Training and Testing Sample Selection

Like other traditional classifiers, providing proper training and testing samples is another important aspect of ANN. In this study, training and testing samples was obtained for three different land-cover classes (vegetation, impervious surface, and soil). A vector file was created with 30 sampling sites for each class; the sites are distributed all over the study area so that they could represent both spatial and spectral variations among the classes under consideration. Training the neural network requires a certain number of training and testing pixels per category (must be between 0 and 200, in NEURALNET). The training and testing pixels used in the analysis are a subset of the total pixels found in the training site file, which was obtained from the IKONOS image. The testing pixels are required for setting an accuracy rate (in this case it is 95%) that is used to terminate the ANN classifier. As it is advised not to use a very small number of training pixels, in this study 100 pixels randomly selected from the 30 sampling sites for each land-cover class were used to train the ANN classifier, and another

100 testing pixels for each land-cover class were used to compute the accuracy rate.

2.4   Results

After finalizing the network structure, various combinations of learning rate and momentum factors were tested to obtain the optimal convergence. By analyzing the sigmoid function curve, it was ascertained that a learning rate of 0.16 and a momentum factor of 0.57 produce a better result. The ANN classifier was allowed to automatically decide the sigmoid function constant (14.48). After setting all these parameters, the network was trained and then activation-level maps were obtained for three different land-cover classes. The output in the form of an activation-level map for impervious surface is shown in Figure 2.4. The values in the map are in the range of 0–1 representing the degree of membership to imperviousness. The higher the value, the more likely that the pixel contains a larger amount of impervious surfaces. As the activation-level maps represent the likelihood of a pixel belonging to a particular land-cover class, it could be inferred that this likelihood should closely relate to the percentages of that land-cover class within the pixel.

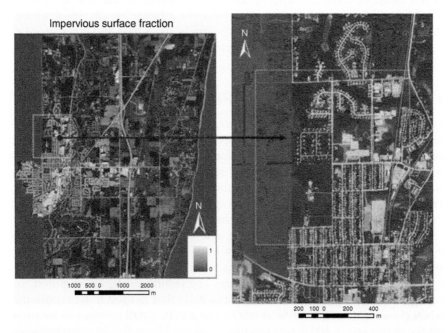

FIGURE 2.4
Impervious surface fraction imagery obtained through the ANN model.

2.4.1 Output Transformation

The results generated from the ANN classifier indicate the degree of membership that a pixel belongs to impervious surfaces, not exactly the percentage of impervious surfaces within that pixel. To compare with the "real" impervious surface percentage on the ground, several transformation methods have been applied to the ANN results. In the first method (method 1), the average activation values of a 5×5 pixel window of impervious surface activation-level map were converted into percentage by simply multiplying the original value with 100. The fraction values obtained from activation-level maps were converted to percentage so that the real imperviousness, which is in percentage, could be compared with the estimated one. In the second method (method 2), at first a normalization technique was applied on the estimated mean activation values of impervious surfaces so that it ranges exactly between 0 and 1. Then this normalized fraction value of impervious surface (x_j) was converted into percentage for further comparison with the real imperviousness. The normalization was carried out following the equation mentioned below

$$x_j = \frac{x_i - x^{min}}{x^{max} - x^{min}},$$

where

x_j is the estimated value of x_i after normalization
x^{min} and x^{max} are the minimum and maximum values of a given set of data

The third method (method 3) takes into account all three land-cover classes for estimating the percentage of impervious surfaces. As the VIS model of Ridd (1995) was followed for unmixing of the IKONOS pixel composition, the average activation values of three different land-cover classes (vegetation, impervious surface, and soil) were added to get the cumulative activation value for a particular sampling unit. Once the total activation value is calculated for a sampling unit, considering this cumulative activation value as 100%, the proportions for three land-cover classes were computed. However, as the present study focuses only on the estimation of impervious surface fraction, the calculation was restricted to impervious surface.

2.4.2 Result Verification

To evaluate the effectiveness of the established methods, it is necessary to compare these results obtained through ANN classifier with the reality on the ground. One hundred and fifty random samples with a sampling unit of 5×5 pixels were created to compare the results. For each sample, a $20 \text{ m} \times 20 \text{ m}$ sampling window was used to digitize the real impervious surface over color aerial photograph manually. The real percentage of imperviousness was calculated by dividing digitized impervious surface area to the total area of a sampling unit (400 m^2) and then multiplying the result by 100. The imperviousness of IKONOS pixels was obtained from the activation-level maps. A zonal statistics tool available in ERDAS imagine was used to estimate the mean activation value for the sampling units.

Two quantitative estimators were used to compare the estimated percentage imperviousness with the real percentage imperviousness. The estimators are the Pearson's correlation coefficient (R) and mean average error (MAE). The correlation coefficient is a measure of reliability and describes the strength of relationship between the estimated and real imperviousness. The MAE is about the relative prediction error and is estimated as

$$\text{MAE} = \frac{1}{N} \sum \left| I_i - \hat{I}_i \right|,$$

where
I_i is the real imperviousness for sampling unit i obtained from color aerial photograph
$\hat{I}_i$ is the estimated imperviousness for the same sampling unit i
N is the total number of samples

The estimated results obtained through the previously mentioned methods (methods 1, 2, and 3) are reported in Table 2.1. The best result was obtained through method 2 while comparing the estimated imperviousness to the real imperviousness. The MAE for method 2 was 7.87 whereas for methods 1 and 3 it is 9.14 and 9.99, respectively. This suggests the effectiveness of normalization of activation values in attaining a lesser prediction error. The Pearson's correlation coefficients (0.94, 0.94, and 0.93 for methods 1, 2, and 3, respectively) reveal that there is a strong correspondence between the predicted percent imperviousness and the real percent imperviousness. The estimated percentage imperviousness for all the 150 sampling units was plotted against the real percentage imperviousness (Figure 2.5a through c). In methods 1 (Figure 2.5a) and 3 (Figure 2.5c), it is observed that there is an underestimation of impervious surface fraction. Figure 2.5b (method 2) reveals that there are certain sampling units where the real imperviousness is 0% but in the result obtained through ANN it shows a

TABLE 2.1

Accuracy Assessments of Impervious Surface Estimation

	R	MAE
Method 1	0.94[a]	9.14
Method 2	0.94[a]	7.87
Method 3	0.93[a]	9.99

[a] Correlation is significant at the 0.01 level (one tailed).

very high percentage of imperviousness. A crossverification of these sampling units with the help of aerial color photograph reveals that in reality these sampling units are located on recently ploughed agricultural fields. It is also observed that misclassification of soil as impervious surface is enhanced in method 2 although the best MAE (7.78) and correlation coefficient (0.94) are achieved through this method.

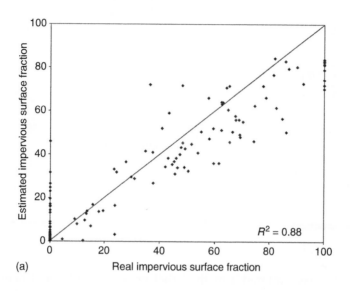

(a)

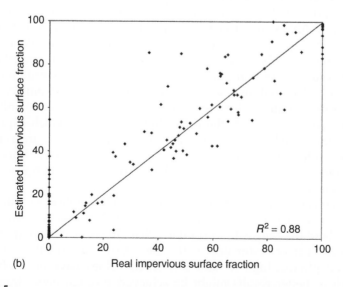

(b)

FIGURE 2.5

Results of impervious surface estimation accuracy assessment (a) method 1, (b) method 2, and

(continued)

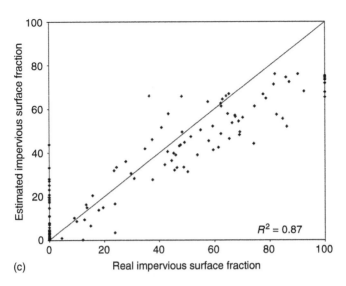

FIGURE 2.5 (continued)
(c) method 3.

2.5 Conclusion

This chapter has explored a method that uses a three-layered neural network classifier to extract subpixel impervious surface information from high-resolution satellite imagery. The study successfully demonstrated the usefulness of ANNs in extracting subpixel impervious information from pixels of 4 m × 4 m IKONOS image with very few training data. Results show that the soft ANN classification method has potential in estimating impervious surface fraction in an urban/ex-urban area. The result is encouraging while differentiating most of the impervious surfaces from vegetation, although some confusion between soil and impervious surface still exists. The study also proved that the activation-level maps, which explain the degree of membership of various underlying land covers, are very well correlated with subpixel information. Therefore, the spectral information of pure land-cover classes, instead of the fraction information from mixed pixels, can be utilized to train the ANN.

The established method has been successfully applied to extract the subpixel imperviousness in the fast-growing Grafton area near Milwaukee where the land use ranges from rural to urban. In particular, comparing the impervious surface estimates with the "real" information digitized from aerial photos, high R^2 and low MAE values have been obtained. Nevertheless, better results might be achieved through better training of the neural network as it is helpful in improving the classification results and by addressing the issues relating to the misclassification of ploughed soil as impervious surface. The future developments to the established method could be achieved by incorporating ancillary information such as the increased textual information available in IKONOS image, and moisture and temperature information obtained from medium-resolution satellite data. As the obtained information provides meaningful data to urban planners and policy makers, there is a need for further examination of this method in other urban/ex-urban areas to prove its global worthiness. Although one of the disadvantages of the ANN algorithm is the black-box nature of functioning, where the expert does not intervene in defining the relation between input and output, further research in enhancing the method is needed as the method provides a practical way to extract subpixel imperviousness information on a fine scale.

References

Arnold, C.A. Jr. and Gibbons, C.J. (1996). Impervious surface: the emergence of a key urban environmental indicator. *American Planning Association Journal*, 62(2), 243–258.

Carlson, T. (2003). Applications of remote sensing to urban problems. *Remote Sensing of Environment*, 86, 273–274.

Civco, D.L. and Hurd, J.D. (1997). Impervious surface mapping for the state of Connecticut. *ASPRS 1997 Annual Convention*, Seattle.

Clarke, K.C. and Gaydos, L.J. (1998). Loose-coupling a cellular automaton model and GIS: long-term urban growth prediction for San Francisco and Washington/Baltimore. *International Journal of Geographical Information Science*, 12(7), 699–714.

Dougherty, M., Dymond, R.L., Goetz, S.J., Jantz, C.A., and Goulet, N. (2004). Evaluation of impervious surface estimates in a rapidly urbanizing watershed. *Photogrammetric Engineering & Remote Sensing*, 70(11), 1275–1284.

Dwivedi, R.S., Kandrika, S., and Ramana, K.V. (2004). Comparison of classifiers of remote-sensing data for land-use/land-cover mapping. *Current Science*, 86(2), 328–335.

Eastman, J.R. (2003). *IDRISI Kilimanjaro*, Guide to GIS and Image Processing (Manual Version 14). Clark University Press, Massachusetts.

Elvidge, C.D., Milesi, C., Dietz, J.B., Tuttle, B.T., Sutton, P.C., Nemani, R., and Vogelmann, J.E. (2004). U.S. constructed area approaches the size of Ohio. *EOS Transactions*, 85(24), 233–240.

Flanagan, M. and Civco, D.L. (2001). Subpixel impervious surface mapping. *ASPRS 2001 Annual Convention*, St. Louis.

Goetz, S.J., Wright, R.K., Smith, A.J., Zinecker, E., and Schaub, E. (2003). IKONOS imagery for resource management: tree cover, impervious surfaces, and riparian buffer analyses in the mid-Atlantic region. *Remote Sensing of Environment*, 88, 195–208.

Herold, M., Roberts, D.A., Gardner, M.E., and Dennison P.E. (2004). Spectrometry for urban area remote sensing-development and analysis of a spectral library from 350 to 2400 nm. *Remote Sensing of Environment*, 91, 304–319.

Ingram, J.C., Dawson, T.P., and Whittaker, R.J. (2005). Mapping tropical forest structure in southeastern Madagascar using remote sensing and artificial neural networks. *Remote Sensing of Environment*, 94, 491–507.

Ji, C.Y. (2000). Land-use classification of remotely sensed data using kohonen self-organizing feature map neural networks. *Photogrammetric Engineering & Remote Sensing*, 66(12), 1451–1460.

Kavzoglu, T. and Mather P.M. (2003). The use of backpropagating artificial neural networks in land cover classification. *International Journal of Remote Sensing*, 24(23), 4907–4938.

Lee, S. and Lathrop, R.G. (2006). Subpixel analysis of Landsat ETM+ using Self-Organizing Map (SOM) neural networks for urban land cover characterization. *IEEE Transactions on Geoscience and Remote Sensing*, 44(6), 1642–1654.

Li, X. and Yeh, A.G. (2002). Neural-network-based cellular automata for simulating multiple land use changes using GIS. *International Journal of Geographical Information Science*, 16(4), 323–343.

Linderman, M., Liu, J., Qi, J., An, L., Ouyang, Z., Yang, J., and Tan, Y. (2004). Using artificial neural networks to map the spatial distribution of understorey bamboo from remote sensing data. *International Journal of Remote Sensing*, 25(9), 1685–1700.

Liu, W. and Wu, E.Y. (2005). Comparison of non-linear mixture models: sub-pixel classification. *Remote Sensing of Environment*, 94, 145–154.

Mertens, K.C., Verbeke, L.P.C., Westra, T., and De Wulf, R.R. (2004). Sub-pixel mapping and sub-pixel sharpening using neural network predicted wavelet coefficients. *Remote Sensing of Environment*, 91, 225–236.

Mesev, V. (1997). Remote sensing of urban systems; hierarchical integration with GIS. *Computer Environment and Urban systems*, 21(3/4), 175–187.

Pal, M. and Mather, P.M. (2003). An assessment of the effectiveness of decision tree methods for land cover classification. *Remote Sensing of Environment*, 86, 554–565.

Ridd, M.K. (1995). Exploring a V-I-S (vegetation-impervious surface-soil) model for urban ecosystem analysis thorough remote sensing: comparative anatomy for cities. *International Journal of Remote Sensing*, 16(12), 2165–2185.

Small, C. (2003). High spatial resolution spectral mixture analysis of urban reflectance. *Remote Sensing of Environment*, 88, 170–186.

Song, C. (2005). Spectral mixture analysis for subpixel vegetation fractions in the urban environment: how to incorporate endmember variability? *Remote Sensing of Environment*, 95, 248–263.

Song, C., Woodcock, C.E., Seto, K.C., Pax-Lenney, M., and Macomber, S.A. (2001). Classification and change detection using Landsat TM data: when and how to correct atmospheric effects? *Remote Sensing of Environment*, 75, 230–244.

Southeastern Wisconsin Regional Planning Commission (SEWRPC). (2004). The population of Southeastern Wisconsin, Technical Report #11. http://www.sewrpc.org/publications/techrep/tr-011_population_southeastern_wisconsin.pdf

Space Imaging. (2005). IKONOS relative spectral response and radiometric cal coefficients, 2005. http://www.spaceimaging.com/products/ikonos/spectral.htm

United Nations. (2001). World Urbanization Prospects: the 2001 Revision. United Nations Population Division, U.N. http://www.un.org/esa/population/*population/publications/wup2001/WUP2001report.htm*

United States Census Bureau. (2000). Census 2000 Summary File 1 (SF 1) 100-Percent Data Geographic Comparison Tables. http://factfinder.census.gov

United States Environmental Protection Agency (USEPA). (2003). Draft Report on the Environment 2003. United States Environmental Protection Agency, Washington, D.C. http://www.epa.gov/indicators/roe/pdf/EPA_Draft_ROE.pdf

Ward, D., Phinn, S.R., and Murray, A.T. (2000). Monitoring growth in rapidly urbanizing areas using remotely sensed data. *Professional Geographer*, 52(3), 371–386.

Wu, C. (2004). Normalized spectral mixture analysis for monitoring urban composition using ETM+ imagery. *Remote Sensing of Environment*, 93, 480–492.

Wu, C. and Murray, A.T. (2003). Estimating impervious surface distribution by spectral mixture analysis. *Remote Sensing of Environment*, 84, 493–505.

Xian, G. and Crane, M. (2005). Assessments of urban growth in the Tampa Bay watershed using remote sensing data. *Remote Sensing of Environment*, 97, 203–215.

Yang, X. (2006). Estimating landscape impervious index from satellite imagery. *IEEE Geoscience and Remote Sensing Letters*, 3(1), 6–9.

Yuan, F. and Bauer, M.E. (2006). Mapping impervious surface using high resolution imagery: a comparison of object-based and per pixel classification. *ASPRS 2006 Annual Conference*, Reno, Neveda.

3

Mapping Impervious Surfaces Using Classification and Regression Tree Algorithm

George Xian

CONTENTS

3.1 Introduction

Urban development is usually associated with the conversion of land in rural areas to residential and commercial land use. As the extent of built-up land increases, further development generally raises concerns about the impacts of land use and land cover (LULC) change on urban and rural environmental conditions and on quality of life. Spatial distributions and patterns of LULC often affect socioeconomic (Douglass, 2000), environmental (Gillies et al., 2003), and regional climatic conditions (Arnfield, 2003; Kalnay and Cai, 2003; Voogt and Oke, 2003). The influences of urban environments on the global population and the monitoring of spatial–temporal changes in large urban and suburban areas are both becoming increasingly important (Small, 2001). The ability to monitor urban LULC changes is highly desired by local communities and by urban management to help provide a more detailed picture of the human-influenced landscape (Carlson, 2003).

Many previous studies focused on qualitative descriptions of conditions in administrative regions, rather than defining regions quantitatively based on urban growth and urbanization. Using administrative definitions of urban extent has many benefits, but physical measurements of urban areas from remotely sensed imagery can provide a self-consistent metric and enable comparative analysis of urban extent in transboundary regions (Davis and Schaub, 2005; Small et al., 2005). Physical measurement metrics overcome such difficulties as capturing changes in the exurban rural fringe, often characterized by scattered, low-density development, and distinguishing between a moderate- and a fast-growth city. Studies have indicated that increased availability and improved quality of multispatial and multitemporal remote sensing data, as well as new analytical techniques, make it possible to monitor urban LCLU changes and urban sprawl in a timely and cost-effective way (Weng, 2001; Seto and Liu, 2003; Wu, 2004).

One successful approach that was designed to map and measure the area of impervious surfaces using remote sensing information was applied to urban land-use estimation at a local or regional scale (Ward et al., 2000; Gillies et al., 2003; Carlson, 2004; Xian and Crane, 2005, 2006; Lu and Weng, 2006). Impervious surface area (ISA) is considered a key indicator of environmental quality and can be used to address complex urban environmental issues, particularly those related to the health of urban watersheds (Schueler, 1994). ISA is also an indicator of non-point source pollution or polluted runoff (Slonecker et al., 2001). ISA is highly connected to urban land-use condition, such as the size and density of built-up areas (Plunk et al., 1990; Morgan et al., 1993; Hebble et al., 2001).

However, urban landscapes are highly heterogeneous, and most urban image pixels in remotely sensed imagery, such as those from Landsat and other similar sensors, are composed of a mixture of different surfaces. Nearly every pixel in an urban area represents a mixture of different land cover types including grass, trees, sidewalks, driveways, roads, and buildings. Pixel-level analysis often creates considerable spectral confusion, especially in residential areas where impervious surfaces are usually mixed with tree canopy and other vegetation coverage (Clapham, 2003). To quantify spatial extents and distribution patterns of urban LULC by using satellite remote sensing data, subpixel analysis is needed (Ji and Jensen, 1999; Yang et al., 2003; Xian et al., 2006). Subpixel techniques break down the mixed pixel into percentages of its components based on spectral characteristics and provide quantifiable measurements of ISA. By selecting different ISA threshold values, subpixel percent ISA data derived from medium-resolution satellite images, such as Landsat imagery, have been used to quantitatively determine the spatial extent and development density of large urban areas (Xian, 2006). Physical measurements of urban areas from remotely sensed data provide a self-consistent metric for urban LULC analysis. Studies of urban LULC have benefited from widely available data sources, which have allowed researchers to overcome data inconsistencies across administration boundaries. The current National Land Cover Database (NLCD) developed by the US Geological Survey (USGS) provides a one-time per-pixel percent imperviousness dataset for the entire United States (NLCD, 2001). This dataset is available for investigations of nation-wide impervious surface distribution.

This chapter investigates the spatial distribution and changes over time of impervious surfaces and associated urban land-use conditions by applying a recently developed subpixel imperviousness assessment model (SIAM) approach (Xian, 2006). Landsat satellite data were used to quantify multitemporal variations of urban spatial extents and development intensities in two geographic locations—Seattle-Tacoma, Washington and Las Vegas, Nevada. Landsat Thematic Mapper (TM) and Enhanced Thematic Mapper Plus (ETM+) images, in combination with high-resolution aerial photographs, were used as the primary source for the estimation of multitemporal subpixel ISA distribution in these two areas. Accuracy has been analyzed by classifying high-resolution orthoimages to calculate true ISA extent and comparing that with model-derived impervious surfaces. In addition, the study summarized general characteristics of ISA and associated urban LULC features, including housing densities and spatial extents of development. Results supported practical applications and were of interest to urban planners and management communities, as well as to natural resource and natural hazards researchers.

3.2 Selection of Study Areas

Two geographically different regions—Seattle-Tacoma, Washington on the western coast and Las Vegas, Nevada in the western inland—were selected as the study areas (Figure 3.1). Both regions have a history of long-term urban development and both have experienced tremendous growth in the past 20 years. Growth management policies have been implemented to prioritize urban development and urban land uses in different zonings in these regions. In addition, high-resolution aerial photos have been routinely collected by the local governments in Seattle-Tacoma and Las Vegas for use in planning and monitoring.

3.2.1 Seattle-Tacoma Area

The first study area was the Seattle-Tacoma region, which lies in the northwest corner of the continental United States, on Puget Sound in western Washington. This area extends about 140 km north to south and 60 km east to west. It encompasses 6700 km^2 of land in Island, King, Kitsap, and Pierce

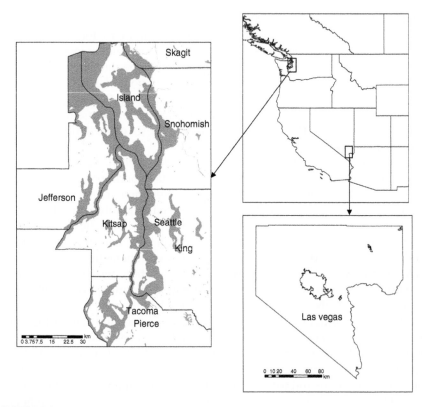

FIGURE 3.1

Seattle-Tacoma metropolitan area (*left*) in western Washington State. The Las Vegas Valley (*low right*) is in the southern part of Nevada.

counties. Major cities include Seattle, Tacoma, Bremerton, and Marysville. More than 3 million people reside in this region.

The Seattle–Tacoma area's temperate climate and growing economy have led to the cities being ranked as some of the most livable in the United States. The corridor of Interstate five connects most major cities in the region. The total population of these cities has almost doubled since 1965. Associated with this population growth is an increase in urbanization and sprawl. In King County between 1970 and 2000, for example, the population increased 44%, from 1.2 to 1.7 million, whereas the number of households increased by 72%, from 400,000 to 680,000 (KCORPP, 2006). Growth management efforts started in 1964 when King County introduced a comprehensive plan. Serious growth management efforts were implemented in the 1985 and 1994 comprehensive plans to manage new growth while meeting economic needs and providing affordable housing (Robinson et al., 2005). The main goals introduced by the growth management plan included encouraging development in urban areas and discouraging inappropriate low-density development. Residential developments in urban areas were zoned for higher residential densities, usually 1–12 dwelling units per acre, whereas rural areas were zoned for lower residential densities, generally 1 dwelling unit per 2.5–10 acres (KCORPP, 2006). Forest and agricultural areas were also zoned for very low residential densities of 1 dwelling unit per 10–80 acres. As a result, urban growth planning in King County had a dual approach: most new growth desired over a 20-year planning period was to be within urban growth boundaries (UGBs), while low-density residential zoning and long-time resource production lands were planned to reduce the potential for new development outside the UGBs.

3.2.2 Las Vegas Valley

The Las Vegas Valley is located in southern Nevada and encompasses about 1320 km^2, including the cities of Las Vegas, Henderson, North Las Vegas, and Boulder City. The area's dry and hot desert climate and its gaming and entertainment facilities attract many people to visit and live in the region. Outside the city, natural vegetation consists of desert flora; inside the city, landscaping with grass, shrubs, and trees is common. However, gravel and bare sandy soils that, on satellite images, appear similar to concrete are also found throughout the urban area (Xian and Crane, 2006). The region has experienced a remarkable increase in urban land use over the past 50 years. According to Census data, the population of Clark County increased from <50,000 in 1950 to slightly >740,000 in 1990 and >1.37 million in 2000 (Clark County, 2006). The population in the Las Vegas Valley urban area reached 1.36 million in 2000 and increased to >1.68 million in 2004. This population increase makes Las Vegas the fastest-growing metro area in the United States (Frey, 2005). Tremendous housing developments have been built to meet the needs of population growth. Total housing units reached 680,897 in 2005 in the Las Vegas Valley urban area. Single-family detached housing and apartments made up 53.3% and 27.6% of total housing units, respectively. Comprehensive planning that focused on zoning-based districts developments has been introduced to manage growth in Clark County. In rural areas of Clark County, residential density is designated as 0.5–3 dwelling units per acre. In the suburban and compact single-family residential districts, dwelling units per acre range from 5 to 14. In the multifamily residential districts, dwelling units per acre range from 18 to 50 (Clark County, 2006).

3.3 Impervious Surface Estimations

The anthropogenic impervious surfaces associated with urban development have been used to assess spatial and temporal variations of urban land use. This study used SIAM, which was developed to map multitemporal subpixel ISA estimates. The method requires high-resolution imagery to create a training dataset and medium-resolution imagery to estimate ISA in a large area through regression tree models. Details of the method have been described in Xian (2006).

3.3.1   Data

Two types of remote sensing data are required to map ISA. One type consists of high-resolution images that usually have at least 1-m spatial resolution, such as 1 m digital orthophoto quarter quadrangles (DOQQ), which are generated from aerial photography. Other high-resolution image sources include QuickBird or 0.3 m orthoimagery. Another type of required data is medium-resolution satellite imagery, such as Landsat, Advanced Land Imager (ALI), and Advanced Spaceborne Thermal Emission and Reflection Radiometer (ASTER), which cover larger areas than DOQQ. High-resolution images are normally used to build training and validation datasets for regression models and to conduct accuracy assessments. Medium-resolution images are used for ISA estimation in a large area.

Generally, the number of high-resolution images selected for a training dataset depends on the size of the study area. We usually select images that can represent major urban LULC features in different locations. However, there is no specific requirement for the maximum number of images. Eight 0.3 m orthoimages acquired from eight different locations in the Seattle metropolitan area were selected. Each Seattle orthoimage covers $\sim$1.6 km $\times$ 1.6 km. Similarly, eight 0.3 m orthoimages from eight different locations were selected for the Las Vegas Valley. Each Las Vegas orthoimage covers $\sim$1.5 km $\times$ 1.5 km. All orthoimages were downloaded from the USGS Seamless Data Distribution System (http://seamless.usgs.gov) as simulated natural color composites in a Universal Transverse Mercator (UTM) projection (zone 10 for Seattle, zone 11 for Las Vegas), referenced to the North American Datum of 1983.

For the Seattle area, four Landsat scenes were selected for path 46, rows 26 and 27: two scenes of Landsat TM from August 26, 1986 and two scenes of Landsat ETM+ from August 14, 2002. For the Las Vegas area, two Landsat scenes were selected from 2002 for path 39, row 35: one Landsat TM scene from April 13, 1984 and one ETM+ scene from June 10. Images for both Seattle and Las Vegas were acquired in clear skies for the entire area to minimize atmospheric scatter effects. All images were preprocessed by the USGS Center for Earth Resources Observation and Science (EROS) to correct radiometric and geometric distortions of the images. Terrain correction was applied using a digital elevation model to correct errors caused by local topographic relief. No atmospheric corrections were made to the Landsat images. Visual inspection showed that the coregistration uncertainty between the Landsat images and orthoimages was within 0.1 m. Slopes derived from USGS 30 m DEM for both Seattle and Las Vegas areas were also used for helping to distinguish imperviousness from other types of land cover.

FIGURE 3.2
A 0.3 m orthoimage from the south of Tacoma (*left*), classification of urban and nonurban (*middle*), and 30 m percent impervious surface (*right*).

3.3.2 Classifications of High-Resolution Images

The orthoimages used to build the training dataset had to be first processed into urban and nonurban land-use classes. This procedure can be done manually through photo interpretation, which is very labor intensive, or through the use of automated feature extraction software. We used Feature Analyst* software, which is able to identify several samples and then run a supervised classification based on them. Once samples were identified and the classification run, the results were all merged into one high-resolution raster dataset. Nonimpervious areas were removed through manual interpretation. The classification accuracy of high-resolution images was over 99%. Pixels classified as urban were then totaled to calculate impervious surface as a percentage and the result was rescaled to 30 m to match Landsat pixels. Figure 3.2 presents results from processing one orthoimage to obtain a 30 m imperviousness training dataset near downtown Tacoma. All buildings and roads were included in measurements of impervious surfaces in each 30 m pixel.

3.3.3 Landsat Imagery

Landsat images in digital number (DN) were first converted to spectral radiance at the sensor and then to at-satellite reflectance using procedures provided by the Landsat Science Data Users Handbook (Landsat Project Science Office, 2006). Reflectance bands 1 through 5 and 7 were used at a spatial resolution of 30 m. Reflectance values from the visible and near-infrared bands of Landsat images were used to compute the Normalized

---

* Use of any trade, product, or company names is for descriptive purposes only and does not imply endorsement by the U.S. Government. Limited information on this program can be found at http://www.featureanalyst.com/.

Difference Vegetation Index (NDVI) values. The Landsat thermal bands had their original pixel sizes of 120 m for TM and 60 m for ETM+ images resampled to 30 m using the nearest neighbor algorithm to match the pixel size of the other spectral bands.

3.3.4   Regression Tree Models

To estimate ISA in a large area, SIAM uses close-to-true percent imperviousness datasets derived from high-resolution imagery as dependent variables in the regression tree models. Landsat reflectance, thermal bands, and derived information such as the NDVI, together with other geographic information such as slope, are then used as independent variables to build regression tree models.

The regression tree model is a machine-learning algorithm. Regression trees are constructed using a partitioning algorithm, which builds a tree by recursively splitting the training sample into smaller subsets. In the partitioning process, each split is made such that the model's combined residual error for the two subsets is significantly lower than the residual error of the single best model. A set of rules are produced for predicting a target variable (percent imperviousness) based on training data. Each rule set defines the condition under which a multivariate linear regression model is established for prediction (Breiman et al., 1984; Quinlan, 1993). Each rule includes three parts: statistical descriptions of the rule, conditions that determine if the rule can be used, and a linear model. The statistical descriptions present the number of cases covered by the rule, the mean range of the dependent variable, and a rough estimate of the error to be expected when this rule is used for new data. The condition for each rule controls the values of independent variables by different thresholds. The linear model is a simplified equation to fit the training data covered by the rule. Models based on the regression tree provide a proposition logic representation of these conditions in the form of number tree rules. Generally, the model can be expressed as

Rule *i*: If conditions for $x_1, x_2, x_3, \ldots, x_n$ are true, then

$$y_i = a_i + \sum_{j=1}^{m} b_j x_j, \tag{3.1}$$

where
  i is the *i*th rule
  x_n are independent variables
  y_i is the dependent variable (percent ISA)
  a_i and b_j are constants
  m is the number of independent variables used in the *i*th rule

The value of *i* ranged from 10 to 20 at different times. Each rule was formed according to the conditions generated from evaluating the training cases.

The main advantages of the regression tree algorithm include simplifying of complicated nonlinear relationships between predictive and target variables into a multivariate linear relation and accepting both continuous and discrete variables as input data for continuous variable prediction. The Cubist regression tree algorithm was used to estimate percent impervious surfaces for the Seattle and Las Vegas areas.

3.3.5 Imperviousness Estimates

Percent impervious surface distributions in 1986 and 2002 in the Seattle area, and 1984 and 2002 in the Las Vegas Valley, were mapped at a subpixel level. Landsat reflectance, NDVI derived from Landsat reflectance and thermal bands, and slope information were all input as independent variables to build regression tree models. NDVI helped to discriminate urban residential land use from rural land in areas where housing was mixed with trees and other vegetation canopy. The Landsat thermal bands, however, were helpful for eliminating nonimpervious areas, especially at the urban fringe of Las Vegas. Slope layers helped eliminate steep areas that were misclassified as urban in the mountain ranges surrounding Las Vegas because most urban areas have developed in valleys or on the lower alluvial flanks of the mountains. Slope also improved ISA estimates for the Seattle area. Due to the uncertainty of the regression model prediction for imperviousness, a 10% threshold was selected for capturing almost all developed land, including low-, medium-, and high-density residential and commercial areas and eliminating most uncertain pixels from urban land use. Pixels with a 30 m spatial resolution were classified as urban when their ISA values were ≥10%. Pixels with ISA values of <10% were classified as nonurban. Furthermore, urban development densities were also defined by different ISA thresholds such as 10%–40% for low-density urban, 41%–60% for medium-density urban, and 60% or higher for high-density urban.

Figure 3.3 represents the spatial distribution of ISA for both regions. The Puget Sound metro region is the most developed in the Seattle area. Total areas with imperviousness ≥10% rose from ~1285 km^2 in 1986 to 2007 km^2 in 2002, representing a 56% increase in urban land use. Many high-density urban areas (ISA >60%) were found within the metro regions. Most new developments occurred on the eastern side of Seattle and the southern part of Tacoma in King and Pierce counties. Table 3.1 presents pixel numbers of three ISA categories and their proportions to the total land area. The low-, medium-, and high-density urban areas took about 9.0%, 6.3%, and 3.7%, respectively, of the total land area in 1986. These percentages increased to 14.8%, 9.4%, and 5.6% in 2002.

In the Las Vegas Valley, the wide distribution of impervious surface is a reflection of how urban development expanded in almost all directions (Figure 3.3). The spatial extent of urban land delineated by the 10% ISA threshold measured ~290 km^2 in 1984 and increased 113% to about 620 km^2

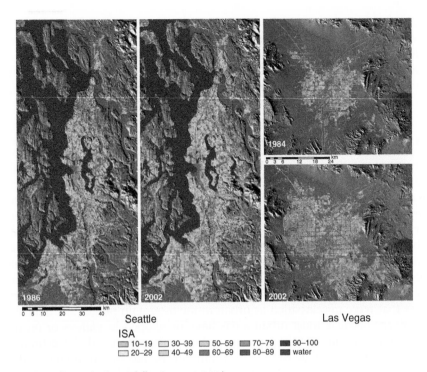

FIGURE 3.3 (See color insert following page 292.)
Impervious surfaces in 1986 and 2002 in Seattle, 1984 and 2002 in Las Vegas.

in 2002. During the 1980s, most medium- to high-percentage ISA were located in the downtown and Las Vegas strip areas. More recently, high-percentage ISA has expanded to the southeast and northwest portions of Las Vegas. Table 3.2 represents numbers of pixels for each imperviousness category and their percentages of total area of the valley. The low-, medium-, and high-density impervious surfaces were summarized as 2.3%, 4.2%, and 3.8%, respectively, of the total land area in 1984. The proportions increased to 4.9%, 7.4%, and 9.5% for the three imperviousness categories in 2002.

TABLE 3.1

Impervious Surface Estimation by Pixel Numbers in the Seattle Area

Year	Low 10%–40% (% of Total Land)	Medium 41%–60% (% of Total Land)	High >60% (% of Total Land)
1986	676,439 (9.0)	470,603 (6.3)	280,187 (3.7)
2002	1,104,363 (14.8)	705,676 (9.4)	419,917 (5.6)

TABLE 3.2

Impervious Surface Estimation by Pixel Numbers in the Las Vegas Area

Year	Low 10%–40% (% of Total Area)	Medium 41%–60% (% of Total Area)	High >60% (% of Total Area)
1984	70,827 (2.3)	130,284 (4.2)	120,038 (3.8)
2002	155,043 (4.9)	233,114 (7.4)	298,609 (9.5)

To evaluate temporal variations of ISA, percent imperviousness was summed in 10% increments and are displayed in Figure 3.4 and Figure 3.5. Figure 3.4 shows that the 20%–49% imperviousness class has the largest numbers of pixels between 1986 and 2002 in Seattle. In addition, the largest ISA increase took place in the 20%–59% category, or low- to medium-development densities, during that period, indicating most new urban growth occurred as low to medium density. Figure 3.5 shows that ISA categories from 50% to 70% comprise ~44% of all pixels between 1984 and 2002 in Las Vegas. However, the dominant fraction of ISA was in the 50%–59% class in 1984 and 60%–69% in 2002. The increase in ISA density reflected a shift toward higher housing development density in the Las Vegas region.

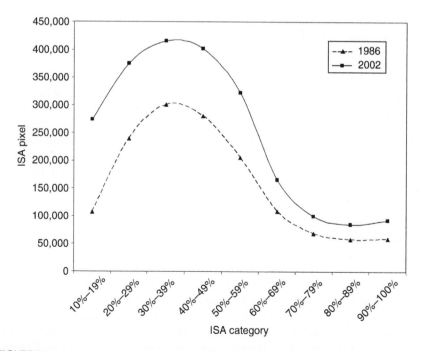

FIGURE 3.4

Percent impervious surface and changes of ISA in different categories between 1986 and 2002 in the Seattle area.

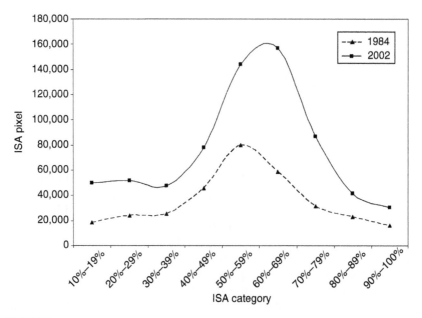

FIGURE 3.5
Percent impervious surface and changes of ISA in different categories between 1984 and 2002 in
the Las Vegas Valley.

3.3.6   Accuracy of ISA Estimations

The regression tree model implemented for the large-area ISA estimation
used several parameters to measure the model's predictive accuracy. These
included average error, which is the average of absolute difference between
model-predicted value and true value, and the correlation coefficient, which
measures the agreement between the actual values of the target attribute
and those values predicted by the model. Evaluation of the training dataset
for the 2002 Seattle ISA estimate calculated an average error of 9.9 and a
correlation coefficient of 0.88. The same parameters measured from the 2002
ISA estimate for Las Vegas were 11.2 and 0.85, respectively. Seattle results
had a higher accuracy than those of Las Vegas.

   To perform accuracy assessment using true ISA data, we compared modeled ISA with digital close-to-true high-resolution orthoimages for selected
locations. In each randomly selected orthoimage, one 5×5 30 m sampling
unit was classified into urban and nonurban land use. The area classified as
urban was divided by the total area of the unit, or 900 m^2, and taken as the
true fraction of imperviousness for that unit. After all selected orthoimages
were processed, comparisons were made between ISA interpreted from
orthoimages and modeled ISA. Details of this method can be found in our
previous study (Xian, 2006). Two statistics—root-mean-square error (RMSE)
and systematic error (SE)—were used to summarize differences between
modeled and true impervious surfaces. RMSE and SE are defined as

TABLE 3.3

Percent ISA and Accuracies for Each Sample Site

Sample Site	Seattle	Las Vegas
Mean ISA measured from DOQQ ISA	33.8	62
Modeling ISA% from Landsat	34	57
Mean SE	0	-6.0
Mean RMSE	15.10	16.10

$$\text{RMSE} = \sqrt{\frac{\sum\limits_{i=1}^{N}\left(\hat{U}_i - U_i\right)^2}{N}}, \tag{3.2}$$

$$\text{SE} = \frac{1}{N}\sum\limits_{i=1}^{N}\left(\hat{U}_i - U_i\right), \tag{3.3}$$

where
$\hat{U}_i$ is the modeled ISA for sample i
U_i is the true ISA data for sample i
N is the total number of samples

The values of RMSE and SE for seven and six randomly selected ortho-images were calculated for the Seattle and Las Vegas areas, respectively.

Table 3.3 represents accuracy assessment results for two regions. The modeled ISA for Seattle area had values of 0.0 for mean SE and 15.10 for mean RMSE. Values of mean SE and mean RMSE were −6.0 and 16.10 for Las Vegas. The ISA estimate in Seattle had a relatively higher accuracy than that in Las Vegas, where ISA was slightly underestimated. The reason for this lower estimate for Las Vegas might be caused by the region's landscaping gravel and rocks, bare alluvial soils, and the surrounding gravel in rural areas. These materials have a similar appearance to concrete and had brighter reflectance than some urban buildings. This might cause significant confusion in regression tree models and produce lower imperviousness values for the region.

3.4 Urban Land-Use Density and Percent Imperviousness

Urban development typically converts natural land into impervious surfaces. The urban landscape can be characterized in terms of ISA, vegetation covered area, and others such as bare soil and water. The impervious surface at the subpixel level can represent the spatial sizes of most dwelling units in residential areas. Furthermore, the fraction of imperviousness in a parcel can be used to represent the density of built-up areas in urban zones.

By choosing thresholds of subpixel impervious surfaces to define urban land-use boundaries and then calculate changes in built-up land, the characteristics of urban extent and land-use change can be objectively quantified for any metropolitan region. Many of the specific decisions that influence urban development, such as built-up density, extent, and building form, are often implemented at the development, tract, or individual parcel level. The physical measurement of ISA from remote sensing data provides a quantitative and self-consistent metric, which can avoid the considerable difficulties that result from different administrative and political definitions of urban and nonurban land use, as well as associated land-use changes for multimunicipal areas.

The differences in impervious surface density displayed in Figures 3.4 and 3.5 suggest apparent and distinct patterns of urban development density and temporal variations in these two metropolitan areas. In Seattle, the predominant development densities were between 30% and 50% imperviousness. However, the major development densities were higher in the Las Vegas area, ranging between 50% and 70%. These imperviousness densities are determined by dwelling unit densities and represent intensities of built-up land for the area. For an area with the same size of one Landsat 30 m $\times$ 30 m pixel, which has 900 m^2 or 9687.5 ft^2, 40% imperviousness coverage, for instance, represents 3874.8 ft^2 of built-up land. For an area zoned as 1–12 dwelling units per acre and each lot is 8100 ft^2 (which is common in the Seattle region), the maximum built-up area would take ~45% (1 acre/8100/12) of each lot.

To demonstrate the percent of built-up area in each planning lot and its possible connection to percent imperviousness, Figure 3.6 presents a high-resolution aerial photo of a parcel on the northeastern side of Seattle, where developments were zoned for high-density residential land use as at least 1–12 dwelling units per acre. Each lot designated for single-family housing is 8100 ft^2. Most housing units and associated sidewalks consumed about 40%–50% of each lot. As a result, the total built-up land in this section was estimated as <50% as interpreted from the photography. In addition, the 2002 impervious surface map obtained from Landsat imagery was precisely subset for this section and mean imperviousness was calculated as 42.6%, which was very close to the estimate from zoning information. Housing units shown in the photography represented a medium density of ISA category in the region.

Local growth management also zoned urban areas with different density developments for the Seattle area (Robinson et al., 2005). In King County, for example, besides high-density residential areas zoned for urban area, there were rural areas zoned for low-density residential area as 1 dwelling unit per 2.5–10 acres. New developments were prioritized to increase housing density within UGBs and to limit housing densities outside these boundaries. Data showed that land committed to construction of residential housing ranged from 72% to 85% in the 1985–1994 and 1999–2001 periods,

FIGURE 3.6
Aerial photography from Seattle urban area. The lots on the patch are outlined by white lines. The mean ISA value of the patch is 42.6%.

respectively (Robinson et al., 2005). Seattle's growth management strategy resulted in widespread, low-density single-family residential development outside the UGBs in the region. The change pattern measured from the subpixel ISA estimate also indicated that the maximum increase was in the 30%–40% ISA category from 1986 to 2002. This increase was believed to correspond to the land-use trend resulting from growth management of the region.

Las Vegas urban land use was highly restricted by local geographic conditions where most lands suitable for urban development were located in the valley. Urban planning was implemented through the comprehensive plan that divided the county into zoning districts, including rural residential, suburban and compact residential, multiple-family residential, commercial, manufacturing and industrial, and special districts (Clark County, 2006). Within the suburban and compact single-family residential district, the

FIGURE 3.7
Aerial photos from Las Vegas urban area. *Left*: Older housing built in the 1980s. *Right*: New developments built in the 2000s. Mean ISA values for left and right are 57.7% and 61.5%, respectively.

development densities were designated as 5 dwelling units per acre with lot area of 5200 ft^2, 8 dwelling units per acre with lot area of 3300 ft^2, and 14 dwelling units per acre with lot area of 2000 ft^2. This development policy determined that urban area development density in Las Vegas was higher than that in Seattle.

The spatial distribution and temporal change patterns of estimated sub-pixel ISA in Figure 3.5 suggest that imperviousness categories of 50%–59% in 1984 and 60%–69% in 2002 contain the largest parts of the region. The increase in ISA category during this period resulted in a higher modeled built-up density in 2002. To illustrate features of urban development density in the area, two aerial photos from the central and southeast sectors of the Las Vegas metropolitan area are displayed in Figure 3.7. Photos of two residential areas show two types of housing buildings in the 1980s (*left*) and 2000s (*right*). Calculation from mapped ISA indicated that the 1980s housing sector had a mean imperviousness of 57.7%. The mean imperviousness for the 2000s built-up sector increased to 61.6%. The percent imperviousness estimated from Landsat TM and ETM+ imagery quantitatively revealed both the extent of urban development and its temporal transition patterns, which were affected by local and regional growth management.

3.5 Discussion

The spatial extents, density distributions, and temporal variations of sub-pixel percent impervious surface estimates from remote sensing and slope information revealed several important characteristics of urban land use and growth in two metropolitan areas.

The imperviousness mapping depicted a clear link between urban development policy and real land-use patterns. Developments in the Seattle area have been designated for low-density residential development in rural and wildland areas outside of UGBs while keeping high-density housing growth within UGBs. In Puget Sound between 1990 and 2000, 55% of new growth— about 253,000 new residents—settled in low-density areas with fewer than 12 people per acre (Davis and Schaub, 2005). This zoning policy has resulted in widespread, low-density single-home residential developments in the region. Associated with this land-use pattern were predominantly low- to medium-density impervious surfaces distributed in the region. The ISA temporal change pattern, in which the percentage of low-density ISA increased from 47.4% to 49.5% from 1986 to 2002, represented the characteristics of urban land use and reflected the implementation of the growth policy in the region.

The ISA mapping for the Las Vegas Valley reflected a high-density development policy in the region. In 1984, the largest proportion (40.6%) of ISA was in the medium-density category. However, in 2002, the largest ISA category (43.5%) was in the high-density category. The encouragement of high-density development led to a dominant high-density category of imperviousness in the region.

The impervious surface metric, which relies on satellite imagery, provides a uniform data source, which is unchanged across political jurisdictions. The method and concept are useful for regional scale analyses and should be applicable to other forms of urban landscape delineation, such as individual developments and neighborhoods. Planners could monitor the physical manifestations of planning policies within cities by integrating quantitative analysis of remote sensing data with zoning information. In addition, urban land-use conditions derived from ISA status can be verified without accessing local administrative data, which may not be available for many analyses. This advantage is very important when implementing a multicity or a regional analysis, which usually involves different city- or county-level administrations and when data availability and reliability are major concerns.

Building an optimistic regression tree model is very important for accurately estimating impervious surfaces in large areas. A large effort was needed for the development of good training data from high-resolution imagery. Appropriate numbers of high-resolution images that represent major features of urban land use in the study area are needed. Another important procedure is to select independent variables to construct a good regression tree model. Currently, spectral reflectance and several ancillary data are used as independent variables to build the model. Other data such as the socioeconomic information including population data may also be valuable for a better regression tree model.

3.6 Conclusion

The characteristics of urban land use are highly influenced by local geographic conditions, social and economic forces, and land management policies. Policies that encourage levels of development density within urban or rural areas have clear implications for the urban ecosystem. This study demonstrates that subpixel impervious surfaces estimated from satellite remote sensing data and regression tree models provide a regionally consistent and reliable source to assess the size, spatial distribution, and temporal change of urban land use. Quantitative analysis of ISA distributions for two geographically distinct areas reveals that subpixel impervious surfaces are closely related to urban land-use patterns and variation trends. Urban extent in the Seattle area had a moderate growth rate of 56% from 1986 to 2002. In the Las Vegas Valley, the growth rate was much faster and reached 113% from 1984 to 2002. Subpixel ISA values derived from satellite data captured distinct patterns of development density for these areas, in which two different development policies have been implemented. Variations quantified from subpixel impervious surfaces also revealed urban development patterns that have been influenced by different public policies in these two metropolitan regions.

Acknowledgments

This research was performed for SAIC under USGS contact O3CRCN0001. Thanks to Drs. David Meyer and Gabriel Senay for their comments and suggestions, which improved the manuscript. I also thank Brian Granneman for processing some of the high-resolution images.

References

Arnfield, A.J., Two decades of urban climate research: A review of turbulence, exchanges of energy and water, and the urban heat island, *International Journal of Climatology*, 231, 1, 2003.

Breiman, L., Friedman, J., Olshen, R., and Stone, C., *Classification and Regression Trees*, Wadsworth International Group, Belmont, CA, 1984, p. 358.

Carlson, T.N., Applications of remote sensing to urban problems, *Remote Sensing of Environment*, 86, 273, 2003.

Carlson, T.N., Analysis and prediction of surface runoff in an urbanization watershed using satellite imagery, *Journal of the American Water Resources Association*, 40, 1087, 2004.

Clapham, W.B. Jr., Continuum-based classification of remotely sensed imagery to describe urban sprawl on a watershed scale, *Remote Sensing of Environment*, 86, 322, 2003.

Clark County, *Comprehensive Planning, Clark County, Nevada*. Accessed on 16 August 2006 at http://www.co.clark.nv.us/Comprehensive-planning.

Davis, C. and Schaub, T., A transboundary study of urban sprawl in the Pacific Coast region of North America: The benefit of multiple measurement methods, *International Journal of Applied Earth Observation and Geoinformation*, 7, 268, 2005.

Douglass, M., Mega-urban regions and world city formations: Globalization, the economic crisis and urban policy issues in Pacific Asia, *Urban Studies*, 37, 2315, 2000.

Frey, W.H., *Metro America in the New Century: Metropolitan and Central City Demographic Shifts since 2000*, Brookings Institution, Washington, 2005. Accessed on September 16, 2005 at http://www.brookings.edu/metro/pubs/.

Gillies, R.R., Box, J.B., Symanzik, J., and Rodemaker, E.J., Effects of urbanization on the aquatic fauna of the Line Creek watershed, Atlanta—a satellite perspective, *Remote Sensing of Environment*, 86, 411, 2003.

Hebble, E.E., Carlson, T.N., and Daniel, K., Impervious surface area and residential housing density: A satellite perspective, *Geocarto International*, 16, 13, 2001.

Ji, M.H. and Jensen, J.R., Effectiveness of subpixel analysis in detecting and quantifying urban imperviousness from Landsat Thematic Mapper imagery, *Geocarto International*, 14, 31, 1999.

Kalnay, E. and Cai, M., Impact of urbanization and land-use change on climate, *Natural*, 423, 528, 2003.

KCORPP (King County Office of Regional Policy and Planning), *The Annual Growth Report 2000*, King County, Washington. Accessed on August 10, 2006 at http://www.metrokc.gov/budget/agr/agr00/.

Landsat Project Science Office, *Landsat 7 Science Data Users Handbook*, 2006. URL: http://ltpwww.gsfc.nasa.gov/IAS/handbook/handbook_toc.html; last accessed June, 2006.

Lu, D. and Weng, Q., Use of impervious surface in urban land-use classification, *Remote Sensing of Environment*, 102, 146, 2006.

Morgan, K.M., Newland, L.W., Weber, E., and Busbey, A.B., Using spot satellite data to map impervious cover for urban runoff predictions, *Toxicological and Environmental Chemistry*, 40, 11, 1993.

NLCD 2001, National Land Cover Database 2001. Accessed on August 16, 2006 at http://www.mrlc.gov/mrlc2k.nlcd.asp.

Plunk, D.E., Morgan, K., and Newland, L., Mapping impervious cover using Landsat TM data, *Journal of Soil and Water Conservation*, 45, 589, 1990.

Quinlan, J.R., *C4.5: Programs for Machine Learning*, Morgan Kaufmann, San Mateo, CA, 1993, p. 302.

Robinson, L., Newell, J.P., and Marzluff, J.M., Twenty-five year of sprawl in the Seattle region: Growth management responses and implication for conservation, *Landscape and Urban Planning*, 71, 51, 2005.

Schueler, T.R., The importance of imperviousness, *Watershed Protection Techniques*, 1, 100, 1994.

Seto, K. and Liu, W., Comparing ARTMAP neural network with the maximum-likelihood classifier for detecting urban change, *Photogrammetric Engineering and Remote Sensing*, 69, 981, 2003.

Slonecker, E.T., Jennings, D.B., and Garofalo, D., Remote sensing of impervious surfaces: A review, *Remote Sensing Review*, 20, 227, 2001.

Small, C., Estimation of urban vegetation abundance by spectral mixture analysis, *International Journal of Remote Sensing*, 22, 1305, 2001.

Small, C., Pozzi, F., and Elvidge, C.D., Spatial analysis of global urban extent from DMSP-OLS night lights, *Remote Sensing of Environment*, 96, 277, 2005.

Voogt, J.A. and Oke, T.R., Thermal remote sensing of urban climates, *Remote Sensing of Environment*, 86, 370, 2003.

Ward, D., Phinn, S.R., and Murray, A.T., Monitoring growth in rapidly urbanized areas using remotely sensed data, *Professional Geographer*, 52, 371, 2000.

Weng, Q., A remote sensing-GIS evaluation of urban expansion and its impact on surface temperature in the Zhujiang Delta, China, *International Journal of Remote Sensing*, 22, 1999, 2001.

Wu, C., Normalized spectral mixture analysis for monitoring urban composition using ETM+ imagery, *Remote Sensing of Environment*, 93, 480, 2004.

Xian, G., Assessing urban growth with sub-pixel impervious surface coverage, *Urban Remote Sensing*, Weng and Quattrochi, Eds., Taylor & Francis Group, Boca Raton, FL, 2006, p. 179.

Xian, G. and Crane, M., Assessments of urban growth in the Tampa Bay watershed using remote sensing data, *Remote Sensing of Environment*, 97, 203, 2005.

Xian, G. and Crane, M., An analysis of urban thermal characteristics and associated urban land cover in Tampa Bay and Las Vegas using Landsat satellite data, *Remote Sensing of Environment*, 104, 2, 147, 2006.

Xian, G., Yang, L., Klaver, J.M., and Hossain, N., Measuring urban sprawl and extent through multi-temporal imperviousness mapping, *Rates, Trends, Causes, and Consequences of Urban Land-Use Change in the United States, U.S. Geological Survey Professional Paper*, 1726, 2006, p. 59.

Yang, L., Xian, G., Klaver, J.M., and Deal, B., Urban land-cover change detection through sub-pixel imperviousness mapping using remotely sensed data, *Photogrammetric Engineering and Remote Sensing*, 69, 1003, 2003.

4

Mapping Urban Impervious Surfaces from Medium and High Spatial Resolution Multispectral Imagery

Dengsheng Lu and Qihao Weng

CONTENTS

4.1   Introduction

Research on impervious surface extraction from remotely sensed data has attracted interest since the 1970s. During the 1970s and 1980s, much research on impervious surface extraction was based on aerial photographs [1,2]. In the past decades, research was shifted to develop more advanced approaches for quantitative impervious surface extraction from satellite images. The approaches include per-pixel image classification [3–5], sub-pixel classification [6–9], and decision tree modeling [10–13]. Extractions of impervious surface have also been conducted by the combination of high-albedo and low-albedo fraction images [14–16] and by establishing the relationship between impervious surfaces and vegetation cover [17,18]. Because of the complexity of impervious surface materials and the spectral confusion between impervious surfaces and other land covers, extraction of impervious surface from remotely sensed data is still a challenge. The object-ive of this research is to extract impervious surface from Landsat ETM+ image through the integration of land surface temperature (LST) and fraction images. Specifically, the combination of the LST and low-albedo fraction is used to extract the dark impervious surface contents existing in the low-albedo fraction images. Another objective is to explore the approach to extract impervious surface areas from IKONOS data, through the use of a hybrid approach based on the combination of decision tree classifier (DTC) and unsupervised classification to extract the dark impervious surface and other shadowed impervious surface areas.

4.2   Study Area and Dataset

Marion County/the City of Indianapolis, Indiana, United States, is chosen as the study area (Figure 4.1). Indianapolis, the state capital, is a key center for manufacturing, warehousing, distribution, and transportation in the state. Indianapolis has almost 800,000 population, and the city is the nation's 12th largest one. It possesses several other advantages that make it an appropriate choice for this study. It has a single central city and other large urban areas in the vicinity have not influenced its growth. The city is located in a flat plain and is relatively symmetrical, with possibilities of expansion in all directions. Like most American cities, Indianapolis is increasing in both population and extent. The areal expansion is through encroachment into the adjacent agricultural and nonurban land. Certain decision-making

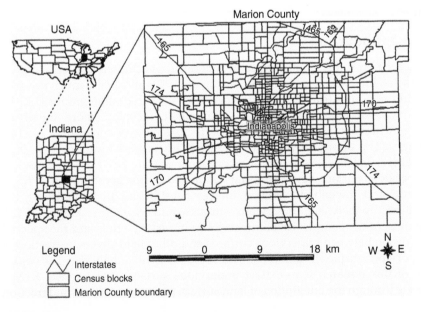

FIGURE 4.1
The study area—Marion County, Indiana State, United States.

forces such as density of population, distance to work, property value, and income structure encourage some sectors of the metropolitan Indianapolis to expand faster than others.

A Landsat 7 ETM+ image (path 21/row 32) of Marion County, Indiana, which was acquired on June 22, 2000 under clear weather conditions, was used in this research. The ETM+ image has one panchromatic band with 15 m spatial resolution, six reflective bands with 30 m spatial resolution, and one thermal infrared band with 60 m spatial resolution. The ETM+ data were geometrically rectified with 1:24,000 topographic maps. The root-mean-square error (RMSE) during image rectification was <0.2 pixels. A nearest-neighbor algorithm was used to resample the ETM+ image (including the six reflective bands and the thermal infrared band) to a pixel size of 30 m $\times$ 30 m during image rectification. The ETM+ reflective bands were used to develop fraction images with SMA, and the thermal infrared band was used to compute LST. The ETM+ panchromatic band was not used for the research. No atmospheric calibration was conducted for the ETM+ image, because previous research had demonstrated that atmospheric calibration did not have an effect on fraction images when image endmembers were used [19].

Orthophotographs were used in this research for validation of impervious surface estimation results. The color orthophotographs were provided by the Indianapolis Mapping and Geographic Infrastructure System, which was acquired in April 2003 for the entire Marion County. The orthophotographs have a spatial resolution of 0.5 ft (i.e., 0.14 m). The coordinate system belongs to Indiana State Plane East, Zone 1301, with North American Datum of 1983. The orthophotographs were reprojected and resampled to 1 m pixel size for the sake of quicker display and shorter computing time.

In order to explore the extraction of impervious surface using IKONOS data, a typical urban study area in Indianapolis, Indiana, was selected. Different urban land uses, such as commercial areas, different intensities of residential areas, forest, grass, and rivers can be found in this study area. It is an ideal area for research on extraction of urban impervious surface areas. The IKONOS images were acquired on October 6, 2003 in a cloud-free condition. The IKONOS images have four bands in blue, green, red, and near-infrared wavelengths with 4 m spatial resolution and they have one panchromatic band with 1 m spatial resolution. The 4 m IKONOS multi-spectral image was used in this research.

4.3   Mapping of Impervious Surface with Landsat ETM+ Data

Spectral mixture analysis (SMA) is regarded as a physically based image-processing tool, which supports repeatable and accurate extraction of quantitative subpixel information [20]. The SMA approach assumes that the spectrum measured by a sensor is a linear combination of the spectra of all components (endmembers) within the pixel and the spectral proportions of the endmembers represent proportions of the area covered by distinct features on the ground [20,21]. The mathematical model of SMA can be expressed as

$$R_{il} = \sum_{k=1}^{n} f_{kl} \, R_{ik} + \varepsilon_{il}, \tag{4.1}$$

where
$i = 1, \ldots, m$ (number of spectral bands)
$k = 1, \ldots, n$ (number of endmembers)
$l = 1, \ldots, p$ (number of pixels)
R_{il} is the spectral reflectance of band i of a pixel, which contains one or more endmembers
f_{kl} is the proportion of endmember k within the pixel
R_{ik} is known as the spectral reflectance of endmember k within the pixel on band i
ε_{il} is the error for band i in pixel l

For a constrained unmixing solution, f_{kl} is subject to the following restrictions:

$$\sum_{k=1}^{n} f_{kl} = 1 \text{ and } 0 \leq f_{kl} \leq 1, \tag{4.2}$$

One critical step in the SMA approach is to select suitable endmembers. In practice, image-based endmember selection approaches are often used, because image endmembers can be easily obtained and they represent the spectra measured at the same scale as the image data. Image endmembers can be derived from the extremes of the image feature space, assuming that they represent the purest pixels in the images [22]. In this research, image endmembers were selected from the feature spaces of minimum noise fraction (MNF) components. The MNF transform approach was applied to transform the ETM+ six reflective bands into a new dataset (Figure 4.2). The first three components account for the majority of the information (~99%) and were used for assisting the selection of four endmembers: vegetation, high albedo, low albedo, and soil (Figure 4.3). A constrained least-square solution was then used to unmix the six ETM+ reflective bands into the four fraction images (Figure 4.4). The high-albedo fraction image mainly reflects the bright impervious surfaces in the urban landscape. The low-albedo fraction image is more complex than other fraction images because it contains different objects, such as water, building shadows in the urban landscape, vegetation canopy shadows in forested areas, and dark impervious surfaces. The soil fraction image reflects the soil contents, which is mainly located in the agricultural areas. The vegetation fraction image represents the vegetation information from forest, pasture, grass, and crops. The fraction images indicate that the impervious surfaces are concentrated on

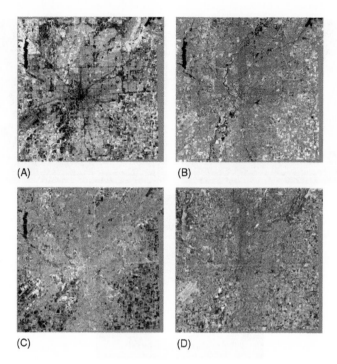

(A)   (B)

(C)   (D)

FIGURE 4.2
Four components derived from a minimum noise fraction transformation of Landsat ETM+ image. (From Lu, D. and Weng, Q., *Remote Sensing Environ.*, 102, 146, 2006. With permission.)

the high- and low-albedo fraction images. It is important to remove the nonimpervious surfaces from both high- and low-albedo fraction images to accurately extract the impervious surfaces.

LST was developed from the Landsat ETM+ thermal band, and the approach for LST extraction was described in our previous work [23]. The temperature is gradually decreased from the highest values in commercial areas, to medium values in high- and medium-intensity residential areas, to the lowest values in nonurban areas such as in forested lands and water (Figure 4.5). The cooling effects from water and forests can effectively reduce the temperature, whereas large impervious surface proportions in urban areas have the heating effects resulting in urban heat island effects. The different LST features between impervious surface and other land covers provide the fundamental basis for the distinction of impervious surface from other land covers. In order to remove nonimpervious surfaces (i.e., water, forest, pasture, grass, and crops) from low-albedo and high-albedo fraction images, it is necessary to develop some criteria, such as "If LST < t, this pixel in low-albedo or high-albedo fraction images is assigned to 0 because it is water or vegetation (such as forest, crops), otherwise, keep the value in low-albedo or high-albedo fraction." The threshold t can be identified based on the reference data of water and

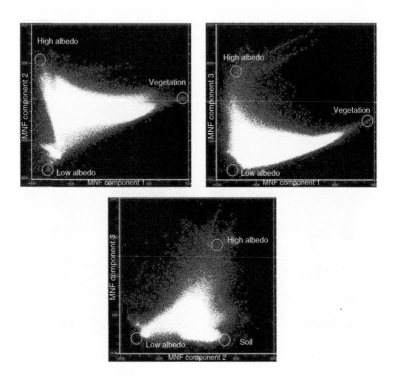

FIGURE 4.3
Feature spaces between the MNF components, illustrating potential endmembers of Landsat ETM+ image. (From Lu, D. and Weng, Q., *Remote Sensing Environ.*, 102, 146, 2006. With permission from Elsevier.)

vegetation. After nonimpervious surfaces were removed from the low-albedo and high-albedo fraction images, the impervious surface image was then derived based on the addition of the adjusted low-albedo and high-albedo fraction images (Figure 4.6). It indicated that the gradient impervious surface change is obvious, from highest in commercial areas in the center of this study area, to medium in the high-intensity residential areas, to the lowest in the low-intensity residential areas.

In order to compare the result obtained in this research with previous work, the same accuracy assessment approach was used here [15], that is, the RMSE and system error (SE).

$$\text{RMSE} = \sqrt{\frac{1}{n} \sum_{i=1}^{n} (v_e - v_r)^2}, \tag{4.3}$$

$$\text{SE} = \frac{1}{n} \sum_{i=1}^{n} (v_e - v_r), \tag{4.4}$$

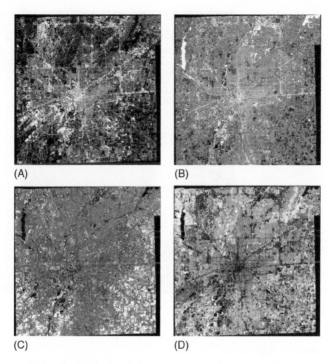

FIGURE 4.4
Four fraction images from spectral mixture analysis of six ETM+ reflective bands (the fraction values range from 0 to 1, with the lowest values in black and the highest values in white in the fraction images) (A: High albedo; B: Low albedo; C: Soil; and D: Vegetation). (From Lu, D. and Weng, Q., *Remote Sensing Environ.*, 102, 146, 2006. With permission from Elsevier.)

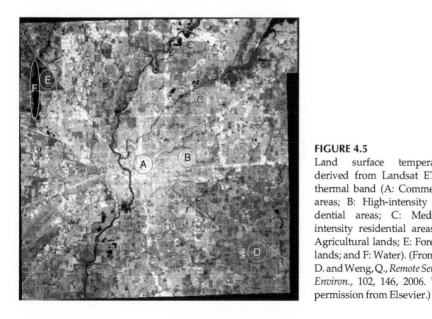

FIGURE 4.5
Land surface temperature derived from Landsat ETM+ thermal band (A: Commercial areas; B: High-intensity residential areas; C: Medium-intensity residential areas; D: Agricultural lands; E: Forested lands; and F: Water). (From Lu, D. and Weng, Q., *Remote Sensing Environ.*, 102, 146, 2006. With permission from Elsevier.)

FIGURE 4.6
Impervious surface image developed from ETM+ data based on the combination of land surface temperature and fraction images.

where

  n is the number of samples

  V_e and V_r are estimated values from the sample plot on ETM+-derived impervious surface image and the reference values from digitizing impervious surface polygons on the same sample plot of the orthophotographs.

A total of 76 samples with 3×3 pixel size (90 m $\times$ 90 m in ETM+ data) were designed and digitized on the orthophotographs. Because of the 3-year difference between ETM+ data and orthophotographs, a careful examination of each sample plot between ETM+ color composite and orthophotographs found that seven sample plots have obvious changes and thus they were removed from the accuracy assessment. Thus, 69 samples were used for assessment of impervious surface estimation image. Based on the 69 samples, the overall RMSE with 9.22% and SE with 5.68% were obtained. When impervious surface is <30%, the RMSE becomes 9.98% and SE becomes 8.59%, but when impervious surface is ≥30%, RMSE is 8.36% and SE is 2.77%. This result has been much improved compared with previous research [15].

4.4 Mapping of Impervious Surface with IKONOS Data

Different colors or kinds of impervious surfaces have their specific reflectance values. Based on spectral features, impervious surfaces can be grouped as low- and high-reflectivity impervious surface (LRIS and HRIS). LRIS is defined here as dark impervious objects with low reflectance, especially in visible and near-infrared bands, thus LRIS objects appeared as dark gray or black in IKONOS color composite. HRIS is defined here as the impervious objects with high reflectance, thus they appeared white in the composite. HRIS usually has significantly different spectral features when compared with other land covers. Thus, it can be easily extracted. The challenge is to extract LRIS from other land covers because the LRIS, such as some dark building roofs and roads, has similar spectral features with water bodies and shadows cast by buildings or tree canopy. Previous research has shown the difficulty in separating them based on spectral signatures with traditional classification approaches such as maximum likelihood classifier (MLC) [24,25].

Classification approaches, such as MLC [26] and DTC [11,27], are often used to extract impervious surface areas. The MLC is a parametric classifier, which assumes normal distribution for each feature of interest, associated with an equal prior probability among the classes. Hence, training samples that have insufficient or nonrepresentative features of interest or with multimode distributions often lead to poor classification results because of inaccurate estimation of the mean vector and the covariance matrix used in the MLC algorithm. In urban landscapes, the assumption of normal distribution was often violated and the selection of training sample plots was often difficult because of the complex urban environments. Much previous research has indicated that nonparametric classifiers provide better classification results than parametric classifiers in complex landscapes [28]. The nonparametric classifiers, such as DTC and neural network, have attracted increased attention in urban land-cover classification. Hence, the DTC is used in this research for extraction of impervious surface areas. Many previous works have detailed the description of DTC [11,29,30].

The strategy for extraction of impervious surface areas is illustrated in Figure 4.7. A DTC is used to classify the IKONOS spectral signatures into two reflectance levels of impervious surface areas, and the confusion between some water bodies and shadows cast by tall buildings is further analyzed with an unsupervised ISODATA approach. Vegetation is first separated from nonvegetation types based on NDVI, and HRIS from others based on the visible bands. The confused pixels (some dark impervious surface, shadows from tall buildings, and water) are masked out and classified using unsupervised ISODATA. The clusters are then merged into water or shadowed impervious surfaces based on the analysis of overlaying the classified image on a color composite. Finally, LRIS, HRIS, and shadowed impervious surface are recoded as impervious surface class (Figure 4.8).

Accuracy assessment is an important part in image classification. Different elements for accuracy assessment, such as overall accuracy and producer's and user's accuracy, can be used and they can be calculated from an

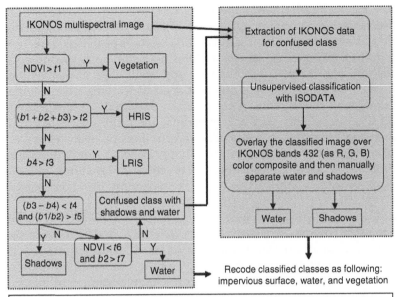

Note: *b1*, *b2*, *b3*, and *b4* represent IKONOS bands 1, 2, 3, and 4; NDVI is the normalized difference of vegetation index; HRIS and LRIS are high- and low-reflectivity impervious surfaces; t1, ..., t7 are thresholds developed from training samples

FIGURE 4.7
Strategy of impervious surface extraction based on a combination of decision tree classifier and unsupervised classification on IKONOS spectral signatures.

FIGURE 4.8
Extraction of impervious surfaces from IKO-NOS data with a hybrid approach based on a combination of decision tree classifier and unsupervised classification.

Land covers
☐ Impervious surface
■ Vegetation
■ Water

N

1 0 1 2 3 km

TABLE 4.1

Comparison of Different Approaches for Extracting Impervious Surface Areas

Method	Type	ISA	WAT	VEG	PA%	UA%	OA%
MLC	ISA	77	2	3	93.9	89.5	89.3
	WAT		13		100.0	76.5	
	VEG	9	2	44	80.0	93.6	
DTC	ISA	83	1	2	96.5	96.5	96.0
	WAT		16		100	94.1	
	VEG	3		45	93.8	95.7	

Note: ISA, WAT, and VEG represent impervious surface area, open water bodies, and vegetation, including forest and associated tree crown shadows, and grass; PA, UA, and OA represent producers' accuracy, user's accuracy, and overall classification accuracy; MLC and DTC represent maximum likelihood classifier and decision tree classifier.

error matrix. Many previous literatures have detailed the approaches for accuracy assessment [31–33]. In this study, a total of 150 sample plots were randomly selected for accuracy assessment. The land-cover class for each sample plot was visually interpreted on the color composite of IKONOS bands 4, 3, and 2. An error matrix was then developed for each classifier, that is, MLC and DTC. Finally, producer's and user's accuracy for each class and overall accuracy for each classifier were calculated based on the error matrix (Table 4.1). In urban environments, the land-use/cover classification with a traditional per-pixel classifier such as MLC is a challenge, even using high spatial resolution images such as IKONOS data. Use of nonparametric classifier—DTC improved classification accuracy. In the MLC approach, some dark impervious surface or shadows cast by tall buildings in the impervious surface areas were confused with water or with shadows cast by tree crowns in the vegetation areas. In the DTC approach, this confusion was significantly reduced. Both producer's and user's accuracies for DTC approach were 96.5%, compared with 93.9% and 89.5% for MLC. This research indicated that selection of a suitable classification approach was important for producing accurate classification results.

___

4.5 Discussions

This research showed that impervious surface was overestimated in the less-developed areas, but was underestimated in the well-developed areas. This situation is the same as previous research [15]. In the less-developed areas, such as medium- and low-intensity residential lands, impervious surfaces are often mixed with vegetation and soil. However, bright impervious surface objects have higher reflectance values than vegetation, especially in the visible and shortwave infrared bands, exaggerating their area proportions in the mixed pixels. Another factor is that dark impervious surface objects and water or shadows have similar spectral characteristics and similar values in the low-albedo fraction image. Dark impervious surfaces from the low-albedo fraction image may be overextracted because the extraction is solely based on the difference in LST. The relatively low spatial resolution of Landsat ETM+ thermal band smooths boundaries between impervious surfaces and other cover types. In the well-developed areas, such as high-intensity residential lands, similar spectral responses in some bright impervious surfaces and dry soils created confusion. Tree crowns that cover portions of some impervious surfaces, such as roads, may be another factor causing underestimation of impervious surfaces. This problem is much improved in high spatial resolution IKONOS image because of less influence from the mixing pixel. However, many impervious surface areas may be in the shadows caused by tall buildings and large tree crowns. The shadow problem that occurred in the high spatial resolution image is a challenge in impervious surface extraction. To date, there is lack of an effective approach to remove the shadow impacts. The shadow problem also affects the extraction of impervious surface areas. In medium spatial resolution images, such as Landsat TM/ETM+, or Terra ASTER, use of a thermal band is an effective way of separating dark impervious surface from water, or shadows on impervious surface areas from the shadows on vegetated area because of the different LSTs [16]. However, high spatial resolution images often lack thermal bands and have limited spectral bands. This research used an unsupervised classification to separate the dark impervious surface, water, and shadows because an analyst can effectively use his knowledge in the study area during merging the clusters into meaningful classes and can use different features, such as textures and patterns of land-use distributions, inherent in the high spatial resolution images.

The importance of thermal image in assisting impervious surface extraction has not attracted sufficient attention yet. One limitation of the thermal image is its relatively low spatial resolution. For example, the spatial resolution of thermal image is 120 m in TM and 60 m in ETM+ data, whereas their reflective images are 30 m. The low spatial resolution image contains much mixed information, which cannot provide a clear boundary of the temperature differences between impervious surface and other land covers.

Selection of a suitable image acquisition date is also important for extraction of impervious surface. We had explored to extract impervious surface areas in different seasons using ASTER images acquired in October 2000, June 2001, March 2002, and ETM+ images acquired in June 2000 and April 2003 (unpublished results). We found that the images acquired in vegetation-growing seasons are suitable for impervious surface extraction, but the images acquired in spring or maybe in winter seasons are not. In the vegetation-growing season, vegetation and crops in agricultural areas have significantly different spectral features with impervious surfaces, whereas in other seasons, the nonphotosynthetic vegetation (such as branches and stems after leaves fall down) and bare soils (in agricultural areas) are confused with impervious surfaces.

The spatial and spectral resolutions of remote sensing data are important factors affecting the selection of approaches for extraction of impervious surface. We had explored urban land-use/cover classification based on Landsat TM/ETM+ and Terra ASTER images with different classification methods and different image-processing procedures and found that urban land-use/cover classification is very difficult because of the complex urban landscape resulting in large numbers of mixed pixels. The distinction between commercial/industrial lands and high-intensity residential areas, between low-intensity residential areas and forests, and between dark imperious objects and water or shadows is especially difficult [24,25]. In this case, an SMA-based approach is more suitable for urban land-use/cover classification or extraction of impervious surfaces. For high spatial resolution images such as IKONOS, the use of classification approaches such as the DTC used in this research can successfully extract impervious surface areas with high accuracy. Due to high spatial resolution, the mixed pixel problem is not an important factor affecting classification accuracy. However, the high spectral variation within the same land-cover class may be the major factor. Thus, the use of textures in the image classification procedure or object-oriented classification approach is helpful in improving high spatial resolution image classification performance.

4.6 Conclusions

The important role of impervious surface images in urban-related studies has been recognized. However, impervious surface extraction from remotely sensed data is still a challenge. The integration of LST- and SMA-derived fraction images used in this paper has shown promise in improving the quality of mapping impervious surface distribution. Compared with previous research, this study has shown much improvement in mapping impervious surface distributions. The overall RMSE of 9.22% and SE of 5.68% were obtained in this research. In particular, the impervious surface in well-developed areas can be better estimated than that in less-developed areas.

High spatial resolution images such as IKONOS are important data sources for the extraction of urban impervious surface areas, which can be used as reference data for validation of the results developed from medium or coarse spatial resolution data. One critical step is to extract the dark impervious surface areas, which are often confused with water and shadows. A hybrid approach based on DTC and unsupervised ISODATA classifier can effectively extract the impervious surface areas, which provide significantly better results than MLC.

References

1. Slonecker, E.T., Jennings, D., and Garofalo, D., Remote sensing of impervious surface: a review, *Remote Sensing Reviews*, 20, 227, 2001.
2. Brabec, E., Schulte, S., and Richards, P.L., Impervious surface and water quality: a review of current literature and its implications for watershed planning, *Journal of Planning Literature*, 16, 499, 2002.
3. Hodgson, M.E., et al., Synergistic use of Lidar and color aerial photography for mapping urban parcel imperviousness, *Photogrammetric Engineering and Remote Sensing*, 69, 973, 2003.
4. Dougherty, M., et al., Evaluation of impervious surface estimates in a rapidly urbanizing watershed, *Photogrammetric Engineering and Remote Sensing*, 70, 1275, 2004.
5. Jennings, D.B., Jarnagin, S.T., and Ebert, C.W., A modeling approach for estimating watershed impervious surface area from national land cover data 92, *Photogrammetric Engineering and Remote Sensing*, 70, 1295, 2004.
6. Ji, M. and Jensen, J.R., Effectiveness of subpixel analysis in detecting and quantifying urban imperviousness from Landsat Thematic Mapper, *Geocarto International*, 14, 31, 1999.
7. Civico, D.L., et al., Quantifying and describing urbanizing landscapes in the northeast United States, *Photogrammetric Engineering and Remote Sensing*, 68, 1083, 2002.
8. Phinn, S., et al., Monitoring the composition of urban environments based on the vegetation-impervious surface-soil (VIS) model by subpixel analysis techniques, *International Journal of Remote Sensing*, 23, 4131, 2002.
9. Rashed, T., et al., Measuring the physical composition of urban morphology using multiple endmember spectral mixture models, *Photogrammetric Engineering and Remote Sensing*, 69, 1011, 2003.
10. Yang, L., et al., An approach for mapping large-area impervious surface: synergistic use of Landsat 7 ETM+ and high spatial resolution imagery, *Canadian Journal of Remote Sensing*, 29, 230, 2003.
11. Yang, L., et al., Urban land cover change detection through sub-pixel imperviousness mapping using remotely sensed data, *Photogrammetric Engineering and Remote Sensing*, 69, 1003, 2003.
12. Goetz, S.J., et al., Integrated analysis of ecosystem interactions with land use change: the Chesapeake Bay watershed, in *Ecosystems and Land Use Change*, DeFries, R.S., Asner, G.P., and Houghton, R.A., Eds., American Geophysical Union, Washington, D.C., p. 263, 2004.
13. Jantz, P., Goetz, S.J., and Jantz, C.A., Urbanization and the loss of resource lands within the Chesapeake Bay watershed, *Environmental Management*, 36, 808, 2005.
14. Wu, C. and Murray, A.T., Estimating impervious surface distribution by spectral mixture analysis, *Remote Sensing of Environment*, 84, 49, 2003.
15. Wu, C., Normalized spectral mixture analysis for monitoring urban composition using ETM+ imagery, *Remote Sensing of Environment*, 93, 480, 2004.
16. Lu, D. and Weng, Q., Use of impervious surface in urban land use classification, *Remote Sensing of Environment*, 102, 146, 2006.
17. Bauer, M.E., et al., Impervious surface mapping and change monitoring using

Landsat remote sensing, *ASPRS Annual Conference Processings*, Denver, Colorado, American Society for Photogrammetry and Remote Sensing, Bethesda, MD, May 23–28, 2004.

18. Gillies, R.R., et al., Effects of urbanization on the aquatic fauna of the Line Creek watershed, Atlanta—a satellite perspective, *Remote Sensing of Environment*, 86, 411, 2003.

19. Small, C., The Landsat ETM+ spectral mixing space, *Remote Sensing of Environment*, 93, 1, 2004.

20. Smith, M.O., et al., Vegetation in Deserts: I. A regional measure of abundance from multispectral images, *Remote Sensing of Environment*, 31, 1, 1990.

21. Adams, J.B., et al., Classification of multispectral images based on fractions of endmembers: application to land cover change in the Brazilian Amazon, *Remote Sensing of Environment*, 52, 137, 1995.

22. Mustard, J.F. and Sunshine, J.M., Spectral analysis for earth science: investigations using remote sensing data, in *Remote Sensing for the Earth Sciences: Manual of Remote Sensing*, 3rd ed., Vol. 3, Rencz, A.N., Ed., John Wiley & Sons, New York, NY, p. 251, 1999.

23. Weng, Q., Lu, D., and Schubring, J., Estimation of land surface temperature–vegetation abundance relationship for urban heat island studies, *Remote Sensing of Environment*, 89, 467, 2004.

24. Lu, D. and Weng, Q., Spectral mixture analysis of the urban landscapes in Indianapolis with Landsat ETM+ imagery, *Photogrammetric Engineering and Remote Sensing*, 70, 1053, 2004.

25. Lu, D. and Weng, Q., Urban classification using full spectral information of Landsat ETM+ imagery in Marion County, Indiana, *Photogrammetric Engineering and Remote Sensing*, 71, 1275, 2005.

26. Fankhauser, R., Automatic determination of imperviousness in urban areas from digital orthophotos, *Water Science and Technology*, 39, 81, 1999.

27. Goetz, S.J., et al., IKONOS imagery for resource management: tree cover, impervious surface, and riparian buffer analyses in the mid-Atlantic region, *Remote Sensing of Environment*, 88, 195, 2003.

28. Foody, G.M., Hard and soft classifications by a neural network with a non-exhaustively defined set of classes, *International Journal of Remote Sensing*, 23, 3853, 2002.

29. Friedl, M.A. and Brodley, C.E., Decision tree classification of land cover from remotely sensed data, *Remote Sensing of Environment*, 61, 399, 1997.

30. Pal, M. and Mather, P.M., An assessment of the effectiveness of decision tree methods for land cover classification, *Remote Sensing of Environment*, 86, 554, 2003.

31. Congalton, R.G., A review of assessing the accuracy of classification of remotely sensed data, *Remote Sensing of Environment*, 37, 35, 1991.

32. Smits, P.C., Dellepiane, S.G., and Schowengerdt, R.A., Quality assessment of image classification algorithms for land-cover mapping: a review and a proposal for a cost-based approach, *International Journal of Remote Sensing*, 20, 1461, 1999.

33. Foody, G.M., Status of land cover classification accuracy assessment, *Remote Sensing of Environment*, 80, 185, 2002.

Part II

Technological Advances in Impervious Surface Mapping

5

The SPLIT and MASC Models for Extraction
of Impervious Surface Areas from Multiple
Remote Sensing Data

Yeqiao Wang, Yuyu Zhou, and Xinsheng Zhang

CONTENTS

5.1 Introduction

Impervious surface is defined as any impenetrable material that prevents infiltration of water into the soil. Urban pavements, such as rooftops, roads, sidewalks, parking lots, driveways, and other man-made concrete surfaces, are among impervious surface types that feature the urban and suburban landscape. Impervious surface has been identified as a key environmental indicator due to its impacts on water systems and its role in transportation and concentration of pollutants [1]. Urban runoff, mostly through impervious surface, is the leading source of pollution in a nation's estuaries, lakes, and rivers [1,2]. A watershed-planning model predicted that most stream-quality indicators would decline when impervious surface areas (ISA) in a watershed exceeded 10% [3]. ISA have been recognized as an indicator of the intensity of urban environment. With the advent of urban sprawl, ISA have been identified as a key factor affecting the health of habit [4]. Quantification of the percentage and derivation of spatial distribution of ISA in landscapes have become increasingly important with the growing concern over water quality in this country [5–8].

Extraction of ISA from remotely sensed data has been challenging because of the nature of spectral mixing within pixel data from different sensors. In suburban areas, ISA are always mixed with tree canopies and other types of land cover at Landsat data level. Although high spatial resolution data are readily available, the volumes and costs of data prohibit their extensive usage for large areas. Therefore, modeling approaches that can integrate high spatial resolution remote sensing data with Landsat data for extraction of subpixel proportions of ISA are in demand. On the other hand, high spatial resolution true-color orthophoto data are becoming popular and available in planning agencies and different user groups. Although conventional pixel-based methods can be used for classification of high spatial resolution data, their shortcomings are evident. The most noticeable is that spectral classifications will produce dramatic salt-and-pepper effects due to increased spatial resolution and level of classification complexity. In addition spatial information, such as neighborhood, proximity, and homogeneity, cannot be used sufficiently in these methods [9]. Therefore, effective approaches that can extract precise ISA information from digital orthophoto data are in demand.

In this chapter, we introduce two tested models to extract ISA information from different types of remote sensing data. The subpixel proportional land-cover information transformation (SPLIT) model can be used to extract proportions of ISA with Landsat TM data as a base and high spatial resolution airborne digital multispectral videography data as a subpixel information provider [8]. With established relationship through training samples, the SPLIT model can be extrapolated to areas beyond the coverage of finer spatial resolution data. It enables effective and efficient use of limited coverage of high spatial resolution remote sensing data to extended areas in extraction of proportions of ISA through Landsat data. The modeling mechanism of SPLIT allows other types of high spatial resolution sensor data to be used as subpixel information providers as well. Multiple agent segmentation and classification (MASC) is an object-oriented modeling that can be used to extract ISA from fine spatial resolution true-color digital orthophoto data. It facilitates precise ISA mapping for purposes of planning and resource management. The two models present different perspectives in ISA extraction with multiple remote sensing data options.

5.2 Methods

5.2.1 SPLIT Model

The architecture of the SPLIT model includes a modular artificial neural network (MANN) and a control unit. MANN is a global artificial neural network (ANN), which consists of a group of simple-structured sub-ANNs or subnets (Figure 5.1). MANN decomposes a complex task into multiple subtasks through the use of subnets. The subnets are assigned to learn different patterns or proportions of ISA through the control unit. The number of subnets is the same as the number of output-processing elements in an ANN. The control unit is designed to perform multiple functions including task assignment, inverse simulation of spectral features, and classification decisions.

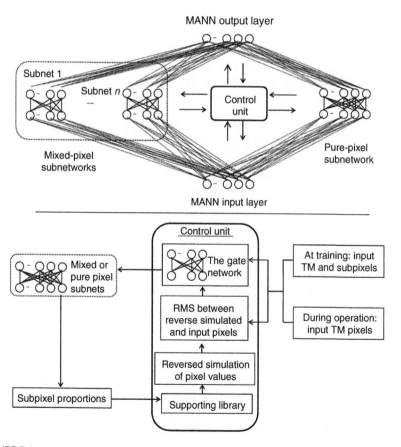

FIGURE 5.1
The SPLIT model consists of a modularized artificial neural network and a control unit. (From Wang, Y. and Zhang, X., *Photogramm. Eng. Remote Sensing*, 70, 821, 2004. With permission.)

The control unit includes a gate network and a supporting library, which records ANN parameters and training patterns. During ANN training, the control unit and supporting library learn patterns of land-cover compositions, proportions of ISA, and corresponding spectral features of TM pixels and the corresponding finer spatial resolution subpixel values. Once trained, the control unit will be able to screen input data and subpixel patterns, dissect and distribute input to suitable subnets for pattern recognition. This design allows complex cases of subpixel patterns to be decomposed and effectively handled.

Let L be the number of output-processing elements of the control unit, which is the same as the number of subnets, N the number of MANN output, which represents the level of proportions of ISA, $y_k = (y_{k1}, \ldots, y_{kN})$ the activation vector of the kth subnet output layer, $y = (y_1, \ldots, y_N)$ the MANN output, and $c = (c_1, \ldots, c_N)$ the activation vector of the output of the control unit. The output of the lth subnet is adjusted by c_l:

$$y = \sum_{l=1}^{L} c_l y_l. \tag{5.1}$$

MANN is trained by updating the weights connecting each of the processing elements. Updating of weights is derived by maximizing the objective function:

$$J = \ln\left(\sum_{l=1}^{L} c_l e^{-0.5(d-y_l)^T(d-y_l)}\right), \tag{5.2}$$

where $d = (d_1, \ldots, d_N)$ is a desired output vector for the MANN. The quantity of h_k is used to describe learning equations:

$$h_k = \frac{c_k e^{-0.5(d-y_k)^T(d-y_k)}}{\sum_{l=1}^{L} c_l e^{-0.5(d-y_l)^T(d-y_l)}}. \tag{5.3}$$

Standard back-propagation attempts to minimize a global error function E by

$$-\frac{\partial E}{\partial I} = -\frac{\partial E}{\partial y}\frac{\partial y}{\partial I} = (d-y)\frac{\partial y}{\partial I}, \tag{5.4}$$

where $I_k = (I_{k1}, \ldots, I_{kN})$ represents the preactivation vector of the kth subnet output layer, and a standard quadratic error function is assumed. The corresponding values to back-propagation include [10].

- Back-propagate error for the kth subnet:

$$-\frac{\partial J}{\partial I_k} = -\frac{\partial J}{\partial y_k}\frac{\partial y_k}{\partial I_k} = h_k(d-y_k)\frac{\partial y_k}{\partial I_k} \tag{5.5}$$

- Back-propagate error for the kth output-processing element of the control unit:

$$\frac{\partial J}{\partial s_k} = \sum_{l=1}^{L} \frac{\partial J}{\partial c_l} \frac{\partial c_l}{\partial s_k} = \sum_{l \neq k} \left(-\frac{h_1}{c_l} c_k c_l \right) + \frac{h_k}{c_k} \left(c_k - c_k^2 \right)$$

$$= -c_k \sum_{l \neq k} h_l + h_k - c_k h_k = h_k - c_k \sum_{l=1}^{L} h_k = h_k - c_k. \qquad (5.6)$$

Equation 5.5 and Equation 5.6 indicate that the error at each subnet is weighted by its control unit.

In the training stage, processing elements in the input layer accept multi-spectral TM data and the corresponding subpixel values from training samples. Patterns of mixed land-cover compositions, proportions of ISA, and corresponding characteristics of spectral features between the two datasets are recognized. The output layer of MANN represents proportions of ISA in TM pixels, which are defined by subpixel patterns. In model operation, input-processing elements accept TM data as input data. The control unit evaluates spectral patterns of the input and determines which subnet or a group of subnets should be assigned to handle the input data. For example, if a TM pixel contains a mixture of two land-cover types, one subnet that was trained to handle this composition would be assigned to process the input data. If three or more land-cover types mixed in one TM pixel, the subnet that was trained to handle the likely composition would be assigned to extract land-cover proportions of the composition. Besides, the supporting library would record spectral relations and similarities among land-cover compositions. Therefore, closely related and confusing compositions can be better identified.

A pixel is considered "pure" if there is a dominant land-cover type that counts for over 80% of the TM pixel area as determined by the corresponding subpixel information. A pure-pixel subnet (Figure 5.1) handles the pixels that have close to homogeneous subpixel spectral features. The number of subnets depends on the number of considered compositions of mixed land-cover types and that can be calculated [8].

After initial extraction of subpixel proportions, the control unit performs an inverse transformation from the proportion domain to the spectral domain. The control unit supports the inverse simulation (Figure 5.1). The inverse transformation simulates spectral features of the TM pixel based on obtained proportions of given land-cover composition and on spatial aggregation pattern of subpixels. The purpose of inverse transformation is to examine the degree of similarity between the original spectral feature of the TM pixel and the simulated spectral feature from proportions of land-cover by the subpixel information provider. The most likely composition and proportion of land-cover types are determined by the root-mean-squared (RMS) error derived from inverse transformation.

$$\text{RMS} = \sqrt{\frac{\sum (b_i - \hat{b}_i)^2}{N}}, \tag{5.7}$$

where
  N is the number of output-processing elements in an inverse subnet
  b_i is the spectral value of the ith spectral band derived by the inverse
  simulation from the extracted proportions and land-cover combinations from pixels of the subpixel information provider
  $\hat{b}_i$ is the pixel value observed from TM data

RMS is a measure of closeness between simulated spectral features and the spectral value of the TM pixels. If RMS error is greater than a threshold, the control unit will make an adjustment and reassign the job to another subnet that has the most relevant land-cover composition for a new round of proportion extraction. Thresholds come from selected training samples in which the proportions of land-cover types are known. The proportion extraction is accepted if the RMS error is lower than the threshold. If there is no subnet that can achieve RMS lower than the threshold, the output of the subnet that has the lowest RMS will be saved.

Digital multispectral videography data have four spectral bands that are identical to the spectral coverage of the first four bands of TM sensors. The spatial relationship between videography and TM data determines that 225 videography pixels (15×15) cover the same ground area as one TM pixel. The videography data in this study were acquired on October 10, 1997 for four selected sites in DuPage and Cook counties in the west suburb of Chicago. The landscape of the county is dominated by urban and suburban settings. As a Landsat TM scene (Path 023/Row 031) on the same day of videography data acquisition was available, the almost simultaneous data acquisition and identical spectral bandwidth between the two sensor systems made the videography data an ideal source of subpixel spectral information provider to team up with TM data (Figure 5.2). However as the videography data were available only for four selected sites, an extrapolation process was necessary to obtain proportional ISA information for the area beyond videography data coverage by the SPLIT model.

We geometrically rectified and georeferenced videography data based on the Landsat TM data. The registration error (RMS) between the two datasets was 0.0043 of a TM pixel, which was about one pixel of videography data. In order to differentiate subpixel ISA from other land-cover types, we conducted initial classifications on both TM and videography data to identify patterns of subpixel proportions and to establish spectral relations between the two datasets. We classified the videography and TM data into six general land-cover categories including impervious surface, deciduous trees, coniferous trees, agricultural land/grassland, wetland/water, and urban grass.

We conducted GPS-guided ground verification and recorded proportions of different land-cover types in selected sampling sites. The obtained ground-referencing locations served as references for selection of training

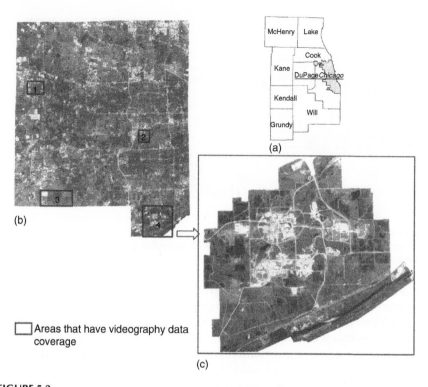

FIGURE 5.2
The study area and locations within the DuPage county where digital videography data were available (a, b) and an example of digital videography data (c).

data between videography and TM pixels. The locations of initial candidate training samples were selected from classified TM and videography data based on their land-cover types (Figure 5.3). It is necessary to point out that the purpose of initial classifications was to provide background data to facilitate training data selection. The real training data came from selected TM pixels and the corresponding subpixels, which established the spectral relations between TM and the corresponding videography data. We applied five generalized proportion intervals, that is, 0%–20%, 21%–40%, 41%–60%, 61%–80%, and 81%–100%, for extraction of proportional ISA.

5.2.2 MASC Model

The MASC model consists of four submodels including segmentation, shadow effect, MANOVA-based classification, and postclassification (Figure 5.4). The segmentation submodel employs a parameter of heterogeneity change for merging regions (Figure 5.4a). We introduced shape information in the segmentation submodel to enhance the performance of ISA extraction. The shadow-effect submodel uses a split-and-merge process to separate shadows and objects that cause shadows (Figure 5.4b). The classification submodel

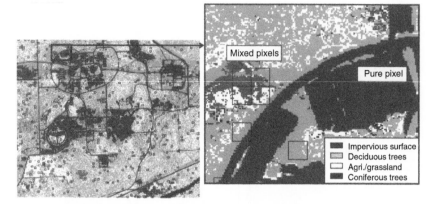

FIGURE 5.3
Locations of training samples (small squares) and examples of pure and mixed subpixel information within TM pixels.

uses a MANOVA-based classifier so that the variability within the object and the relationship between spectral bands can be taken into account (Figure 5.4c). We employed GIS-based postclassification to improve ISA extraction for those shadow areas that were impossible to be separated (Figure 5.4d).

The multiple agent segmentation considers spectral, texture, and shape information. In an object-oriented approach, the segmentation technique makes it possible to generate a hierarchical net of image segments on several levels of scale. In the segmentation process, the size and shape of desired objects can be defined by the calculation of heterogeneity between adjacent pixels. After a preliminary classification, objects are merged by classification-based segmentation. In MASC modeling, we incorporated shape information by heterogeneity change in place of spectral difference as the cost function for merging of two regions. There are different possibilities to describe change of heterogeneity before and after a merge [11,12]. A common method for heterogeneity change is described as follows.

The overall heterogeneity change h_{change} includes spectral, texture, and shape agents.

$$h_{\text{change}} = \sum_{A=1}^{m+1+2} w_A \left(n_{\text{obj1}} \left(h_{A,m} - h_{A,\text{obj1}} \right) + n_{\text{obj2}} \left(h_{A,m} - h_{A,\text{obj2}} \right) \right), \qquad (5.8)$$

where

h_{change} is the overall change of heterogeneity when two regions are merged

$h_{A,m}$ is a heterogeneity of merged region for agent A

$h_{A,\text{obj1}}$ and $h_{A,\text{obj2}}$ are the heterogeneities of two regions merged for agent A

n_{obj1} and n_{obj2} are the number of pixels in each of the two regions merged

w_A is the weight of each heterogeneity measure for agent A

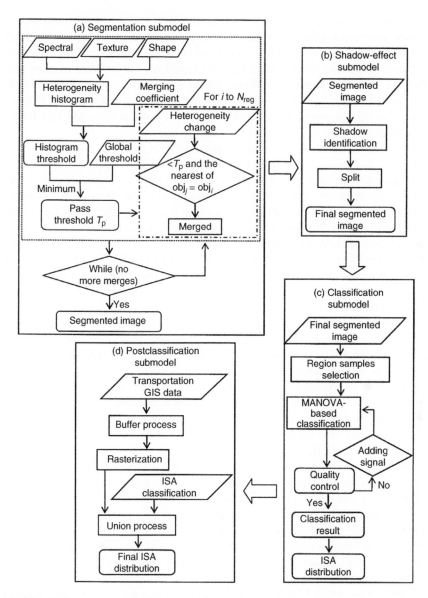

FIGURE 5.4
The flowchart of MASC modeling, which includes a segmentation submodel (a); a shadow-effect submodel (b); a classification submodel (c); and the postclassification submodel (d).

We used three spectral channels from the true-color orthophoto as $A = 1, 2, 3$ (i.e., $m = 3$) for the heterogeneity measures described in Equation 5.8. We used the fourth channel ($A = 4$) for the texture component and the fifth and sixth channels ($A = 5, 6$) for the shape components.

The shadows caused by tall vertical objects, such as tree crowns, are unavoidably associated with high spatial resolution remote sensing imageries. As the segmentation submodel has difficulty in differentiating shadow areas, we added a split-and-merge method in the shadow-effect submodel to deal with those types of mixed regions. This shadow-effect submodel identifies mixed regions based on the spectral feature and separates mixed regions into single pixels and then applies the multiple pass algorithm [13]. The submodel takes a small value as the global threshold for this new segmentation. Finally, it takes a process of region constraint in the multiple pass algorithm in the segmentation.

As we obtained a segmented image after initial and additional segmentation processes, we implemented a MANOVA-based classifier on the segmented image. Based on this classifier, we took into account the relationship between spectral bands, the variability in training objects, and the objects to be classified. We used a repeating algorithm based on a quality index in this submodel to obtain appropriate numbers of land-cover classes. When the mean of quality index for an image was lower than the previous mean, a new signal of object, which had the largest quality index in the image, was added into the training clusters. The classification stopped when the mean of quality index was larger than the previous one.

We selected a subset area from the township of East Greenwich, Rhode Island, as a test area. The landscape of this test area is characterized by extensive suburban development (Figure 5.5a). The true-color digital orthophoto data possess 1 m spatial resolution with red, green, and blue spectral bands and are distributed in GeoTIFF format (Figure 5.5b). We used a 3×3 window to extract the texture information of variance as one of the features in the segmentation process, as texture information can be helpful for definition of regions that have different levels of internal variance [13].

In MASC modeling, we first established training samples for five main classes including urban land, agricultural land, forest land, barren land, and water. Each class contained several subclasses. In order to obtain better classification results for the testing site, we employed a preclassification stratification method to divide the study area into five main categories as listed here before applying the MASC algorithm. The repeating algorithm added new subclasses into training samples and combined these new samples into existing subclasses or assigned new labels. We employed 1997 land-use dataset from Rhode Island Geographic Information System (RIGIS) database to build boundaries of five generalized categories. As some of the objects from the segmented image covered more than one category, this method assigned those to appropriate categories based on the percentage of its pixels within the five categories. We performed classifications based on this method and recorded results into ISA and non-ISA only.

We performed the MASC among each category with specified training object samples. We also used a rasterized GIS transportation data as a reference to identify road networks and integrated road data with output from the classification submodel to obtain final ISA coverage.

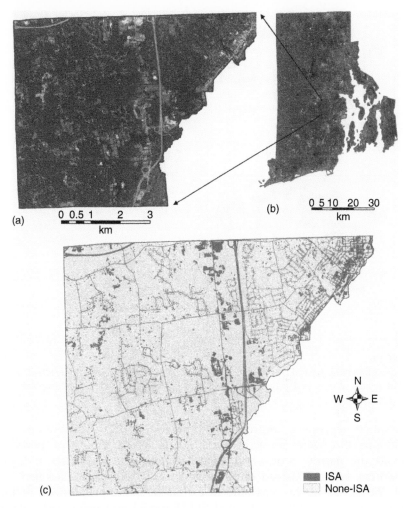

FIGURE 5.5 (See color insert following page 292.)
An example of true-color digital orthophoto data with 1 m spatial resolution for the East
Greenwich, Rhode Island (a, b) and the ISA extracted by MASC modeling (c).

5.3   Result

5.3.1   SPLIT Modeling Result

The trained SPLIT model was first used to derive proportions of impervious surface and other types of land covers for the four sites that had
coverage of videography data. A comparison of conventional pixel-based
classification of TM image (Figure 5.6a) and the proportions of ISA
obtained from the SPLIT model (Figure 5.6b) indicates that proportional

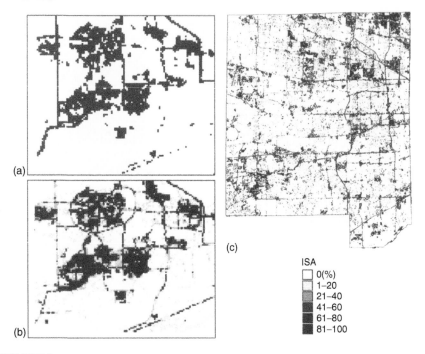

FIGURE 5.6
A comparison of ISA derived by conventional classification of TM data (a) and by SPLIT model
(b). The results of extracted proportions of ISA for the DuPage County by extrapolation of
SPLIT model (c).

information represents improved information of ISA distribution. The
continuous ISA were correctly identified at 80%–100% in proportions,
whereas the narrow roads and edges of ISA were identified at lower
proportions of ISA. Instead of using one fixed label for a classified TM
pixel, the SPLIT model was able to extract subpixel proportions of ISA
within TM pixels. With the trained SPLIT model and using TM data as
the input, we were able to extrapolate the extraction of proportions of ISA
to the areas beyond the coverage of videography data, in this case the
entire DuPage county (Figure 5.6c).

The SPLIT model performed well when the proportional level of ISA was
between 41% and 60%. The reason could be that there was no dominant
proportion of land-cover types at the middle ranged mixing proportions
(41%–60%). Therefore, the relationships between TM pixel values and the
mixing patterns of subpixels could be more accurately identified and
extracted. When the proportion intervals were off the middle ground, that
is, at 21%–40% and 61%–80% intervals, the influence from one dominant
land-cover type could reduce the effectiveness of extracting proportions of
other types within TM pixels. The 81%–100% interval was considered pure
pixels and these pixels were handled by the pure-pixel subnet. Accuracy
assessment result in Table 5.1 indicates that the higher the proportions of

TABLE 5.1

Accuracy Assessment of SPLIT-Derived Proportions of ISA

Reference	SPLIT Model–Derived Proportions of ISA					Row Total	Omission Error (%)	Accuracy (%)
	0%–20%	21%–40%	41%–60%	61%–80%	81%–100%			
0%–20%	154	30	7	1		192	19.79	80.21
21%–40%	28	182	10	3	1	224	19.91	81.25
41%–60%	16	7	173	10	2	208	16.83	83.17
61%–80%		2	24	164	6	196	16.33	83.67
81%–100%				10	88	98	10.20	89.79
Column total	198	221	214	188	97	918		
Commission error (%)	22.22	17.67	19.16	12.77	9.28			Overall 82.90

TABLE 5.2

Accuracy Assessment of MASC-Derived ISA

	MASC Model–Derived ISA				
Reference	ISA	Non-ISA	Row Total	Omission Error (%)	Accuracy (%)
ISA	89	11	100	11	89
Non-ISA	1	99	100	1	99
Column total	90	110	200		
Commission Error (%)	1.11	10			Overall 94

ISA within a TM pixel, the better the accuracy achieved. The high-end intervals were either close to or among the pure pixels. The pure pixels at 80%–100% interval achieved the best accuracy.

5.3.2   MASC Modeling Result

An example of extracted ISA using MASC modeling for the test site is illustrated in Figure 5.5c. The MASC modeling treated groups of pixels as classification targets and obtained detailed spatial distribution of ISA using high spatial resolution digital orthophoto data.

We used random point sampling method to evaluate the accuracy of ISA extracted by MASC. We selected 200 test samples and examined the classification accuracies for ISA and non-ISA only. The confusion matrix indicated that the MASC modeling achieved 94% overall accuracy using orthophoto (Table 5.2). The producer's and user's accuracies were 89% and 98.9% for the ISA, and 99% and 90% for the non-ISA categories, respectively. The kappa coefficient was 0.88.

5.4   Conclusion and Discussions

With a variety of remote sensing data from multiple sensor systems becoming available, different modeling procedures and algorithms need to be developed and tested to meet the requirements in data and information acquisition. The SPLIT model extracted proportions of ISA in a predominately suburban residential and commercial landscape. Instead of using a fixed label of land-cover type for classified TM pixels, proportions of ISA were extracted through integration of TM and finer spatial resolution subpixel information provider. Several factors contributed to the performance of the SPLIT model.

Firstly, the identical spectral coverage between digital videography data and the first four spectral bands of TM established the relationships between TM pixels and subpixel spectral patterns. The same day and almost simultaneous data acquisition of TM and subpixel videography data assured the quality of subpixel information. Otherwise, appropriate image processing needs to be conducted to match coarser resolution sensor data with finer resolution subpixel information providers.

Secondly, the mechanism of SPLIT modeling was able to decompose a complex task, such as mixing scenarios of subpixel proportions, into simplified subtasks with specific targets of land-cover compositions and proportions. The subnets of MANN were capable of obtaining reliable information about proportion compositions.

Thirdly, the inverse simulation reinforced evaluation of extracted land-cover proportions. This process assured the accuracy of subpixel information extraction since it adopted combination of land-cover types and proportion intervals that most closely matched original TM pixel values.

Further, the supporting library imbedded within the control unit recorded the patterns between TM pixel features and the corresponding subpixel compositions during the training process. Once trained, the SPLIT model can be extrapolated to areas that have no coverage of finer spatial resolution data by simulation operations. Therefore, the SPLIT model is an effective way of using limited resources of high spatial resolution data to obtain extended subpixel proportions for a large area.

For obtaining precise ISA information from high spatial resolution true-color digital orthophoto data, object-oriented modeling methods are among the essentials. To this end, we developed the MASC model. Besides segmentation and shadow effect, the MANOVA-based classification exploited correlations of spectral bands to explain the spectral distance between training objects and those to be classified. An established quality index was able to measure the classification performance. With this index and a repeating algorithm, appropriate number of classes can be derived to achieve improved classifications. The MASC-obtained ISA maps confirmed the effectiveness of this object-oriented approach in mapping precise ISA from popularly used true-color digital aerial photography data.

Acknowledgments

The developed models were part of the research projects funded by National Aeronautic and Space Administration (Grant No. NAG5-8829) and Rhode Island Agricultural Experiment Station (Project No. H-330).

References

1. Arnold, C.A. Jr. and Gibbons, C.J., Impervious surface coverage: the emergence of a key urban environmental indicator, *Journal of the American Planning Association*, 62, 243, 1996.

2. Booth, D.B. and Jackson, C.R., Urbanization of aquatic systems: degradation thresholds, stormwater detection, and the limits of mitigation, *Journal of American Water Resources Association*, 35, 1077, 1997.

3. Schueler, T., *Impacts of Impervious Cover on Aquatic Systems*, Center for Watershed Protection (CWP), Ellicott City, MD, 2003, p. 142.

4. Brabec, E., Schulte, S., and Richards, P.L., Impervious surfaces and water quality: a review of current literature and its implications for watershed planning, *Journal of Planning Literature*, 16, 499, 2002.

5. Weng, Q.H., Modeling urban growth effects on surface runoff with the integration of remote sensing and GIS, *Environmental Management*, 28, 737, 2001.

6. Civco, D.L. et al., Quantifying and describing urbanizing landscapes in the northeast United States, *Photogrammetric Engineering and Remote Sensing*, 68, 1083, 2002.

7. Dougherty, M. et al., Evaluation of impervious surface estimates in a rapidly urbanizing watershed, *Photogrammetric Engineering and Remote Sensing*, 70, 1275, 2004.

8. Wang, Y. and Zhang, X., A SPLIT model for extraction of subpixel impervious surface information, *Photogrammetric Engineering and Remote Sensing*, 70, 821, 2004.

9. Burnett, C. and Blaschke, T., A multi-scale segmentation/object relationship modelling methodology for landscape analysis, *Ecological Modelling*, 168, 233, 2003.

10. NeuralWare, *Reference Guide: Software Reference for Professional II/Plus and Neural Works Explorer*, NeuralWare Inc., Pittsburgh, PA, 1993.

11. Shackelford, A.K. and Davis, C.H., A combined fuzzy pixel-based and object-based approach for classification of high-resolution multispectral data over urban areas, *IEEE Transactions on Geoscience and Remote Sensing*, 41, 2354, 2003.

12. Baatz, M. et al., *eCognition User Guide*, URL: http://www.definiens-imaging.com, Definiens Imaging, München, Germany (last date accessed: 25 July 2005), 2004.

13. Woodcock, C.E. and Harward, V.J., Nested-hierarchical scene models and image segmentation, *International Journal of Remote Sensing*, 13, 3167, 1992.

6

Extracting Impervious Surface from Hyperspectral Imagery with Linear Spectral Mixture Analysis

Qihao Weng, Xuefei Hu, and Dengsheng Lu

CONTENTS

6.1 Introduction

Many techniques have been developed to estimate and map impervious surfaces using remotely sensed data. Spectral mixture analysis (SMA), as a subpixel information extraction algorithm, is gaining interest in the remote sensing community in recent years. The linear SMA model assumes that the spectrum measured by a sensor is a linear combination of the spectra of all components within the pixel (Adams et al., 1986). Because of its effectiveness in handling the spectral mixture problem associated with medium-resolution (10–100 m) satellite imagery (such as Landsat TM/ETM+, and Terra's ASTER images), linear SMA has been widely used in the estimation of impervious surfaces (Ward et al., 2000; Madhavan et al., 2001; Phinn et al., 2002; Wu and Murray, 2003; Lu and Weng, 2006a,b). Three distinct methods based on the SMA model have been developed for estimation of impervious surface. These include (1) extraction of impervious surface as one of the endmembers in the standard SMA model (Phinn et al., 2002), (2) estimation by the addition of high-albedo and low-albedo fraction images, both as the SMA endmembers (Wu and Murray, 2003), and (3) the combination of several impervious surface endmembers from a multiple endmember SMA model (Rashed et al., 2003). However, these SMA-based methods have a common problem, that is, impervious surface is often overestimated in the areas with a small amount of impervious surface, but is underestimated in the areas with a large amount of impervious surface.

This problem has to do with the limitation of spatial and spectral resolution in the medium-resolution imagery. These images are regarded as too coarse for use in mapping urban landscape because of its heterogeneity and the complexity of urban impervious surface materials. Identifying one suitable endmember to represent all types of impervious surfaces is often found problematic. Lu and Weng (2004) suggested that three possible approaches may be taken to overcome these problems: (1) by stratification, (2) by use of multiple endmembers, and (3) by use of hyperspectral imagery. In the SMA model, the maximum number of endmembers is directly proportional to the number of spectral bands used. The vastly increased dimensionality of a hyperspectral sensor may remove the sensor-related limit on the number of endmembers available. More significantly, the fact that the number of hyperspectral image channels far exceeds the likely number of endmembers for most applications readily permits the exclusion from the analysis of any bands with low signal-to-noise ratios or with significant atmospheric absorption effects (Lillesand et al., 2004).

Little research has been conducted in impervious surface estimation with hyperspectral imagery based on SMA. The recently launched EO-1 satellite with the Hyperion sensor provides a great opportunity for such a study. Hyperion is a hyperspectral sensor of 242 bands with 30 m spatial resolution. Onboard the same satellite, the ALI sensor is a multispectral one with the same spatial resolution. This satellite platform is ideal for comparative studies between a hyperspectral and a multispectral sensor. The objectives of this chapter are to conduct SMA with various combinations of endmembers by using Hyperion images and to compare impervious surface maps extracted from the Hyperion and ALI images. We intend to address the following questions in this research: (1) Is the hyperspectral data better suited than multispectral data in impervious surface estimation? and (2) Which combination of SMA endmembers would produce the most accurate impervious surface map?

6.2   Data Used

An EO-1 ALI image and a Hyperion image covering Marion County, Indiana, United States (refer to Chapter 4 for a brief description), which were acquired on April 12, 2003, under clear weather conditions, were used in this research. Both images were georectified to a Universal Transverse Mercator coordinate system, using a nearest-neighbor resampling method. The RMSE of <0.5 pixels were obtained from both rectifications. The ALI image had 10 bands ranging from 433 to 2350 nm in wavelength, with a panchromatic band. The Hyperion image had 242 bands covering from 400 to 2500 nm in wavelength. However, there were 38 empty bands and 43 noisy bands, which were removed before data processing. The fourth track of the ALI image overlapped with the Hyperion image (256 pixels in width and 6478 lines in length). Both images were subset into the same study area for the purpose of comparison.

Orthophotographs were used for validation of impervious surface estimation results and for accuracy assessment. The color orthophotographs were provided by the Indianapolis Mapping and Geographic Infrastructure System, which was acquired in April 2003 for the entire county. The orthophotographs have a spatial resolution of 0.14 m. The coordinate system belongs to Indiana State Plane East, Zone 1301, with North American Datum of 1983. The orthophotographs were reprojected into the same coordinated system as the ALI and Hyperion images and resampled to 1 m pixel size for the sake of quicker display and shorter computing time.

6.3   Methodology

6.3.1   Linear Spectral Mixture Analysis

Linear spectral mixture analysis (LSMA) is a physically based image-processing method. It assumes that the spectrum measured by a sensor is a linear combination of the spectra of all components within the pixel (Adams et al., 1986). The mathematical model of LSMA can be expressed as

$$R_i = \sum_{k=1}^{n} f_k R_{ik} + \text{ER}_i \qquad (6.1)$$

where
 $i = 1, \ldots, m$ (number of spectral bands)
 $k = 1, \ldots, n$ (number of endmembers)
 R_i is the spectral reflectance of band i of a pixel that contains one or more endmembers
 f_k is the proportion of endmember k within the pixel
 R_{ik} is the known spectral reflectance of endmember k within the pixel on band i
 ER_i is the error for band i

To solve f_k, the following conditions must be satisfied: (1) selected endmembers should be independent of each other; (2) the number of endmembers should be less than or equal to the spectral bands used; and (3) selected spectral bands should not be highly correlated.

Estimation of endmember fraction images with LSMA involves image processing, endmember selection, unmixing solution, and evaluation of fraction images (Boardman, 1993; Boardman et al., 1995). Of these steps, selecting suitable endmembers is the most critical one in the development of high-quality fraction images. Following the georectification, principal component analysis was performed to reduce data redundancy and correlations between spectral bands. Both ALI and Hyperion images were found to be highly correlated, especially the Hyperion image. It is found that most of the information content was concentrated in the first three principal components (accounting for 94.4% of the total variance in the ALI image and 98.4% in the Hyperion image). These components were retained for use in the LSMA models, whereas the higher-order components were discarded due to the high proportion of noise content (Figure 6.1).

6.3.2 Endmember Selection

Endmembers were initially identified from the ALI and Hyperion images based on high-resolution aerial photographs. Four types of endmembers were selected: green vegetation (vegetation), soils (including dry soil and dark soil), low-albedo (asphalt, water, etc.), and high-albedo surfaces (concrete, sand, etc.). Vegetation was selected from the areas of dense grass, pasture, and forestry. Different types of impervious surfaces were selected from building roofs, airport runway, highway intersections, and so on. Soils were selected from bare grounds in agricultural lands. Next, these initial endmembers were compared with those endmembers selected from the scatter plots of PC1–PC2, PC2–PC3, and PC1–PC3. The endmembers with similar PC spectra located at the extreme vertices of the scatter plots were finally selected. Figure 6.2 shows the selection of the endmembers and their spectral reflectance characteristics.

(a)   PCA Band1(66%) PCA Band2(28%) PCA Band3(0.4%)

(b)   PCA Band1(96%) PCA Band2(3%) PCA Band3(0.4%)

FIGURE 6.1
PCA bands of ALI and Hyperion images.

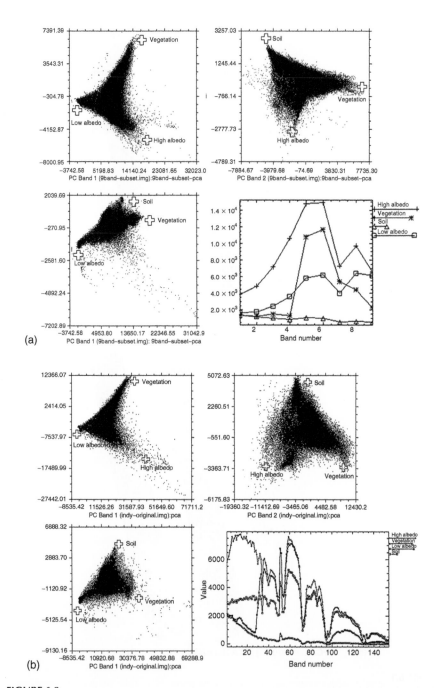

FIGURE 6.2
Scatter plots of the principal components showing the locations of potential endmembers.
Spectral reflectance characteristics of the selected endmembers are also shown.

To find the best quality fraction images for estimation of impervious surfaces, different combinations of endmembers were examined and compared. The combinations included (1) four endmembers of high albedo, low albedo, vegetation, and soil; (2) three endmembers of high albedo, low albedo, and vegetation; (3) three endmembers of high albedo, low albedo, and soil; (4) three endmembers of high albedo, vegetation, and soil; (5) three endmembers of low albedo, vegetation, and soil. Visualization of fraction images, analysis of fractional spectral properties of representative land-cover types, and assessment of error images were conducted to determine which combination provided the best fractions for each image. Because this study was more interested in estimating impervious surfaces in the urban area than in the rural area, the criteria for selecting the most suitable fraction images were based on (1) high-quality fraction images for the urban landscape, (2) relatively low error, and (3) the distinction among typical land-use and land-cover types.

A fully constrained LSMA was then applied to the ALI and Hyperion images. Overall, the SMA results using four endmembers for both ALI and Hyperion were found to be most satisfactory. The RMS errors were small. The fraction images of high albedo, vegetation, and soil were consistent with the land-use and land-cover pattern of the study area (Weng et al., 2004). However, the low-albedo fraction was found to be much harder to interpret. In fact, the low-albedo fraction related to many materials, such as water, shade, dark impervious surfaces, and so on. It is assumed that in the central business district (CBD), low-albedo fraction should be directly related to impervious surface, but for the residential and the surrounding rural areas, water and shade were both shown with high fraction values in the low-albedo images. In addition, LSMA results using three endmembers might produce reasonably good results, especially those of the three-endmember combination of vegetation, low albedo, and soil.

6.3.3 Impervious Surface Estimation

The estimation of impervious surface was implemented by using the relationship between the reflectance of two endmembers (high albedo and low albedo) and the reflectance of the impervious areas. Wu and Murray (2003) developed an estimation procedure based on this relationship. The fraction images of high albedo and low albedo were added together directly and it was found that impervious surface was located on or near the line connecting the low-albedo and high-albedo endmembers in the feature space. The impervious surface image can be computed as

$$R_{imp,b} = f_{low}R_{low,b} + f_{high}R_{high,b} + e_b \qquad (6.2)$$

where

$R_{imp,b}$ is the reflectance spectra of impervious surfaces for band b
f_{low} and f_{high} are the fractions of low albedo and high albedo, respectively

$R_{low,b}$ and $R_{high,b}$ are the reflectance spectra of low albedo and high albedo for band b

e_b is the unmodeled residual

The fitness of this two-endmember linear spectral mixture model has been demonstrated by Wu and Murray for the CBD of Columbus, Ohio, United States. However, some low-reflectance materials (e.g., water and shades) had to be masked out before adding high-albedo and low-albedo fractions together to get an impervious surface image.

In this work, Wu and Murray's method (2003) was applied. The final impervious surface images were considered satisfactory. Figures 6.3 and 6.4 display the final images, with color maps to show the distribution of impervious surface at four categories. Before developing the impervious surface images, the impacts from low-reflectance materials (e.g., water and shade) and high-reflectance materials (e.g., clouds and sand) were isolated and removed. Green vegetation and soil endmembers were considered as not contributing to impervious surface estimation. After removing these

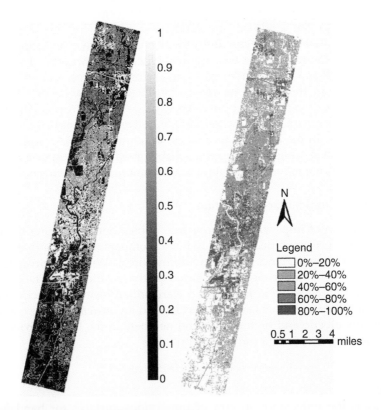

FIGURE 6.3 (See color insert following page 292.)
Impervious surface image derived from ALI image. The color figure to the *right* shows the distribution of impervious surface at four categories.

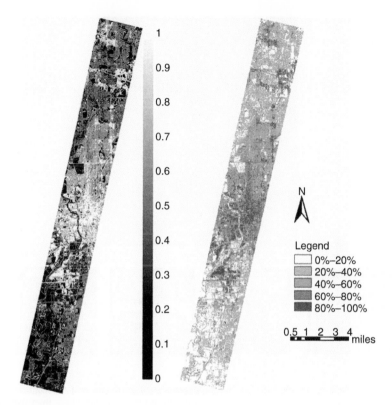

FIGURE 6.4 (See color insert following page 292.)
Impervious surface image derived from Hyperion image. The color figure to the *right* shows the distribution of impervious surface at four categories.

pixels, pure impervious surfaces were estimated with the addition of low- and high-albedo endmembers by a fully constrained linear mixture model.

6.3.4 Evaluation of Impervious Surface Images

Accuracy assessment of impervious surface images was regarded as an important aspect of our method. Selecting sufficient number of reference data through a proper sampling method is crucial. A total of 100 samples with 3×3 pixel size (90 m $\times$ 90 m) were designed using a stratified random sampling scheme. For each sample, impervious surface was digitized on the corresponding DOQQ using ArcGIS. After the digitization, the proportion of impervious surface area was computed by dividing the area of impervious surface by the sampling area. Figure 6.5 illustrates the design of sample plots and the method for obtaining reference data by digitizing impervious surface polygons within selected samples. The root-mean-square error (RMSE), the mean average error, and the correlation coefficient (R^2) were

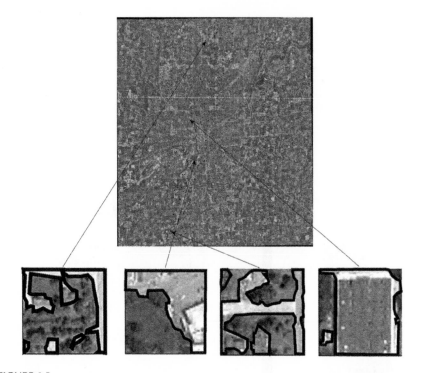

FIGURE 6.5
Method for collecting sample plots on the impervious surface images and for digitizing the impervious surface areas on the orthophotograph. These reference data were then used for assessment of the accuracy of impervious surface images.

then calculated to indicate the accuracy of impervious surface estimation. The following were the equations:

$$\text{RMSE} = \sqrt{\frac{\sum\limits_{i=1}^{N} \left(\hat{I}_i - I_i\right)^2}{N}} \tag{6.3}$$

$$\text{MAE} = \frac{1}{N} \sum\limits_{i=1}^{N} \left|I_i - \hat{I}_i\right| \tag{6.4}$$

$$R^2 = \frac{\sum\limits_{i=1}^{N} \left(\hat{I}_i - \bar{I}\right)}{\sum\limits_{i=1}^{N} \left(I_i - \bar{I}\right)^2} \tag{6.5}$$

where
  $\hat{I}_i$ is the estimated impervious surface fraction for sample i
  I_i is the impervious surface proportion computed from aerial photography
  $\bar{I}$ is the mean value of the samples
  N is the number of samples

6.4 Analysis of Results

6.4.1 Results of Spectral Mixture Analysis

LSMA was performed on both ALI and Hyperion images using five different combinations of endmembers.

6.4.1.1 Four Endmembers of High Albedo, Low Albedo, Vegetation, and Soil

The SMA results derived from both ALI and Hyperion images using the four endmembers were very good. The mean RMS errors were <0.05 in both cases. The fraction images of high albedo, vegetation, and soil showed unique information and fit the general patterns (Figure 6.6). High-albedo fractions had high values in the CBD area, whereas, a large amount of soil and vegetation was concentrated in residential and rural areas. However, low albedo appeared to be harder to interpret because of its complicated spatial patterns.

Correlation analysis was conducted to analyze the relationship between the fraction images acquired from ALI and Hyperion data. Results show that there were strong correlations between the fraction images derived from the two images. Especially, for the vegetation and low-albedo

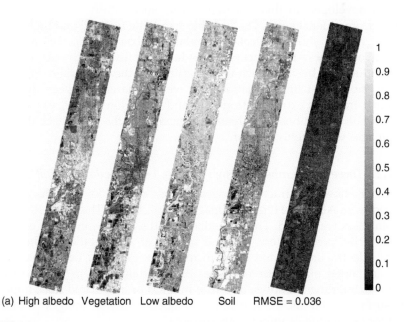

(a) High albedo   Vegetation   Low albedo   Soil   RMSE = 0.036

FIGURE 6.6
Fraction images (high albedo, low albedo, vegetation, and soil) from spectral mixture analysis of the ALI and Hyperion images.

(continued)

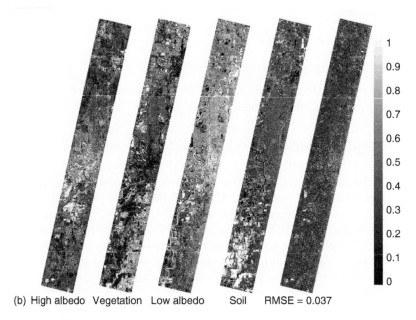

(b) High albedo Vegetation Low albedo   Soil   RMSE = 0.037

FIGURE 6.6 (continued)

fractions, correlation coefficients reached 0.93 and 0.91, respectively. These high corrections indicate that the impervious surface extraction method developed in this research can be successfully used for satellite images from other sensors. On the other hand, the two soil and high-albedo fraction images showed weaker correlations (coefficients: 0.66 and 0.74, respectively). These two fractions may have confused with each other, or were influenced by the spectral properties of other materials. Figure 6.7 further shows that linear relationships were apparent between the corresponding fractions extracted from ALI and Hyperion images, especially in the vegetation fractions. This linearity means that the two vegetation-fraction images were highly similar.

6.4.1.2 Three Endmembers of High Albedo, Low Albedo, and Soil

Figure 6.8 shows the fraction images of high albedo, low albedo, and soil derived from ALI and Hyperion images, respectively. The mean RMS error in both cases was <0.05. The high-albedo and soil fractions were extracted satisfactorily. Nevertheless, because of lack of a vegetation endmember, low-albedo fractions ended up with mixing vegetation. Low-albedo fractions contained many different types of materials, such as water, canopy shadows, building shadows, moisture in grass or crops, and dark impervious surface materials. As a result, low-albedo fraction images cannot be used for estimation of impervious surfaces.

Correction analysis indicates that the two low-albedo fractions correlated highly with a coefficient of 0.96. However, the correlation coefficients

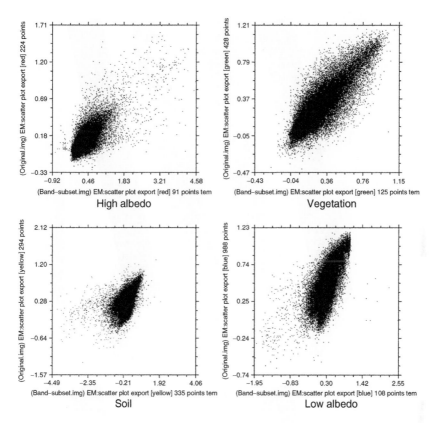

FIGURE 6.7
Correlations between the fraction images of ALI and those of Hyperion (high albedo, low albedo, vegetation, and soil).

between the two high-albedo and soil fractions were 0.72 and 0.74, respectively. Figure 6.9 demonstrates the strong linear relationships again between the corresponding fractions derived from the images of different sensors. Therefore, our SMA method is regarded to be consistent. Low-albedo fraction values estimated from the Hyperion image were higher than those generated from the ALI image.

6.4.1.3 Three Endmembers of High Albedo, Low Albedo, and Vegetation

Figure 6.10 shows two groups of the three fraction images of high albedo, low albedo, and vegetation. The RMS error for SMA of the ALI image is 0.12, and the RMS error for Hyperion is 0.037. This indicates that Hyperion was better than ALI when using three endmembers of high albedo, low albedo, and vegetation for spectral unmixing. The information on high albedo, low albedo, and vegetation can be extracted clearly. Soil ended up with being categorized into residuals, which increased the RMS error as a whole.

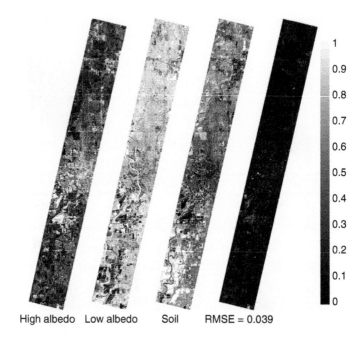

(a)   High albedo Low albedo   Soil   RMSE = 0.039

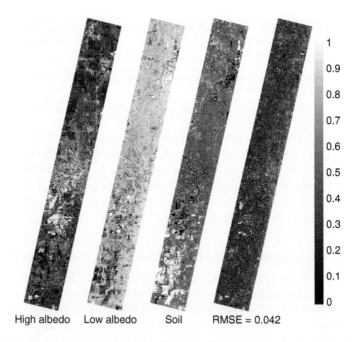

(b)   High albedo Low albedo   Soil   RMSE = 0.042

FIGURE 6.8
Fraction images (high albedo, low albedo, and soil) from spectral mixture analysis of the ALI
and Hyperion images.

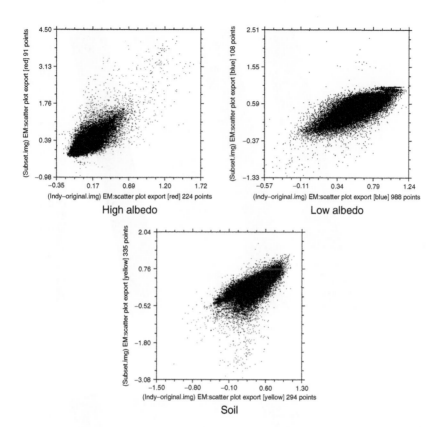

FIGURE 6.9
Correlations between the fraction images of ALI and those of Hyperion (high albedo, low albedo, and soil).

The small RMS error for the SMA results of Hyperion image offers the potential for the fractions to be used in impervious surface estimation and mapping.

Correlation analysis yielded a coefficient of 0.94 between the vegetation fractions and 0.89 between the high-albedo fractions. The low-albedo fractions had a low correlation coefficient of 0.65. Figure 6.11 indicates that the vegetation fractions perfectly matched along the diagonal line, and similarly, with the high-albedo fractions. Low-albedo fraction values from the Hyperion image appeared to be higher than those from ALI in many cases.

6.4.1.4 Three Endmembers of High Albedo, Vegetation, and Soil

Figure 6.12 displays the SMA results with the endmembers of high albedo, soil, and vegetation. The mean RMS error of 0.2 was achieved for both ALI and Hyperion data, denoting a poor fit of the SMA model. This is due to the fact that low-albedo materials were so unique that they cannot be combined into any other category. The combination of low albedo with other

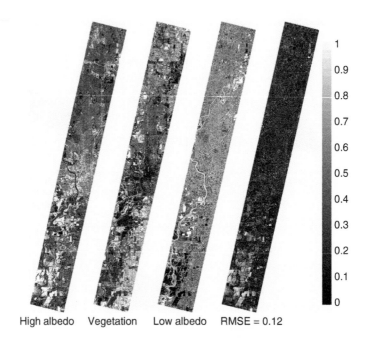

(a)   High albedo   Vegetation   Low albedo   RMSE = 0.12

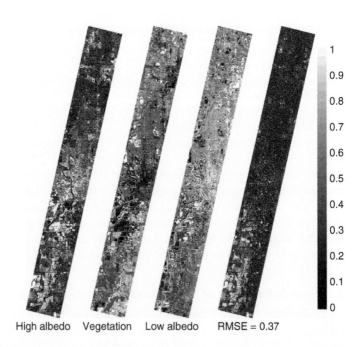

(b)   High albedo   Vegetation   Low albedo   RMSE = 0.37

FIGURE 6.10
Fraction images (high albedo, low albedo, and vegetation) from spectral mixture analysis of the ALI and Hyperion images.

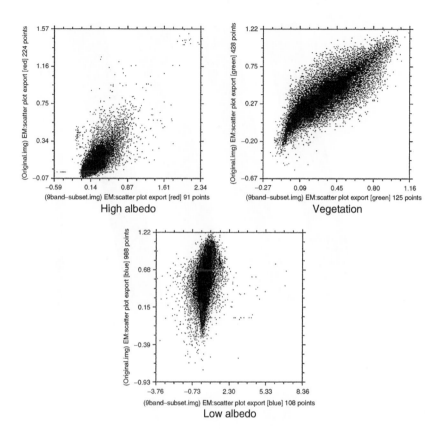

FIGURE 6.11
Correlations between the fraction images of ALI and those of Hyperion (high albedo, low albedo, and vegetation).

endmembers would increase the residuals of the SMA model and the RMS error. Although high-albedo, soil, and vegetation fractions were all extracted satisfactorily, because of the higher RMS error, this combination of the endmembers is regarded unsuitable for further use in impervious surface estimation.

The corresponding fraction images were found significantly correlated too (Figure 6.13). The correlation coefficients for the high-albedo, vegetation, and soil fractions were 0.85, 0.84, and 0.7, respectively. Soil fractions possessed the lowest correlation coefficient value. The scatter plot between the two vegetation fractions showed a diagonal trend, implying that the extractions were agreeable from the ALI and Hyperion images.

6.4.1.5 Three Endmembers of Low Albedo, Vegetation, and Soil

Figure 6.14 displays vegetation, low-albedo, and soil fractions derived from ALI and Hyperion images. The mean RMS error for SMA of ALI image was 0.075 and for Hyperion, 0.2. The vegetation and soil fractions were extracted

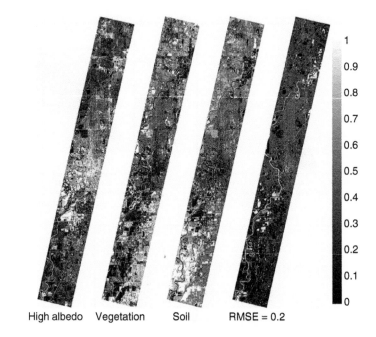

(a)   High albedo   Vegetation   Soil   RMSE = 0.2

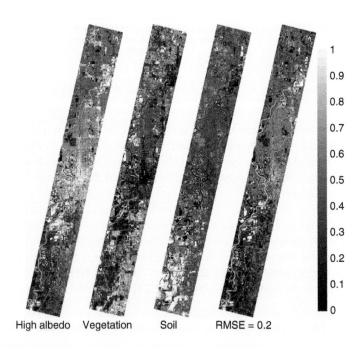

(b)   High albedo   Vegetation   Soil   RMSE = 0.2

FIGURE 6.12
Fraction images (high albedo, vegetation, and soil) from spectral mixture analysis of the ALI and Hyperion images.

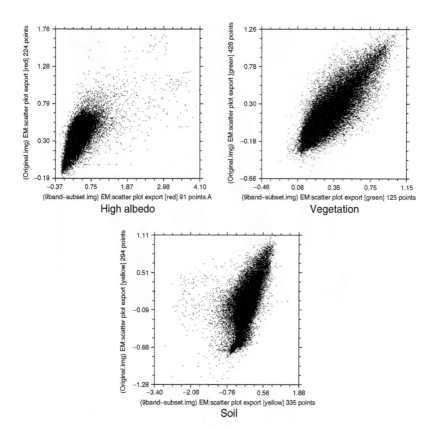

FIGURE 6.13
Correlations between the fraction images of ALI and those of Hyperion (high albedo, vegeta-tion, and soil).

successfully, but low-albedo and high-albedo materials were mixed together. The RMS error for ALI image was small, and the results may be used for further studies in impervious surface estimation. The correlation coefficients between vegetation fractions reached 0.93, between low-albedo fractions, 0.83, and between the soil fractions 0.95. Figure 6.15 shows that the derived vegetation, low-albedo, and soil fractions were all agreeable according to both sensors.

In sum, using four endmembers in both cases generated satisfactory results, with low RMS error and fraction values consistent with land-cover patterns. Some combinations of three endmembers may also give rise to a good extraction of fraction images and to small residuals. The results of correlation analysis between the corresponding ALI and Hyperion fractions were all strongly positively correlated to each other, especially with the vegetation fractions. The fact that vegetation fraction images showed such a perfect diagonal match in the scatter plots suggests that for vegetation studies, Hyperion and ALI images would not have much difference with

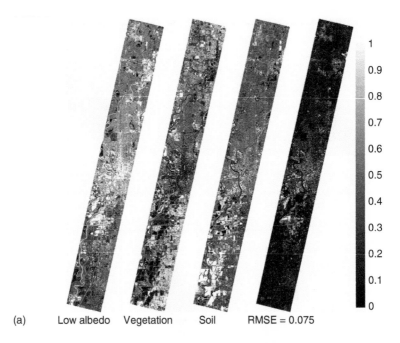

(a)   Low albedo   Vegetation   Soil   RMSE = 0.075

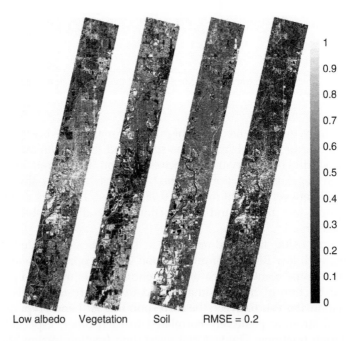

(b)   Low albedo   Vegetation   Soil   RMSE = 0.2

FIGURE 6.14
Fraction images (low albedo, vegetation, and soil) from spectral mixture analysis of the ALI and Hyperion images.

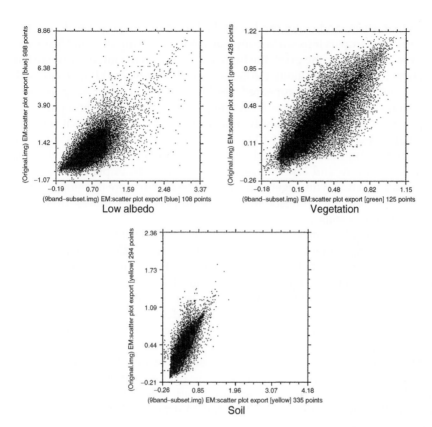

FIGURE 6.15

Correlations between the fraction images of ALI and those of Hyperion (low albedo, vegetation, and soil).

our SMA-modeling approach. However, the correspondent soil and low-albedo fractions had weaker correlations than vegetation and high-albedo fractions. Fractions derived from the Hyperion image were a little better than those from ALI image. This is because Hyperion imagery has a much narrower bandwidth and more data dimensions than ALI imagery, and therefore is more effective in the extraction of fraction images, especially for low-albedo fraction.

6.4.2 Results of Impervious Surface Estimation

In this study, fraction images derived from both ALI and Hyperion images using the four-endmember SMA model were used to generate impervious surface images. Before adding high-albedo and low-albedo fraction images to calculate impervious surfaces, water and shade had to be removed. Because of lower temperatures of water and shade compared to impervious surfaces, it was possible to differentiate shade and water by calculating their temperatures from Landsat thermal bands (Lu and Weng, 2006a).

After that, shade and water can be masked out from the low-albedo fraction images. Two impervious surface images were then computed. In Figures 6.3 and 6.4, the brighter the pixel, the higher the value in the impervious surface fraction images. Impervious surfaces such as roads, building roofs, side walks, and parking lots appeared very bright, whereas forest, grass, and cropland showed a dark tone. Both images quantified the general pattern of impervious surfaces of the study area successfully, meaning that the fraction value was higher in the CBD areas, lower in residential areas, and near zero in the rural and vegetated areas. However, in less-developed areas, such as medium- and low-intensity residential lands, impervious surfaces were found to mix with vegetation and soil. While in more developed areas, such as high-intensity residential areas, impervious surfaces may have mixed with dry soils.

In order to determine which impervious surface image was closer to the reality, an accuracy assessment was performed. RMSE, MAE, and R^2 were calculated for both images. Results indicate that for ALI-derived impervious surface image, $RMSE = 15.3\%$, $MAE = 12.4\%$, and $R^2 = 0.7478$ and for Hyperion-derived impervious surface image, $RMSE = 17.5\%$, $MAE = 14.8\%$, and $R^2 = 0.7302$. It is concluded that both images had a similar accuracy, although the Hyperion-derived impervious surface image outperformed the ALI-derived image in the low-albedo areas. Figure 6.16 illustrates the accuracy assessment results. It indicates that the samples with low impervious surface were overestimated, but the samples with high impervious surface were underestimated. The trend is in agreement with previous researches (Wu and Murray, 2003; Wu, 2004; Lu and Weng, 2006a).

6.5 Discussion

The use of remote sensing techniques to estimate impervious surfaces is still a challenging task due to the characteristics of remotely sensed data, the complexity of urban landscape, and the diversity of impervious surface materials. Remotely sensed data are related to the reflectance of different surface features and materials, but the reflectance of impervious surfaces is very diverse and easy to be confused with other features, such as sand, dry soils, and so on, especially in the seasons when tree canopies are gone. The atmospheric conditions prevailing when satellite images were acquired can also affect the estimation of impervious surfaces. For example, the reflectance of clouds is similar to that of impervious surfaces, which may result in confusion. Another factor is that dark impervious surfaces have similar spectral characteristics with water, moist soils, and shade. Future studies are warranted to differentiate these features from impervious surfaces. Nevertheless, although the urban landscape of Indianapolis city is complex, this research that has demonstrated LSMA using four endmembers

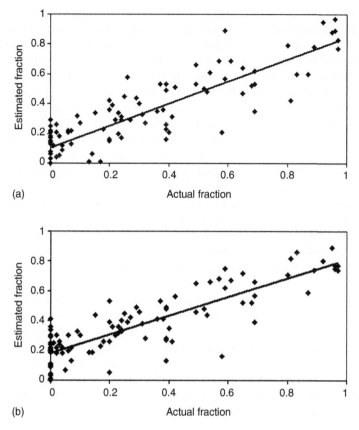

FIGURE 6.16

Accuracy assessment results of estimated impervious surfaces from the ALI and Hyperion images.

(i.e., high albedo, low albedo, vegetation, and soil) is an effective approach for estimating impervious surfaces.

In SMA, endmember selection is the most important step which is directly associated with the accuracy of resultant impervious surface images. The selection of endmembers is a process involving trial and error. It may take much time beginning from selecting initial endmembers, testing the results, and then refining the endmembers until satisfied results are achieved. Moreover, the image-based endmember selection method may not be able to identify different types of impervious surface endmembers. For most SMA studies to date, limited endmembers, that is, three or four endmembers are used because of limited data dimensions and high correlation among the image bands. These limitations lead to the inability of representing the spectral properties of impervious surfaces of all sorts. To solve these problems, hyperspectral imagery may be employed. This study has demonstrated that Hyperion data were more effective than ALI data in estimation of low-albedo surface materials, which include a range of impervious surface features. The combination of hyperspectral imagery and reference data acquired either from in situ field survey or from a spectral library would be promising in future studies of impervious surface estimation and mapping. Another approach is to use a multiple endmember spectral mixture analysis (MESMA). The MESMA allows a large number of endmembers to be extracted in the model and has been proved more effective than a standard SMA method (Painter et al., 1998; Roberts et al., 1998; Okin et al., 2001). This approach starts with a series of candidate two-endmember models, and then evaluates each model based on the criteria of fraction values, RMSE, and residual threshold, and finally produces fraction images with the lowest error (Roberts et al., 1998).

The composition of low-albedo materials is complicated. These materials may contain impervious surfaces, water, building shade, tree canopy shade in forested areas, moist soils, and so on. As a result, before computing an impervious surface image, water, shade, and other pervious materials have to be removed. Although water is easy to be masked out through an unsupervised classification, shade is difficult to be identified and removed. To remove shade, several methods can be applied. For example, topographical shade can be masked out by topographical correction, while vegetation shade may be removed after a detailed study of vegetation canopy structures. In addition, the use of thermal infrared and radar imagery may aid greatly in impervious surface estimation, possibly through data fusion (Slonecker et al., 2001). Lu and Weng (2006a) have successfully used Landsat thermal infrared data to enhance their estimation of impervious surfaces based on the difference in land surface temperature between impervious and pervious surfaces. The land surface temperature image was used as a threshold to remove dry soils from the high-albedo fraction image and to remove water and shadows from the low-albedo fraction image. Radar data have inherent advantages in the identification and estimation of impervious surfaces because of the high dielectric properties of most construction materials and the unique geometry of man-made features (Slonecker et al., 2001).

6.6 Conclusions

In this chapter, impervious surface fraction images were directly extracted from EO-1 Hyperion and ALI images by applying a fully constrained LSMA. This chapter further demonstrates that remote sensing-based impervious surface estimation has the potential to take place of the labor-intensive method of digitizing on aerial photographs, especially for a large study area. Although SMA has been conducted on medium spatial resolution imagery in previous researches, using EO-1 Hyperion and ALI data to estimate impervious surfaces has not been seen in the literature. Because of the integration of hyperspectral and multispectral sensors on the same satellite, it is possible to use images of both sensors for comparative studies. In this chapter, both Hyperion and ALI images have been analyzed by LSMA and the results were compared based on the quality of fraction images and RMS error. Correlation analysis was further performed to make the comparison quantitatively. The results indicate that there was a strong positive linear relationship between the fraction images derived from the Hyperion and from the ALI images. The fraction images were highly similar, especially between the vegetation fractions. This implies that our method of impervious surface estimation was consistent, and thus has the potential to apply to other images. The comparison has also provided some new insights for future impervious surface studies. Our result suggests that the low-albedo endmember was most difficult to identify due to its spectral variation, and that the Hyperion image was more effective. Results from this research offer a foundation for subsequent impervious surface estimation with moderate spatial resolution imagery. Future researches are suggested in the following areas: (1) the combined use of hyperspectral imagery with reference data acquired either from in situ field survey or a spectral library streamlining the selection of endmembers in SMA; (2) the multiple end-member LSMA, instead of the standard LSMA, should be applied for the development of fraction images in impervious surface estimation; (3) different methods may be used to remove shades from satellite imagery, which tend to confuse with low-albedo materials; and (4) impervious sur-face estimation and mapping should utilize the data and techniques of thermal and radar remote sensing.

Acknowledgments

This research is supported by National Science Foundation (BCS-0521734) for a project entitled "Role of Urban Canopy Composition and Structure in Determining Heat Islands: A Synthesis of Remote Sensing and Landscape Ecology Approach" and by the USGS IndianaView program and by the NASA's Indiana Space Grant program for a project entitled "Indiana Impervious Surface Mapping Initiative." We also thank reviewers for their constructive comments and suggestions.

References

Adams, J.B., Smith, M.O., and Johnson, P.E., 1986, Spectral mixture modeling: A new analysis of rock and soil types at the Viking Lander site. *Journal of Geophysical Research*, 91:8098–8112.

Boardman, J.W., 1993, Automated spectral unmixing of AVIRIS data using convex geometry concepts, *Summaries of the Fourth JPL Airborne Geoscience Workshop*, JPL

Publication 93-26, NASA Jet Propulsion Laboratory, Pasadena, California, pp. 11–14.

Boardman, J.M., Kruse, F.A., and Green, R.O., 1995, Mapping target signature via partial unmixing of AVIRIS data, *Summaries of the Fifth JPL Airborne Earth Science Workshop*, JPL Publication 95-1, NASA Jet Propulsion Laboratory, Pasadena, California, pp. 23–26.

Lillesand, T.M., Kiefer, R.W., and Chipman, J.W., 2004, *Remote Sesning and Image Interpretation*, John Wiley & Sons, New York, p. 614.

Lu, D. and Weng, Q., 2004, Spectral mixture analysis of the urban landscape in Indianapolis with Landsat ETM+ imagery. *Photogrammetric Engineering & Remote Sensing*, 70:1053–1062.

Lu, D. and Weng, Q., 2006a, Use of impervious surface in urban land use classification. *Remote Sensing of Environment*, 102(1–2):146–160.

Lu, D. and Weng, Q., 2006b, Spectral mixture analysis of ASTER imagery for examining the relationship between thermal features and biophysical descriptors in Indianapolis, Indiana. *Remote Sensing of Environment*, 104(2):157–167.

Madhavan, B.B., Kubo, S., Kurisaki, N., and Sivakumar, T.V.L.N., 2001, Appraising the anatomy and spatial growth of the Bangkok Metropolitan area using a vegetation-impervious-soil model through remote sensing. *International Journal of Remote Sensing*, 22:789–806.

Okin, G.S., Roberts, D.A., Murray, B., and Okin, W.J., 2001, Practical limits on hyperspectral vegetation discrimination in arid and semiarid environments. *Remote Sensing of Environment*, 77:212–225.

Painter, T.H., Roberts, D.A., Green, R.O., and Dozier, J., 1998, The effects of grain size on spectral mixture analysis of snow-covered area from AVIRIS data. *Remote Sensing of Environment*, 65:320–332.

Phinn, S., Stanford, M., Scarth, P., Murray, A.T., and Shyy, P.T., 2002, Monitoring the composition of urban environments based on the vegetation-impervious surface-soil (VIS) model by subpixel analysis techniques. *International Journal of Remote Sensing*, 23:4131–4153.

Rashed, T., Weeks, J.R., Roberts, D., Rogan, J., and Powell, R., 2003, Measuring the physical composition of urban morphology using multiple endmember spectral mixture models. *Photogrammetric Engineering and Remote Sensing*, 69:1011–1020.

Roberts, D.A., Gardner, M., Church, R., Ustin, S., Scheer, G., and Green, R.O., 1998, Mapping chaparral in the Santa Monica mountains using multiple endmember spectral mixture models. *Remote Sensing of Environment*, 65:267–279.

Slonecker, E.T., Jennings, D., and Garofalo, D., 2001, Remote sensing of impervious surface: A review. *Remote Sensing Reviews*, 20:227–255.

Ward, D., Phinn, S.R., and Murray, A.L., 2000, Monitoring growth in rapidly urbanizing areas using remotely sensed data. *Professional Geographer*, 53:371–386.

Weng, Q., Lu, D., and Schubring, J., 2004, Estimation of land surface temperature–vegetation abundance relationship for urban heat island studies. *Remote Sensing of Environment*, 89:467–483.

Wu, C., 2004, Normalized spectral mixture analysis for monitoring urban composition using ETM+ imagery. *Remote Sensing of Environment*, 93:480–492.

Wu, C. and Murray, A.T., 2003, Estimating impervious surface distribution by spectral mixture analysis. *Remote Sensing of Environment*, 84:493–505.

7

Separating Types of Impervious Land Cover Using Fractals

Lindi J. Quackenbush

CONTENTS

7.1 Introduction

This chapter considers the utility of fractal analysis in separating different types of impervious land cover with a focus on roofs, roads, and driveways in suburban imagery.

7.1.1 Classifying Land Cover and Land Use

Traditional image classification is fundamentally a problem of segmenting images into regions that have similar spectral characteristics. These techniques rely on spectral data alone and consider each pixel independent of its neighbors [1]. While such approaches have proven successful for classification of lower-resolution imagery, the same has not been true for high spatial resolution imagery. While providing unique advantages, high-resolution imagery also presents challenges in terms of the level of variability within a scene [2]. Whereas moderate-resolution imagery, such as Landsat Enhanced Thematic Mapper Plus (ETM+), provides an averaged response within a 30 m pixel, digital aerial sensors can produce imagery with a ground sampled distance (GSD) of a decimeter or less.

Many recent studies have focused on using remote sensing techniques to characterize areas of impervious land cover. Such characterization is often tied to research that has established a direct relationship between the area of impervious surface within a watershed and pollution of its surface waters [3]. However, studies show that not all impervious land uses contribute equally to the levels of contaminants present in runoff [4] and different types of impervious surface vary in their effectiveness at producing runoff [5]—for example, roofs are generally found to contribute less to overall catchment runoff than road surfaces. For input into hydrologic models, it is desirable to consider areas most likely to have a hydraulic connection to the downstream drainage system [6].

Numerous chapters in this book are focused on separating impervious and pervious regions. However, distinguishing between types of impervious areas is challenging and can require different techniques. Because of the similarities in construction materials, separation of different types of impervious cover based on spectral information alone is generally found lacking. Objects in imagery are often discernable to the human eye because of contrast variation and differences in roughness. Texture provides an expression of the local spatial structure in digital images [7]. Jensen [8] suggested that adding a spatial measure, such as texture or complexity, might enhance land-cover classifications. Texture relates to the tonal changes in an image [9] and is often calculated by computing variance, min–max, or standard deviation within a localized window [10]. Fractal dimension is another potential method used to characterize textural differences.

7.1.2 Defining Dimensionality

Images of an urban scene reflect the complex interplay among the features present. Such a scene commonly combines natural and artificial features with wide variation in size, shape, and composition. Both the scene and the features within it are often challenging to describe. Objects in the scene are frequently characterized by their dimensionality; for example, roads are commonly called one-dimensional, fields two-dimensional, and buildings are often considered to be three-dimensional. This intuitive dimension is called topological or Euclidean dimension. Although this appears to be a reasonable characterization of features at some scales, it does not apply broadly. An alternative to an integer-based dimension is to consider dimensionality in terms of a fractional value. A common fractional measure of dimensionality is fractal dimension. Quackenbush [11] summarizes the mathematical foundations of the fractal dimension.

Through its noninteger value, fractal dimension provides a description of the intricacy of curves and surfaces [8,12]: the more spatially complex a feature, the higher its fractal dimension [13]. Many natural features—for example, coastlines or cloud boundaries—can be described as fractals and characterized using fractal dimension [14]. For example, depending on the level of complexity, a river could vary between two extremes: at its simplest it could be considered one-dimensional (essentially a straight line), or at its most complex it could be so tortuous that it entirely fills two spaces and appears two-dimensional [15].

Fractal dimension has been used to measure the spatial variability of geographic features such as coastlines and terrain surfaces [16], to incorporate texture in classifying types of vegetation [17], to segment images [18], and to characterize broad classes of land use within an urban environment [19]. Cultural features typically show lower fractal dimension than natural objects, and studies have used this characteristic to separate natural and artificial features within imagery [20,21]. Since different types of impervious land cover often exhibit visual distinction in texture, fractal dimension may be useful in characterizing these differences. One of the challenging factors is that the visual texture of features in a digital image varies depending on the properties of the image—for example, the size of the feature relative to the pixel size—and it is likely that measurements of texture will also change.

7.1.3   Scale

The term *scale* is used in a variety of applications within the context of spatial, temporal, or spatio-temporal analysis [10]. Two aspects of scale often considered in the spatial context are spatial resolution and spatial extent. The spatial resolution of a digital image is partly a function of the instantaneous field of view of the sensor, which generally correlates to the GSD [9]. Determining the appropriate spatial resolution for a project can be challenging since this is a function of both the scene under analysis and the information sought [1]. While finer spatial resolutions provide additional detail, this is not always advantageous and may lead to increased processing time and storage requirements, with minimal gain in terms of the desired information. Markham and Townshend [22] reported that changing spatial resolution could have a positive or negative impact on classification accuracy depending on the spectral overlap between classes.

A wide variety of spatial analyses, such as measurements of texture, are performed using a localized image window. This relates to the second aspect of scale: spatial extent, considering the appropriate size of the region to analyze. This is often tied to the concept of spatial autocorrelation, which is summarized by Tobler's Law: all places are related but nearby places are more related than distant places [23]. Woodcock et al. [24,25] evaluate the use of variograms to assess the relationship between ground scenes and their corresponding images. Variograms provide a means to quantify spatial variation and establish a range within which points are considered correlated. Woodcock et al. [24,25] found that the range of the variogram was related to the size of the objects in the scene.

Scale-dependent phenomena vary with observation under different scales [10]. Most spatial phenomena are scale-dependent [26] and their characteristics, including texture and dimensionality, may vary based on the spatial resolution or spatial extent studied. A theoretically perfect fractal object will have the same fractal dimension at any scale or level of magnification [27]. However, practical limitations (e.g., in image capture) lead to variation in the measured fractal dimension for imagery of even a perfect object. A real object may have integer dimension when studied over some range of scales rather than at all scales [27]. For example, a road may be considered straight and flat at some scales, and thus can be described using Euclidean geometry, but at other scales, it has complex detail that will be better characterized in noninteger dimensions [12].

7.1.4   Fractal Dimension of the Urban Environment

Urban environments contain a complex mixture of both natural and artificial features. Many authors consider fractal dimension useful for describing natural features and Euclidean (integer) dimensions suitable for characterizing artificial features. Other authors have found that man-made features also display characteristics that make them suitable to characterize using fractal dimension at some scales of analysis [12]. The scale of consideration, in terms of both GSD and local neighborhood size, plays an important role in using fractal dimension to separate types of impervious land cover.

Several studies have used fractal dimension to differentiate types of land use. Generally, these studies looked at broad classes of land use (such as commercial or residential) rather than the very specific land-use types studied in this project. Myint [28,29] used fractal analysis to classify 2.5 m GSD urban imagery collected from the Advanced Thermal Land Applications Sensor (ATLAS) into six classes: single-family homes with <50% tree canopy, single-family homes with >50% tree canopy, commercial, woodland, agriculture, and water bodies. Myint [28] visually identified ten sample plots (with dimension 162.5 m × 162.5 m) for each of the six classes and found that using fractal dimension for classification produced overall accuracies that were generally below 60% for the six-class classification, with some particularly low individual class accuracies. The author theorized that the low accuracies were due to the substantially overlapping ranges for calculated fractal dimension for each class. This is not surprising, since the classes—for example, low- and medium-density residential areas—included common characteristics and thus overlapping texture characteristics. Lam et al. [30] have shown that some fractal algorithms are sensitive to contrast enhancement. Myint [28] did not specify what preprocessing was done before analysis.

Although Batty and Longley [19] considered their results preliminary, their research showed the importance of the scale of analysis in characterizing urban environments. Batty and Longley [19] characterized the varying levels of complexity inherent in the urban environment by calculating the fractal dimension of five broadly classified land uses: residential, commercial–industrial, educational, transport, and open space. Torrens and Alberti [31] applied this information in using fractal dimension as a measure of urban sprawl. Torrens and Alberti [31] found that they could correlate characteristics of development, such as sprawl, with specific signatures calculated using a fractal-based approach. They focused on separating compact areas of development from lower density areas on the fringes.

7.1.5 Calculating Fractal Dimension

7.1.5.1 Overview

A large number of algorithms have been developed to calculate the fractal dimension of imagery. Turner et al. [18] broadly categorize the methods into two groups: size–measure relationships and application of relationships. Size–measure relationships are based on repeated measurement of the area of a surface or the length of a curve using different scales. The techniques for the application of relationship methods are based on fitting a curve or surface to a known fractal function. The majority of techniques reported in the literature fall under the former category. There are differences between the techniques, and two methods will rarely produce equal values for the same object [18].

7.1.5.2 Walking Divider and Isarithm

The walking-divider method is used to determine the fractal dimension of a linear feature. This technique uses a pair of dividers, separated by varying amounts, to measure the length of a curving feature as a series of chords [18]. The length of the curve is calculated as the number of steps, $N(\delta)$, of length δ using varying values of δ. As δ decreases, the amount of detail that can be captured increases and the length of the curve will appear to increase. The dimension of the curve is calculated using a linear regression based on the logarithm of the number of steps, $N(\delta)$, against the logarithm of the step length, δ.

The isarithm method is an extension of the walking-divider method and is used to determine the fractal dimension of surfaces [32]. Isarithms are generated by dividing the range of pixel values into a number of equally spaced intervals and connecting pixels of equal intensity interval. Surface fractal dimension is calculated as the mean fractal dimension of the isolines that characterize the surface. Read and Lam [13] used the isarithm method for characterizing land-use and land-cover change in Landsat TM imagery. Myint [29] evaluated the isarithm method for characterizing texture features of urban land-cover classes in ATLAS imagery. Weng [33] used the isarithm method to consider patterns in urban development by studying the urban heat island effect.

7.1.5.3 Box-Counting Dimensions

Box counting generally involves covering an object with a grid of n-dimensional boxes of side length δ and counting the number of nonempty boxes $N(\delta)$ [34]. As with the walking-divider method, boxes of recursively smaller size are used to cover the object, and a bilogarithmic plot is used to determine the fractal dimension [18]. For gray-scale imagery, the n-dimensional box is a cube (three dimensions) with the third dimension being the gray values. Shen [35] used binary images and simplified the n-dimensional box to a square (two dimensions). Shen calculated the fractal dimension of 20 urban environments in several cities to characterize urban sprawl.

Many versions of the box-counting approach have been developed to produce a reasonably fast and accurate algorithm [18]. A common implementation of the box-counting approach for image processing is the differential box-counting (DBC) method, developed by Sarkar and Chaudhuri [36]. Haering and da Vitoria Lobo [37] calculated fractal dimension using DBC to separate deciduous trees from all other features in a variety of terrestrial images. Chaudhuri et al. [38] also used a DBC method to estimate fractal dimension for a variety of nature textures. Sarkar and Chaudhuri [36] compared their DBC method with several other methods for calculating fractal dimension and found that it provided equivalent results with a substantial reduction in computation.

7.1.5.4 Triangular Prism

Clarke [39] developed a triangular prism surface area (TPSA) method for calculating the fractal dimension of topographic surfaces. Prism counting is a derivation of the box-counting approach; instead of cubes, however, the combined area of four-sided triangular prisms is used to determine fractal dimension. Studies have used the triangular prism method for calculating fractal dimension in a variety of image types, including IKONOS and Landsat ETM+ [13,40,41]. In applying the TPSA method to imagery, intensity variation is treated as the topographic surface used by Clarke [39]. The surface area of the image is estimated using triangular prisms of varying sizes. A triangular prism is formed using five values: the intensity of four pixels that make up the corners of a square and the mean intensity value. An illustration of the triangular prism formed using four adjacent pixels is shown in Figure 7.1. The area of the top of the prism is calculated as the sum of the area of the four triangular faces.

The prism, shown in Figure 7.1, has a base dimension of one pixel (from the center of a pixel to the center of the neighboring pixel). The surface area of the image is estimated using contiguous prisms that cover the entire image. The surface area is repeatedly estimated by stepping through the same area calculation using prisms created by groupings of four pixels separated by increasing the base dimension. Clarke [39] analyzed square images of dimension $2^n + 1$ and determined area using prisms with the base dimension increasing by powers of 2 from 2^0 (i.e., one pixel) up to 2^n. As with previous methods, the fractal dimension is calculated by considering the slope of a bilogarithmic plot, in this case the total area of the triangular faces of the prisms against the base area of an individual prism. Clarke [39] used prism dimensions increasing in powers of 2 to ensure an even distribution of points along the log-base dimension axis of the bilogarithmic plot.

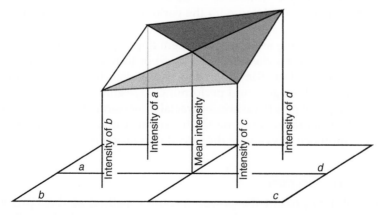

FIGURE 7.1
Illustration of the triangular prism used for calculating fractal dimension.

7.1.5.5 Area–Perimeter Relationship

The area–perimeter method is used to characterize the shape complexity [13] of contiguous regions, for example, areas that have been defined using traditional classification [42]. Each class in an image is assigned a dimension based on the relationship between the area and the perimeter of the regions that belong to the class. Chuvieco [43] used the area–perimeter method to evaluate the impact of fire on landscape patterns. Chuvieco [43] identified clumps by calculating a normalized difference vegetation index (NDVI) and then performing density slicing.

Fractal dimension is based on the principle that a feature displays self-similarity over a variety of scales. Instead of making measurements over various scales, the area–perimeter approach relies on variability in the size of the clumps within a class to act as a substitute for scale variation [44]. Hence, while the area–perimeter method provides a measure of the complexity of a feature class, it does not calculate this measure over a variety of scales and thus is not a true fractal dimension.

7.1.5.6 Fractional Brownian Motion and the Variogram Method

Brownian motion was first used to describe the random motion of pollen grains suspended in water [45]. Water molecules randomly bombard the suspended pollen grains, causing the grains to "walk" about in a random fashion—that is, to move in a series of random steps in random directions. Such motion is fractal in nature: the pattern of the motion observed at 30 s increments will resemble the pattern created if the observation is made at subsecond increments. The fractional Brownian motion (fBm) model is useful for characterizing fractal objects, such as natural features, and also for characterizing images of fractal objects. The fBm model states that there is a statistical relationship between the distance between two pixels and the variance of the difference in their values [29]. Using this model, fractal dimension is calculated with the variogram measure of spatial complexity [32].

Various authors have used the variogram approach in calculating fractal dimension to perform a classification [13,29] or to resample an image [46]. Chen et al. [27] used an fBm approach to calculate fractal dimension for classifying medical imagery and performing edge enhancement. Jaggi et al. [41] used fBm principles to determine fractal dimension for images obtained from NASA's Calibrated Airborne Multispectral Scanner. Rees [15] used the variogram approach to evaluate the fractal dimension of ice sheet surfaces.

7.1.5.7 Power Spectrum

Turner et al. [18] present the power spectrum method as their preferred approach to calculating fractal dimension. In this method, the real-space image is Fourier transformed and the power spectrum is computed. This measured power spectrum is then matched to the formula for an ideal fractal signal using a least-squares fit. The power spectrum approach requires sophisticated preprocessing to estimate the power spectrum of an image sample [47]. Because of the sensitivity of the power spectrum to preprocessing, Soille and Rivest [47] found that the practical applications of this approach were limited. Cox and Wang [48] found that curve-fitting errors could be larger for the spectral method than for other techniques. Despite this, Cox and Wang [48] reported that the technique was popular in geophysics.

7.1.6 Issues in Calculating Fractal Dimension

A variety of factors must be considered in selecting a method to calculate fractal dimension of imagery. Klinkenberg and Goodchild [49] compared seven techniques for calculating fractal dimension of a series of digital elevation models (DEMs). They found that fractal dimension calculated for a single region varied with the seven approaches, but methods tended to be consistent: for example, higher values were typical for one method compared with another. Soille and Rivest [47] compared six methods for computing fractal dimension, using three simulated images with known fractal dimension. Of the methods reviewed by Soille and Rivest [47], the triangular prism and variogram approaches exhibited the least discrepancy from the true fractal dimension. Many studies have found that cultural features generally yield low fractal dimension [20] and Soille and Rivest [47] found that the triangular prism was the most accurate for images with low fractal dimension. Lam et al. [30] compared the isarithm, triangular prism, and variogram methods for calculating fractal dimension and determined that the variogram method was unsuitable for characterizing the fractal dimension of imagery. Myint [28] used both the triangular prism and isarithm methods and found that the triangular prism method performed consistently better.

Cao and Lam [10] discuss many of the issues confronted when considering scale in remote sensing. This consideration is important since the character of a feature varies when imaged using different GSDs or when calculated using different-sized local windows. Selecting an appropriate sized window is partially determined by defining a neighborhood within which pixels show correlation. Woodcock et al. [24,25] used variograms to consider the range of distances in an image where pixels remained correlated. They found that the appropriate window size is a function of the object of interest and the spatial resolution of the imagery. Stein [20] found that separating man-made and natural features required using a local window that was approximately the same size as the objects of interest. Marceau and Hay [50] reported that window size was the most significant factor in accounting for variability in classification results. They found that the accuracy results for different classes varied with different window sizes.

Spatial resolution, often characterized using GSD, also changes the appearance and properties of objects in an image. One of the challenges in image classification is the impact of pixels that contain multiple cover types. Markham and Townshend [22] found that reducing pixel size reduced the overall proportion of mixed pixels and improved the accuracy assessment. Conversely, reducing the pixel size increased the spectral variability within classes and made class separability more challenging. This effect has made classification of high spatial resolution imagery challenging [2].

7.2   Methods and Materials

7.2.1   Imagery

This project used a set of digital aerial imagery acquired using the Emerge DSS model 300 imager with a 4092 × 4077 pixel silicon CCD focal plane array. The CCD array consists of red, green, and blue filter elements with a spectral response similar to color film. The camera was packaged with an integrated GPS/IMU system; thus the imagery was georeferenced without ground control. The imagery was acquired over a study site in the Town of Whitestown, adjacent to the City of Utica in Central New York State (approximate site location: N 43°07.1', W 75°18.2'). The imagery was supplied as three-band normal color orthorectified single frames. The orthorectification was performed using 10 m DEMs. The imagery was processed by Emerge to mitigate the impact of two different components of noise. This included performing a dark signal correction to remove thermal and other dark signals from each pixel and applying individual pixel gain corrections to generate a uniform response throughout the sensor (Kinn, G.J., personal communication, 2004).

The sensor was flown at five altitudes over the study site, producing imagery with GSDs of ~0.1, 0.2, 0.3, 0.4, and 0.5 m. The collection was flown between 1:58 PM and 3:16 PM on a single day in July 2003, beginning with the highest altitude. Two 360 m × 360 m subscenes were visually selected to include primarily residential areas. The first study area included all five altitudes; the second included only the four highest altitudes, generating a total of nine sample images. The multialtitude nature of the dataset provided a unique opportunity for analysis. However, since such datasets are not generally practically available, rather than combine the imagery, this project assessed each image separately with a goal of understanding the potential utility of the different image scales.

7.2.2   Preprocessing

The image data used in this study were acquired using 12 bit radiometry with conversion to 8 bit during the orthorectification process. Studies have found that stretching image data over an 8 bit range produced the most reliable results [13]. To ensure that values in the image subsets covered the available dynamic range, the data were linearly stretched using a model in ERDAS IMAGINE (Leica Geosystems, LLC, Norcross, Georgia). The linear stretch was used to best reflect the radiometric characteristics of the sensor (Kinn, G.J., personal communication, 2004).

7.2.3 Calculating Fractal Dimension

7.2.3.1 Overview of Algorithm

The algorithm used to calculate fractal dimension (FD) was based on the TPSA approach presented by Clarke [39]. As discussed earlier in this chapter, the triangular prism method approximates the surface area of an image using a series of triangular prisms with varying base dimensions. The triangular prism algorithm was selected because it was appropriate for small image windows; it has been shown to be computationally efficient and accurate; and it has provided encouraging results for a variety of applications. The algorithm was coded as a series of modules using Visual Basic. Local fractal dimension was calculated throughout the image based on a neighborhood window.

7.2.3.2 Window Size

Using the triangular prism method to determine fractal dimension involves calculating the area of a series of triangular prisms over several scales. Because this calculation uses a bilogarithmic plot, prism sizes were incremented as powers of 2 to ensure an even distribution of points. Research has suggested that smaller windows better reflect land-cover classes. However, this must be balanced with the number of points needed to fit a line in the linear regression. The minimum window size used was 17×17, which provided five points. The appropriateness of this number of points was confirmed by consideration of the standard error of the slope of the line on the bilogarithmic plot used to determine fractal dimension.

Local fractal dimension was calculated for each image subset using window sizes of 17×17, 33×33, and 65×65 pixels. The corresponding ground area varied based on the GSD of the image. With the smallest GSD, windows of dimension 17×17, 33×33, and 65×65 corresponded to 1.7, 3.3, and 6.5 m, respectively; with the largest GSD, the windows corresponded to 8.5, 16.5, and 32.5 m. The maximum window size was limited in part by the processing power of the computer used.

7.2.3.3 Multiband Modification

The triangular prism algorithm developed by Clarke [39] was originally applied to gray-scale DEMs. Researchers have applied the algorithm to single image bands or from an intensity layer derived from multiband imagery. The TPSA algorithm uses Heron's formula to calculate the area (A) of a triangle with side lengths a, b, and c:

$$A^2 = s(s - a)(s - b)(s - c) \qquad \text{where} \quad s = 1/2(a + b + c) \qquad (7.1)$$

Each side length is calculated based on the coordinates of the triangle vertices. With a single-band image, the coordinates of each vertex are expressed in three dimensions: two based on pixel coordinates, with the third coordinate being the image intensity value. This work used all of the bands in the multispectral imagery by calculating the side lengths of each triangular prism face using the two pixel coordinates and the digital numbers (DNs) from the three bands of the imagery.

7.2.3.4   Pixel Size–Digital Number Ratio

The relationship between pixel size and digital numbers is well defined in DEMs—for example, both are commonly defined in meters—thus the area of the triangular prism has meaningful units. When the algorithm is applied to imagery, the surface area computed mixes both spatial and intensity units. This undefined relationship is one reason why the triangular prism method is sensitive to contrast stretches since only the DNs are affected by such a stretch. In order to assess the impact of such stretching, a factor was incorporated into the algorithm that enabled specification of a ratio between pixel size and digital number before calculating the length of the sides of each triangular face. Processing was performed on each of the image subsets with ratios of 0.1, 1, and 10. This had the effect of stretching the original 8 bit range of the DNs from 0 to 2550, 0 to 255, and 0 to 25.5, respectively.

7.2.4   Assessment

7.2.4.1   Overview

The linear regression used to calculate fractal dimension generates intercept and R^2 values. Several researchers have found that the intercept from the regression provides useful information, and when preliminary analysis suggested that the R^2 values might also provide additional value, all three output variables were considered with a multivariate analysis of variance (ANOVA) using SAS software (SAS Institute, Inc., Cary, North Carolina). Where multivariate differences between the cover types occurred, univariate analysis was performed to determine which variables were significant. The multivariate and univariate analyses considered the three cover types as a group (overall analysis) and where significant differences were found, also considered pairwise groupings of the cover types—that is, road vs. roof, road vs. driveway, and roof vs. driveway.

7.2.4.2   Separating Cover Types

The imagery covered a residential area in Central New York. Roads in the area are asphalt (averaging 7–9 m wide); driveways are asphalt or concrete (averaging 5–6 m wide and 11–14 m long); and roofs are various colors of asphalt shingles (averaging 9–12 m wide and 15–17 m long).

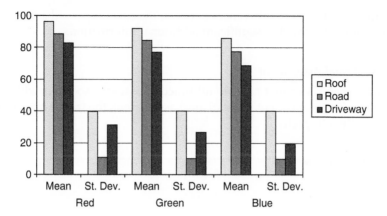

FIGURE 7.2
Mean and standard deviation of digital numbers in bands of 0.1 m GSD image for three impervious cover types.

One sidewalk and small impervious features, such as cars and backyard patios, were present; however, the categorization of roofs, roads, and driveways included the majority of the impervious surface area.

Areas of interest covering all roofs, roads, and driveways within each image were generated within IMAGINE. Figure 7.2 illustrates the mean and standard deviation of the digital numbers for each cover type in the 0.1 m GSD data. The trend in spectral variability was similar for all of the GSDs, with lower variability in roads than in either roofs or driveways. The variability in the driveways was likely related to the presence of both asphalt and concrete construction; variability in the roofs due to the varying colored shingles.

7.2.4.3 Evaluation with SAS Software

Multivariate and univariate ANOVA were used to determine if there were statistical differences between roofs, roads, and driveways based on fractal dimension, intercept, or R^2. Since the generation of the fractal dimension variables is based on a neighborhood, the FD for a single pixel is related to that of its neighbors. Analysis was performed using 100 randomly selected pixels to reduce the likelihood of violating assumptions of statistical independence. Additionally, selecting a very large number of pixels may have led to challenges in distinguishing between statistical and significant differences. With nine input images and output files generated using three window sizes and three pixel–DN ratios, there were 81 output datasets created. Pixel size, window size, and pixel–DN ratios were considered separately to avoid different factors confounding the potential differences in cover type, which generated an enormous number of comparisons. Because the analyses used the same model with the same size datasets, only P-values are reported to simplify the output. Given the purpose of the analysis—that is, determining if there are differences between the cover types—the *F*-statistic would not have provided significant additional information.

7.2.4.4   Visual Assessment

Statistical analysis cannot capture all practical issues associated with data generation. To complement the statistical assessment, the fractal output was also visually interpreted.

7.3   Results

7.3.1   Overview

The triangular prism algorithm to calculate fractal dimension was applied to each of the nine images using three different local window sizes and three different pixel–DN ratios. The output from the algorithm included fractal dimension, as well as *y*-intercept and R^2 values from the linear regression. Because the results attained were similar for the two study sites, the results presented focus on analysis from the study area with five altitudes.

Table 7.1 illustrates the statistics generated using 100 pixels drawn randomly from each of the impervious cover types in the output layers of the 0.2 m GSD imagery, calculated using a 65 × 65 pixel window. The trends shown in this table are similar to those seen in all data layers; for example, the differences between the mean fractal dimension for each cover type were small when compared with the standard deviation. Another trend was that increasing the pixel–DN ratio decreased the derived fractal dimension, while increasing the intercept values. The statistics shown in Table 7.1 also give an indication of the general tendency of the R^2 values. While there

TABLE 7.1

Example Statistics for 0.2 m GSD Image Generated Using a 65 × 65 Pixel Window

	Pixel–DN Ratio	Driveway		Road		Roof	
		Mean	SD	Mean	SD	Mean	SD
Fractal dimension	0.1	2.61	0.14	2.53	0.07	2.64	0.09
	1	2.38	0.12	2.30	0.05	2.43	0.08
	10	2.06	0.04	2.04	0.02	2.09	0.03
Intercept	0.1	9.23	0.57	8.84	0.36	9.74	0.50
	1	11.89	0.89	11.20	0.45	12.46	0.61
	10	13.69	0.44	13.38	0.18	14.00	0.35
R^2	0.1	0.97	0.02	0.97	0.02	0.97	0.02
	1	0.98	0.01	0.99	0.02	0.98	0.03
	10	0.94	0.03	0.95	0.03	0.95	0.03

was some variability depending on the pixel–DN ratio, 99% of the R^2 values were above 0.70 and over 90% of the R^2 values were greater than 0.90.

7.3.2 Statistical Evaluation of Fractal Dimension

Determining if different impervious cover types generated statistically different fractal dimension was addressed using multivariate analysis of variance tools in SAS software. An overall significance level of 0.05 was selected, with the Bonferroni method used to decrease the significance level when considering univariate and pairwise comparisons. In all cases, the overall multivariate analysis of variance showed significant differences ($P < 0.0001$) between the three cover types. Based on this, overall univariate analysis was performed (summarized in Table 7.2). This analysis showed that regardless of pixel size, window size, or pixel–DN ratio, there are generally differences between the three cover types based on fractal dimension or intercept.

Because the overall comparisons showed differences in the three types of impervious cover, pairwise comparisons were performed. Given the trend for the R^2 values not to be significantly different, the pairwise analysis focused on separation based on fractal dimension or intercept values. In only one case (road vs. driveway in 0.1 m imagery with 17×17 window) did the pairwise multivariate analysis of variance show insignificant differences ($P = 0.03$). Tables 7.3 and 7.4 illustrate the univariate pairwise

TABLE 7.2

P-Values for Univariate Comparison of Driveway vs. Roof vs. Road; Significant Differences Where *P*-Value $<0.05/9 = 0.006$ and Pairwise Multivariate $P = 0.000$ Except When Noted

Pixel Size (m)	Window	Ratio = 0.1			Ratio = 1			Ratio = 10		
		FD	Int.	R^2	FD	Int.	R^2	FD	Int.	R^2
0.5	65 × 65	0.000	0.000	0.836	0.000	0.000	0.007	0.000	0.000	0.124
0.5	33 × 33	0.000	0.000	0.185	0.000	0.000	0.003	0.000	0.000	0.000
0.5	17 × 17	0.000	0.000	0.740	0.000	0.000	0.098	0.000	0.000	0.084
0.4	65 × 65	0.000	0.000	0.141	0.000	0.000	0.468	0.000	0.000	0.122
0.4	33 × 33	0.000	0.000	0.818	0.000	0.000	0.179	0.000	0.000	0.000
0.4	17 × 17	0.000	0.000	0.518	0.000	0.000	0.603	0.000	0.000	0.557
0.3	65 × 65	0.000	0.000	0.005	0.000	0.000	0.000	0.000	0.000	0.414
0.3	33 × 33	0.000	0.000	0.920	0.000	0.000	0.124	0.000	0.000	0.130
0.3	17 × 17	0.000	0.000	0.814	0.000	0.000	0.673	0.000	0.000	0.197
0.2	65 × 65	0.000	0.000	0.028	0.000	0.000	0.031	0.000	0.000	0.008
0.2	33 × 33	0.000	0.000	0.014	0.000	0.000	0.006	0.000	0.000	0.004
0.2	17 × 17	0.000	0.000	0.002	0.000	0.000	0.020	0.000	0.000	0.002
0.1	65 × 65	0.000	0.000	0.058	0.000	0.000	0.186	0.000	0.000	0.012
0.1	33 × 33	0.000	0.000	0.014	0.000	0.000	0.047	0.000	0.000	0.835
0.1	17 × 17[a]	0.000	0.000	0.479	0.000	0.000	0.160	0.000	0.000	0.081

[a] *P*-value for driveway vs. road $= 0.03$.

TABLE 7.3

P-Values for Univariate Pairwise Comparison of Cover Type: 0.1 m Pixel

Window	Comparison	Ratio = 0.1		Ratio = 1		Ratio = 10	
		FD	Int.	FD	Int.	FD	Int.
65 × 65	Driveway vs. roof	0.000	0.000	0.000	0.000	0.000	0.000
	Driveway vs. road	0.024	0.000	0.000	0.000	0.000	0.000
	Roof vs. road	0.000	0.000	0.000	0.000	0.000	0.000
33 × 33	Driveway vs. roof	0.000	0.000	0.000	0.000	0.000	0.000
	Driveway vs. road	0.159	0.012	0.259	0.039	0.108	0.093
	Roof vs. road	0.000	0.000	0.000	0.000	0.000	0.000
17 × 17	Driveway vs. roof	0.000	0.009	0.000	0.000	0.000	0.000
	Driveway vs. road	0.044	0.001	0.629	0.011	0.328	0.144
	Roof vs. road	0.004	0.000	0.000	0.000	0.000	0.000

comparisons for the 0.1 and 0.5 m imagery, respectively. These tables show the trend that roofs and roads were generally separable, regardless of the variable considered, while driveways were more frequently confused.

7.3.3   Visual Assessment

With nine image subsets, nine output variables, and three different window sizes, the number of possible combinations for visualization was prohibitively large to display them all. Figure 7.3 illustrates a subset of the results generated using a 17 × 17 window with a pixel–DN ratio of 1. Figure 7.3a shows the intensity from the 0.1 m input file; Figure 7.3b and c show the fractal dimension and intercept output, respectively, for the 0.1 m image; Figure 7.3d and e show the fractal dimension and intercept output, respectively, for the 0.5 m image.

TABLE 7.4

P-Values for Univariate Pairwise Comparison of Cover Type: 0.5 m Pixel

Window	Comparison	Ratio = 0.1		Ratio = 1		Ratio = 10	
		FD	Int.	FD	Int.	FD	Int.
65 × 65	Driveway vs. roof	0.409	0.000	0.007	0.000	0.000	0.000
	Driveway vs. road	0.000	0.000	0.000	0.000	0.000	0.000
	Roof vs. road	0.000	0.000	0.000	0.000	0.000	0.000
33 × 33	Driveway vs. roof	0.005	0.000	0.000	0.000	0.000	0.000
	Driveway vs. road	0.000	0.000	0.000	0.000	0.000	0.000
	Roof vs. road	0.000	0.000	0.000	0.000	0.000	0.000
17 × 17	Driveway vs. roof	0.109	0.000	0.013	0.000	0.002	0.000
	Driveway vs. road	0.000	0.000	0.000	0.000	0.000	0.000
	Roof vs. road	0.000	0.000	0.000	0.000	0.000	0.000

FIGURE 7.3
(a) Emerge 0.1 m GSD image intensity, (b) fractal dimension from 0.1 m image

(*continued*)

(c)

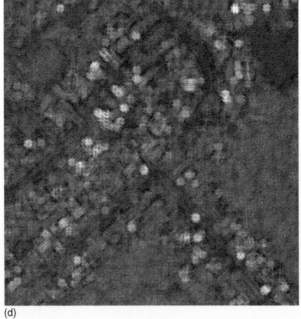

(d)

FIGURE 7.3 (continued)
(c) intercept from 0.1 m image, (d) fractal dimension from 0.5 m image, and

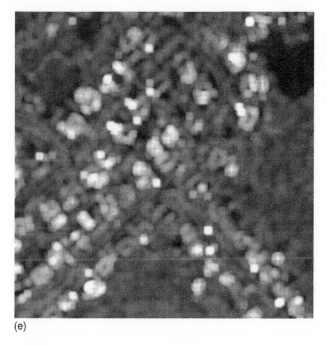

(e)

FIGURE 7.3 (continued)
(e) intercept from 0.5 m image; output generated using a 17 × 17 window with a pixel–DN ratio of 1.

7.4 Discussion

7.4.1 Statistical Evaluation of Fractal Dimension

The overall multivariate analysis of variance demonstrated that there were significant differences between roofs, roads, and driveways considering FD, intercept, and R^2 across the three pixel–DN ratios. Because the overall multivariate analysis showed statistical differences, overall univariate and pairwise multivariate analyses were performed. Table 7.2 summarizes the overall univariate analyses and also indicates the single pairwise multivariate comparison that did not show statistical differences. This table shows that in every example, the cover types were different based on the fractal dimension values and, with the exception of one case, the cover types were also separable based on intercept.

The pairwise univariate analysis showed that statistical distinction between roofs and roads is generally possible regardless of the variables or parameters considered, while separating driveways from roads or roofs is often more difficult. Tables 7.3 and 7.4 illustrate that separation between pairs of cover types was more challenging when calculation used the smallest-sized window. These tables also illustrate the trend that there were some differences in separability of the cover types depending on both the pixel and the window size used. Understanding these differences relied on the visual assessment.

7.4.2   Visual Assessment

The results of the statistical analysis showed that the separation of cover types was weaker when considering the smallest window size. The visual interpretation provides a clue as to the reasoning behind this, as is illustrated in Figure 7.3. With smaller ground extents, the fractal algorithm is a very effective edge detector. In small windows, the edges of both roads and roofs are well defined using either the fractal dimension or intercept variables. In some cases, the edge distinction divided a roof into segments. The fractal layers then saw these segments as sections of road or driveway. Ferro and Warner [7] discuss similar challenges involving edges when incorporating texture in image classifications.

With larger windows, the edges of features such as roads and roofs became less defined in the fractal dimension images, producing clusters of pixels and leading to the objects appearing as a single unit. Studies have shown that there is a correlation between the window size and the feature of interest [1]; the challenge is finding the correct window size for a given feature. Roads in the study area were ~7 m wide and were more defined in the fractal layers in the 0.3 m imagery using a 33 × 33 (9.9 m) window or in the 0.5 m imagery using a 17 × 17 (8.5 m) window.

A potentially confounding factor in the analysis was the seasonality of the imagery. The imagery used in this analysis was collected during the summer with full leaf-out conditions. The spatial characteristics of roads and driveways were complicated by the presence of overhanging trees. In several cases, the fractal dimension calculated using the smallest window sizes led to confusion of roofs when trees were in close proximity. The larger windows seemed less sensitive to this problem.

7.5   Conclusions

This project sought to evaluate if spatial complexity measures might provide a means to separate types of impervious land cover. In particular, the use of fractal dimension calculated using the TPSA method was considered. Fractal dimension is commonly calculated by taking a measure over a series of scales, performing a linear regression, and considering the slope of the line of best fit. Analysis for this project considered fractal dimension as well as the intercept and the R^2 values from the regression.

The primary objective of this study was to consider the separability of three impervious cover types commonly found in residential areas: that is, roads, roofs, and driveways. Overall and pairwise multivariate analysis demonstrated that there were generally significant differences between the three cover types. The visual interpretation provided clues about some of the limiting factors in the separation of these cover types. With smaller windows, the fractal dimension and intercept layers fundamentally detected edges of features such as roads or driveways. Interestingly, it appeared that the edge detection properties appeared to relate more strongly to the number of pixels in the window rather than the GSD of the pixels.

From the results, the following observations are noted:

- The multivariate statistical analysis showed differences in fractal dimension for three impervious cover types.
- When cover types were considered on a pairwise basis, roofs and roads were generally separable, while driveways were more frequently confused.
- In a univariate analysis, samples were generally separable using the derived fractal dimension or intercept values.

The availability of a multialtitude dataset allowed for an assessment of the impact of the image scale on the generation of fractal dimension. While statistical separability of the three cover types under analysis was noted for most of the images, visual analysis showed that there were distinct differences. For example, it was observed that at particular scales, both fractal dimension and intercept values function as edge detectors. This effect was most apparent when considering the roofs using windows that corresponded to a smaller spatial extent. Defining the appropriate pixel and window size for applying such analysis appears to be dependent on the size of the features.

This book describes many different research projects aimed at delineating impervious regions. This project sought to take the next step and looks at separating types of impervious land cover. The analysis described demonstrated that there are statistical differences between fractal dimensions calculated for different classes of impervious land cover. The challenge remains to translate these statistical differences into practical advantage through classification.

References

1. Woodcock, C.E. and Strahler, A.H., The factor of scale in remote sensing, *Remote Sensing of Environment*, 21, 311, 1987.
2. Quackenbush, L.J., Hopkins, P.F., and Kinn, G.J., Developing forestry products from high resolution digital aerial imagery, *Photogrammetric Engineering and Remote Sensing*, 66, 1337, 2000.

3. Sleavin, W. et al., Measuring impervious surfaces for non-point source pollution modeling, in *Proceedings of the ASPRS Annual Convention*, Washington, D.C., 2000.

4. Smith, A.J., Subpixel Estimates of Impervious Surface Cover Using Landsat TM Imagery, M.A. Scholarly Paper, University of Maryland, Maryland, 2000.

5. Arnold, C.L. and Gibbons, C.J., Impervious surface coverage: the emergence of a key environmental indicator, *Journal of the American Planning Association*, 62, 243, 1996.

6. Booth, D.B. and Jackson, C.R., Urbanization of aquatic systems: degradation thresholds, stormwater detection, and the limits of mitigation, *Journal of the American Water Resources Association*, 33, 1077, 1997.

7. Ferro, C.J.S. and Warner, T.A., Scale and texture in digital image classification, *Photogrammetric Engineering and Remote Sensing*, 68(1), 51, 2002.

8. Jensen, J.R., *Introductory Digital Image Processing: A Remote Sensing Perspective*, 2nd ed., Prentice Hall, New Jersey, 1996, chaps. 8, 9.

9. Lillesand, T.M., Kiefer, R.W., and Chipman, J.W., *Remote Sensing and Image Interpretation*, 5th ed., John Wiley & Sons, New York, 2004, chaps. 4, 5.

10. Cao, C. and Lam, N.S.-N., Understanding the scale and resolution effects in remote sensing and GIS, in *Scale in Remote Sensing and GIS*, Quattrochi, D.A. and Goodchild, M.F., Eds., CRC Press, Boca Raton, FL, 1997, chap. 2.

11. Quackenbush, L.J., Classification of Impervious Land Cover Using Fractals, Ph.D. thesis, State University of New York, Syracuse, 2004.

12. Russ, J.C., *The Image Processing Handbook*, 3rd ed., CRC Press, Boca Raton, FL, 1999, chaps. 4, 13.

13. Read, J.M. and Lam, N.S.-N., Spatial methods for characterizing land cover and detecting land-cover changes for the tropics, *International Journal of Remote Sensing*, 23(12), 2457, 2002.

14. Falconer, K., *Fractal Geometry: Mathematical Foundations and Applications*, 2nd ed., John Wiley & Sons, Chichester, England, 2003, chaps. 3, 18.

15. Rees, W.G., Measurement of the fractal dimension of ice-sheet surfaces using Landsat data, *International Journal of Remote Sensing*, 13, 663, 1992.

16. Lam, N.S.-N., Description and measurement of Landsat TM images using fractals, *Photogrammetric Engineering and Remote Sensing*, 56, 187, 1990.

17. De Jong, S.M. and Burrough, P.A., A fractal approach to the classification of Mediterranean vegetation types in remotely sensed images, *Photogrammetric Engineering and Remote Sensing*, 61, 1041, 1995.

18. Turner, M.J., Blackledge, J.M., and Andrews, P.R., *Fractal Geometry in Digital Imaging*, Academic Press, San Diego, 1998, chaps. 3, 5.

19. Batty, M. and Longley, P.A., *Fractal Cities*, Academic Press, London, 1994, chap. 6.

20. Stein, M.C., Fractal image models and object detection, in *Proceedings of the SPIE Visual Communications and Image Processing II*, Cambridge, MA, 1987, 293.

21. Solka, J.L. et al., Identification of man-made regions in unmanned aerial vehicle imagery and videos, *IEEE Transactions on Pattern Analysis and Machine Intelligence*, 20, 852, 1998.

22. Markham, B.L. and Townshend, J.R.G., Land cover classification accuracy as a function of sensor spatial resolution, in *Proceedings of the 15th International Symposium on Remote Sensing of Environment*, Ann Arbor, MI, 1075, 1981.

23. Longley, P.A. et al., *Geographic Information Systems and Science*, 2nd ed., John Wiley & Sons, New York, 2005, chap. 3.

24. Woodcock, C.E., Strahler, A.H., and Jupp, D.L.B., The use of variograms in remote sensing: I. Scene models and simulated images, *Remote Sensing of Environment*, 25, 323, 1988.

25. Woodcock, C.E., Strahler, A.H., and Jupp, D.L.B., The use of variograms in remote sensing: II. Real digital images, *Remote Sensing of Environment*, 25, 349, 1988.
26. Bian, L. and Walsh, S.J., Scale dependencies of vegetation and topography in a mountainous environment of Montana, *Professional Geographer*, 45, 1, 1993.
27. Chen, C.-C., Daponte, J.S., and Fox, M.D., Fractal feature analysis and classification in medical imaging, *IEEE Transactions on Medical Imaging*, 8, 133, 1989.
28. Myint, S.W., Wavelet Analysis and Classification of Urban Environment using High-resolution multispectral image data, Ph.D. thesis, Louisiana State University, Baton Rouge, LA, 2001.
29. Myint, S.W., Fractal approaches in texture analysis and classification of remotely sensed data: comparisons with spatial autocorrelation techniques and simple descriptive statistics, *International Journal of Remote Sensing*, 24, 1925, 2003.
30. Lam, N.S.-N. et al., An evaluation of fractal methods for characterizing image complexity, *Cartography and Geographic Information Science*, 29, 25, 2002.
31. Torrens, P.M. and Alberti, M., Measuring Sprawl, CASA Working Paper 27, presented at Association of Collegiate Schools of Planning Conference, November, Atlanta, GA, 2000.
32. Lam, N.S.-N. and De Cola, L., Eds., Fractal measurement, in *Fractals in Geography*, Prentice Hall, Englewood Cliffs, NJ, 1993, chap. 2.
33. Weng, Q., Fractal analysis of satellite-detected urban heat island effect, *Photogrammetric Engineering and Remote Sensing*, 69, 555, 2003.
34. Keller, J.M., Chen, S., and Crownover, R.M., Texture description and segmentation through fractal geometry, *Computer Vision, Graphics and Image Processing*, 45, 150, 1989.
35. Shen, G., Fractal dimension and fractal growth of urbanized areas, *International Journal of Geographical Information Sciences*, 16, 419, 2002.
36. Sarkar, N. and Chaudhuri, B.B., An efficient approach to estimate fractal dimension of textural images, *Pattern Recognition*, 25, 1035, 1992.
37. Haering, N. and da Vitoria Lobo, N., Features and classification methods to locate deciduous trees in images, *Computer Vision and Image Understanding*, 75, 133, 1999.
38. Chaudhuri, B.B., Sarkar, N., and Kundu, P., Improved fractal geometry based texture segmentation technique, in *IEE Proceedings Part E: Computers and Digital Techniques*, 1993.
39. Clarke, K.C., Computation of the fractal dimension of topographic surfaces using the triangular prism surface area method, *Computers and Geosciences*, 12, 713, 1986.
40. Qiu, H.-L. et al., Fractal characterization of hyperspectral imagery, *Photogrammetric Engineering and Remote Sensing*, 65, 63, 1999.
41. Jaggi, S., Quattrochi, D.A., and Lam, N.S.-N., Implementation of three fractal measurement algorithms for analysis of remote-sensing data, *Computers and GeoSciences*, 19, 745, 1993.
42. De Cola, L., Fractal analysis of a classified Landsat scene, *Photogrammetric Engineering and Remote Sensing*, 55, 601, 1989.
43. Chuvieco, E., Measuring changes in landscape pattern from satellite images: short-term effects of fire on spatial diversity, *International Journal of Remote Sensing*, 20, 2331, 1999.
44. Frohn, R.C., *Remote Sensing for Landscape Ecology: New Metric Indicators for Monitoring, Modeling and Assessment of Ecosystems*, Lewis Publishers, Boca Raton, FL, 1998, chap. 2.

45. Halliday, D. and Resnick, R., *Fundamentals of Physics*, Extended 3rd ed., John Wiley & Sons, New York, 1988, chap. 21.
46. Ramstein, G. and Raffy, M., Analysis of the structure of radiometric remotely-sensed images, *International Journal of Remote Sensing*, 10, 1049, 1989.
47. Soille, P. and Rivest, J.-F., On the validity of fractal dimension measurements in image analysis, *Journal of Visual Communication and Image Representation*, 7, 217, 1996.
48. Cox, B.L. and Wang, J.S.Y., Fractal surfaces: measurement and applications in the earth sciences, *Fractals*, 1, 87, 1993.
49. Klinkenberg, B. and Goodchild, M.F., The fractal properties of topography: a comparison of methods, *Earth Surface Processes and Landforms*, 17, 217, 1992.
50. Marceau, D.J. and Hay, G.J., Remote sensing contributions to the scale issue, *Canadian Journal of Remote Sensing*, 25, 357, 1999.

8

Fusion of Radar and Optical Data for Identification of Human Settlements

Paolo Gamba and Fabio Dell'Acqua

CONTENTS

8.1 Introduction

Urban environment is by far the most complex one that may possibly appear in remotely sensed images, and its analysis requires extracting a wealth of information from the sensed data. On the one hand, identification of very different land-cover classes is required. On the other hand, spatial patterns should be considered to associate land-cover classes to land-use classes and to discriminate between natural and artificial objects. As a result, any single sensor may contribute to urban remote sensing, but no one is in itself sufficient to capture all the available information. In this chapter we therefore deal with data fusion, that is, the idea of using datasets coming from different sensors, to identify impervious surfaces. To limit the scope of our discussion, we focus on synthetic aperture radar (SAR) and optical data, which is at the same time our field of experience in the area and one of the most relevant cases, because of, if nothing else, the sheer quantity of data available.

SAR may be considered as the sensor that conveys the greatest amount of information about two- and three-dimensional structural properties of urban landscape features in addition to dielectric properties of the materials that compose urban objects. Unfortunately, SAR images are obtained at a single wavelength, and thus no spectral discrimination of urban materials is usually possible; some limited classification capabilities may be achieved exploiting polarization [1] and the characterization of different scattering effects, which the latter permits. Thus, SAR data may be useful for spatial (pattern) characterization, but they are almost useless for spectral (composing matter) discrimination. This makes it complicated to discriminate impervious surfaces, or to characterize impervious surface fraction at a subpixel level, relying on SAR data alone.

Impervious surface mapping at the regional or country level is not required only to precisely characterize materials in urban areas, but generally speaking, to identify human settlements all over the earth surface, and to discriminate between artificial and natural land-use classes. Even the simple task of detecting where these settlements are located may be difficult to realize, and sometimes it may even be impossible when using optical sensors alone, for example, in some permanently cloud-occluded tropical areas.

Moreover, impervious surfaces are generally related to man-made structures, like streets and buildings, and other artificial features of the landscape. Because of their properties, radar images appear to be the best candidate for structural characterization of artificial objects and show their potential even at the current, somehow coarse, resolution of satellite SAR sensors. The latter limitation will be relieved by the forthcoming satellite SAR instruments, with significantly improved spatial resolution, down to tens of centimeters [2,3]. This will be certainly appealing where the scope of the analysis is the complex urban environment. Thus, the research on radar sensing of impervious surfaces is striving in a preparatory effort in sight of the new perspectives and issues that higher resolution data will open.

So far, the studies on SAR urban remote sensing have focused on the extraction of city boundaries using texture information [4] and extraction of different land-use classes by means of adaptive approaches [5]. Moreover, with the more recent satellites (ENVISAT and RADARSAT-1) the possibility to change the radar viewing angle was introduced, and some analyses focusing on the variability of this parameter and its influence on impervious surface mapping were carried out as in Refs. [6,7]. However, the results are inferior to optical data analysis.

Therefore, recognizing the limits of SAR data and the fact that many optical sensors are available, and also considering the important structural information that SAR can convey, extensive research has been dedicated to data fusion between SAR and optical data such as those from Landsat TM,

IRS, ASTER, and SPOT sensors for better classification and feature enhancement in urban areas [8]. For example, the higher discriminability of impervious surfaces using spectra in ETM+ images is complementary to the better geometrical characterization of artificial targets and patterns in SAR data. Thus, if SAR data may more easily provide information on structure geometry, stable and oriented features, and patterns of settlement structures, optical images are more suitable for land-cover discrimination and spatial analysis with the finest detail. This has been recognized in a line of research very active today, which jointly exploits spaceborne measurements in the microwave and optical frequency regions for mapping impervious areas [9,10]. The huge amount of archived, past imagery by both sensors is indeed far from being completely analyzed and carries a wealth of information about the evolution of human settlements, still to be fully considered.

This work is aimed at demonstrating that fusing SAR and optical data makes it possible to achieve very interesting results for impervious area mapping, by exploiting spatial patterns on the one side and spectral patterns on the other. In turn, this leads to a substantial improvement in the case of interpretation and accuracy of final maps with respect to using optical or SAR data each considered alone.

8.2 Impervious Area Mapping in Challenging Environments

In order to evaluate the mapping procedure using data fusion in "standard" European or U.S. city environments, an example in this chapter is provided for informal settlement mapping. "Informal settlements" are usually defined as dense settlements comprising communities housed in self-constructed shelters under conditions of informal or traditional land tenure [11]. These areas are characterized by rapid, unstructured, and unplanned development [12]. Detecting informal settlements is the spotlight for various initiatives, for example, European Global Monitoring for Environment and Security (GMES) project and the humanitarian and development aid policies of the United Nations. Unfortunately, for the developing countries, this kind of data is unreliable, obsolete, or just simply nonexistent [13]. Usually the only data available for third world cities are limited to outdated topographic maps and National Census population data, whose accuracy and suitability are widely variable. When no other data are available for this kind of analysis, one solution is to use remote sensing imagery as the primary data source and geographic information systems (GIS) plus secondary data sources in the key role of providing a framework for spatial analysis of remote sensing data [14]. However, due to the microstructure and instability of shape of the informal settlements, the detection is substantially more difficult than in formal settlements [15]. Hence, more sophisticated data and methods of image analysis are necessary.

8.3   Methodology

The methodology used in this work is based on texture analysis of both SAR and optical remotely sensed images for detection and exploitation of spatial patterns. Then, a joint classification of the extracted spatial features, optionally together with the original spectral features, is carried out. The selection of the best feature set for the classifier is ruled by the mapping legend and the training set via the so-called fusion rule base. In order to clarify the overall work flow of the procedure, a graph is presented in Figure 8.1. In Section 8.3.1, first the complete procedure is discussed, then a more detailed description of each processing step is offered.

8.3.1   Overall Procedure

As shown in Figure 8.1, the whole procedure flow may be summarized in four steps.

1. SAR and optical data, properly coregistered, and a training set, with associated mapping legend, are considered.
2. According to the problem considered and thus the most suitable mapping legend level for the problem at hand, a selection of the spatial and spectral features to be classified is performed following the rules in – the fusion rule base (see Section 8.3.2). This rules set implements the a priori knowledge about urban area mapping at different geographical scales and automatically selects the most useful subset of features for the subsequent classification step, some of them already available, others to be provided through automatic preprocessing of the data in the next step.
3. If spatial analysis is required, and texture features are to be computed, then the choice for the best parameter set for texture

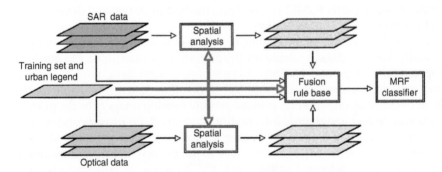

FIGURE 8.1
Conceptual work flow of the overall procedure proposed in this chapter.

computation is automatically performed starting from the training set, using a semiautomatic approach proposed in Section 8.3.3.

4. Finally, the spatial features (if any), the spectral features, and the original data (if considered by the selected rules) are fed into a spatially aware classifier, based on a Markov random field (MRF) structure, proposed in Section 8.3.4.

8.3.2 Fusion Rule Base

Urban areas are required to be analyzed, as discussed in Section 8.1, by a suitable mix of spectral and spatial features. For instance, texture features may capture the spatial patterns of the scene, but they generally can be determined only at a coarser resolution than the original data. This means that boundaries between neighboring objects are blurred in texture maps, and patterns are more relevant if land-use rather than fine-scale land-cover mapping is considered. For instance, residential and industrial areas may be easily discriminated at tens or hundreds of meters, examining building clusters, while a finer resolution may be misleading. On the other hand, land-cover maps (with classes such as roof, pavement, grass) are better achievable using the finest ground resolution available, that is, considering the original data, alone or together with textures.

There should therefore exist a set of relations, or *rules*, to determine the best set of features to be used in a mapping procedure, according to the map legend to be considered. This rule set, susceptible to formalization, constitutes a fusion rule base that may be constantly updated in further researches while allowing to conserve the overall structure of the procedure proposed in this work.

The currently implemented rules reflect the very basic consideration in the preceding paragraphs, which in turn depends on the choice of the data studied in this work. It should be stressed that these rules may become questionable if very high-resolution airborne SAR and optical sensors are considered, while they are acceptable for data recorded by current, coarse resolution satellite sensors. Such rules are

1. Land-cover mapping does not require spatial analysis and exploits the original data, as long as the "salt and pepper" classification noise is reduced by the use of a "spatially aware" classifier, like the one described in Section 8.3.3 and employed here.

2. Land-use mapping requires spatial analysis alone, once a way to select the best feature set is implemented, as described in Section 8.3.1.

8.3.3 Spatial Analysis with Multiple-Width Textures

When dealing with urban areas in different parts of the world, one point that is really interesting is that they have a very different structure, and this means, from a remote sensing standpoint, a strong need for a resolution-dependent approach. Formal and informal settlements have different spatial patterns, and the ability to somehow "guess" the best feature for mapping a given area is therefore one of the most challenging and interesting research themes. Since, as recalled earlier, SAR images are more related to geometric properties of the scene than any other sensor, their analysis is likely to allow the extraction of more information about human settlement patterns than using any other source of information generated from satellites.

Even though several methods have been proposed in technical literature to provide a measure for the spatial relationships among neighboring pixels, the most widely successful is the co-occurrence texture analysis [16]. Co-occurrence texture measures are computed starting from the co-occurrence matrix. A few parameters determine this process, namely the distance between the two locations to be jointly considered the direction of such distance and the width of the window used for computation. The distance provides an important means to discriminate among textures based on the element spacing. Direction is especially important in case of anisotropy in the texture [17]. Finally, the window width is usually neglected.

Remote sensing images, however, usually reveal only very compact patterns. Inside an urban area, for instance, most common distances are on the order of meters, so that basic texture elements are located in adjacent pixels for most satellite images. Moreover, although many environments do have preferred directions, such anisotropy is often immaterial for very fine textures and small-scale texture patches. Sometimes no clear texture segment can be seen, and patterns are continuously, smoothly changing. So, the co-occurrence window width remains the only really important parameter. The window width defines the area around a pixel where we assume that texture patterns are statistically stable. In turn, this number needs to be tied to the mean physical dimension of the textured areas we are looking for. This explains why this parameter has been found to be the most important one in urban remote sensing [18] and the only parameter that is strongly related to the spatial patterns of the image.

A methodology for textures and multiple widths has been proposed in the same work and was labeled *multiscale texture fusion*. The approach is based on a supervised neural classification, fed by a feature extraction step. This step exploits the same training set as the classifier, and it is based on the computation of a discrimination index, the histogram distance index (HDI) [19]. The whole set of textures computed from the co-occurrence matrix for different values of the parameters are considered, and the subset that is highest in HDI ranking is used as an input to a fuzzy ARTMAP multiband classifier [20]. Adaptive resonance theory (ART) networks, basically introduced for solving pattern recognition problems, have indeed shown to be very efficient in multiband remote sensing data analysis (see also Ref. [21]). This is particularly so when we deal with bands whose statistical properties are very different, as is the case with texture measures. ART networks store in their memories information about the training samples and compare test patterns with these memories. Any match assigns the pattern to an output category, for example, a land-use class. We proved the existence of a strong correlation between the HDI rank order and the progression of the overall accuracy values after classification. In particular, the major aim of this step is to provide the minimum number of texture measures with fixed or variable width, which maximizes the classification performance.

8.3.4 Markov Random Field Classifier

In this work an MRF approach is used, because of its capability of dealing with multiple images, even with very distinct statistics, and the possibility to easily adapt it to different inputs. This is why MRF is widely used in data-fusion methodologies, where more datasets, coming for instance from radar and optical sensors, are used to classify the same scene [22]. MRFs are also capable of considering spatial relationships among neighboring pixels in the classification framework. This makes an MRF classifier a "spatially aware" one, capable of fully exploiting the ground resolution of the data while greatly reducing the typical salt and pepper classification noise of per-pixel classification algorithms.

To briefly summarize the MRF framework, let us consider a set of features or images coming from n sensors; then, let us consider the $M \times N$ image acquired by sensor r as made up of MN pixels or feature vectors $X_r(1, 1)$, ..., $X_r(M, N)$, $r = 1, 2, \ldots, n$, where $X_r(i, j) = (x_r(i, j, 1), \ldots, x_r(i, j, B_r))$ and B_r is the number of spectral bands or features for sensor r. We assume that K classes $c_1, c_2, \ldots, c_K$ are present in the images with prior probabilities $P(c_1), P(c_2),$ $\ldots, P(c_K)$. Let us denote with $C(i, j)$ the class for pixel (i, j); we call X_r the set of pixels of the whole image $X_r = \{X_r(i, j); 1 \leq i \leq M, 1 \leq j \leq N\}$ and with $C = \{C(i, j), 1 \leq i \leq M, 1 \leq j \leq N\}$ the set of labels for the same scene; in practice for a given pixel (i, j), $C(i, j) \in \{c_1, c_2, \ldots, c_K\}$.

If we call $P(X_1, \ldots, X_n|C)$ the conditional probability density of feature vectors $X_1, X_2, \ldots, X_n$ given the scene label set C, and with $P(C|X_1, \ldots, X_n)$ the posterior probabilities, the classification task consists of assigning each pixel to the class that maximizes the posterior probabilities. Naturally a relation exists between the data (measurements or features) and the prior information, which can be represented in a Bayesian formulation as

$$P(C|X_1, \ldots, X_n) = \frac{P(X_1, \ldots, X_n|C) \, P(C)}{P(X_1, \ldots, X_n)} \tag{8.1}$$

where $P(C)$ represents the prior model for the class labels.

Thus, we want to maximize the likelihood function $L(X_1, \ldots, X_n|C) = P(X_1|C)^{\alpha 1}, \ldots, P(X_n|C)^{\alpha n} \, P(C)$ where α_r, $0 \leq \alpha_r \leq 1$ is the reliability factor for sensor r.

Furthermore, denoting G_{ij} the local neighborhood of pixel (i, j) we can write

$$P(C(i, j)|C(k, l);\{k, l\} \neq \{i, j\}) = P\big(C(i, j)|C(k, l);\{k, l\} \in G_{ij}\big)$$

$$= \frac{1}{Z}e^{-U(C)/T} \tag{8.2}$$

where

U is the so-called energy function
Z is a normalizing constant factor
T is a temperature term often used in statistical physics

In our model, we always consider a second-order neighborhood, that is, the eight pixels closer to each single pixel of the image. If we want to maximize $P(C(i, j)|C(k, l); \{k, l\} \in G_{ij})$, we find that we need to minimize $U(C)$, where $U(C(i, j)) = \Sigma_{\{k,l\} \in G_{ij}} \beta I(C(i, j), C(k, l))$ and $I(C(i, j), C(k, l)) = -1$ if $C(i, j) = C(k, l)$, 0 if $C(i, j) \neq C(k, l)$.

To perform the classification, we need to minimize $U(X_1, \ldots, X_n, C) = \alpha_s U_{\text{spect}}(X_s) + U_{\text{sp}}(C)$, where U_{sp} is given by the previous equation and U_{spectr} is defined as follows.

$$U_{\text{spectr}}(X_s(i, j), C(i, j)) = \frac{B_s}{2} \ln|2\pi \Sigma_k| + \frac{1}{2}(X_s(i, j) - \mu_k)^T \Sigma_k^{-1}(X_s(i, j) - \mu_k) \tag{8.3}$$

where

Σ_k and μ_k are, respectively, the class-conditional covariance matrix and mean vector for class k
B_r is the number of spectral bands or features for source r

Many algorithms are available for MRF implementation, but they are often demanding in terms of CPU-time. A simple but still effective one is iterated conditional mode (ICM), used in this work, which allows to reach a local minimum of the energy function very quickly. The problem of ICM is that it may get trapped in a local minimum, but since in this work the starting point is an already good neural network classification, convergence is empirically assured.

8.4   Experimental Results

Experimental results are offered in this section for two very different examples of human settlement mapping. The first one is a problem of formal vs. informal settlement mapping, which implies a land-use legend, where spatial patterns are more relevant than spectral features. The second one is instead a more standard impervious area mapping on a European town, with limited stress on spatial patterns and more importance on built-up area/vegetation discrimination. Figure 8.2 shows how the general framework depicted in Figure 8.1 applies to the two different mapping scenarios.

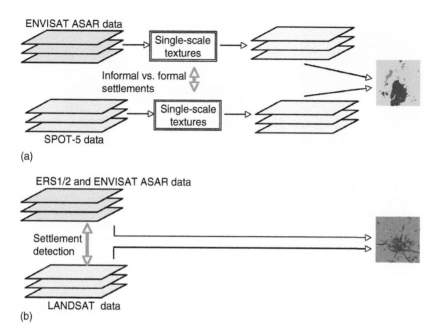

(a)

(b)

FIGURE 8.2

Instances of the framework of Figure 8.1 applied to: (a) informal vs. formal settlement mapping; (b) settlement detection and impervious surface mapping.

8.4.1 Al-Fashir (Sudan)

As outlined in Section 8.2, SAR and optical data analysis for impervious surface mapping in informal settlements stretch the scope of the current methodologies for urban mapping, developed for industrialized countries and mostly European and U.S. urban areas. The test area for this work is the area around the town of Al Fashir, the largest human settlement in the North Darfur region in Sudan. SAR data on Al Fashir were recorded by ASAR on 26 July and 13 August 2004. In that period, the war in progress in the area and the consequent famine produced the sudden raising of a huge tent camp located northwest of the main town area. The area was therefore the target of a wide effort of humanitarian aid, and vast amount of remotely sensed data were recorded to help rescuers and NGO officers.

Figure 8.3 shows the mapping results obtained using both ASAR datasets and applying to them the previously described texture analysis. Best performances were obtained using two sets of textures (mean, second moment, and variance or mean, entropy, variance, and dissimilarity), computed using a 21×21 window, as it was found viable in many similar cases [5,23]. The results of the second choice are provided in Figure 8.4b, next to either of the two original images. The tent camp is the low brightness area northwest of the main urban core, and it is actually partitioned into two main tent blocks.

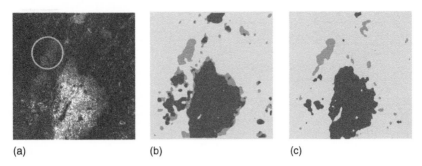

(a)   (b)   (c)

FIGURE 8.3 (See color insert following page 292.)
Classification maps for SAR data classification of human settlements in the area around the town of Al Fashir (Sudan). Three land-use classes are used: informal settlements (light green), formals settlements (red), rocks and bare soil (yellow). (a) Original SAR data with the tent camp highlighted by a green circle; (b) map from classification of SAR textures; (c) map from joint classification of SAR and SPOT textures.

Figure 8.3b shows that texture analysis may provide a first discrimination between formal and informal settlements, but it shows also that spatial analysis at a single frequency does not allow discarding uninteresting areas outside human settlements. There are indeed rock formations and a bare soil area with a radar backscattering very similar to that of some parts of these settlements. A certain amount of misclassification is therefore reported, testified by the red and green "blobs" all around the map, and especially in the top right area, where no settlement can be found, only rocky hills.

However, acquisitions from the SPOT sensor were also available on this area, in particular SPOT-5 images with 2.5 m spatial resolution was acquired on 14 November 2004. According to the fusion rule base, texture features from both SAR and SPOT data were considered and jointly classified. A comparison between Figure 8.3b and c allows understanding the advantage of using both information sources. As a matter of fact, the misclassifications have been greatly reduced and better delineation of both the formal and the informal settlement areas is achieved. Misclassifications with rock soil still persist, but they have been dramatically reduced.

86.79%	154,970	7,159	16,431	74.61%	133,219	42,446	2,895
89.91%	29,913	554,474	32,320	97.65%	3,711	602,221	10,775
86.90%	0	1,930	12,803	80.67%	0	2,848	11,885
	83.82%	98.39%	20.80%		97.29%	93.00%	46.51%

Overall accuracy: 89.17%   Overall accuracy: 92.26%

(a)   (b)

FIGURE 8.4
Confusion matrices for the maps. From top: structured settlements, desert, unstructured settlements.

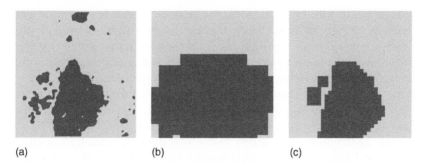

(a)   (b)   (c)

FIGURE 8.5
Urban extent using ASAR data, from the map in Figure 8.3(b), compared with (b) GRUMP and
(c) AFRICOVER urban extents for the town of Al-Fashir (Sudan).

Quantitatively speaking, the confusion matrices for the two maps in
Figure 8.3 are reported in Figure 8.4. In both cases the overall accuracy
values are good (around 90%), and the class accuracy values show a really
good discrimination between structured and unstructured human settlements (following the legend in Figure 8.4, they are represented by red and
green colors, respectively). One may comment that a suitable texture analysis, coupled with a spatially aware classifier, allows exploiting spatial patterns to delineate different human settlements. However, a better
discrimination is obtained by a joint exploitation of optical and radar data.

Another interesting comparison for this particular example may be
obtained comparing the urban area extent found by means of the procedure
adopted in this research work with the urban extents in global datasets,
such as those by the Global Urban-Rural Mapping Project (GRUMP, [24])
and the AFRICOVER initiative [25], proposed in Figure 8.5.

Apparently, the urban extent is widely overestimated in the GRUMP
database, while our result matches very well the AFRICOVER map, obtained
using Landsat data, that is, optical data with a ground spatial resolution
comparable with the ASAR data used in this work. This in turn means that,
as far as urban extent delineation is concerned, SAR data allow to obtain
results similar to those obtained using optical data, which is good news for a
better mapping of urban areas in the African continent. The comparison was
made to show that existing spatial databases contain information that is either
too coarse for real human settlement mapping or not sufficiently detailed as
far as the settlement typology is concerned. Both kinds of information might
instead be extracted from the joint use of SAR and optical data, as shown here.

8.4.2   Pavia (Italy)

The second test site for this work is the town of Pavia, northern Italy, which has
already been widely analyzed in many works, not only by our research
group [26]. In this area, a relevant number of SAR scenes were acquired by

sensors on board ERS-1/2, RADARSAT-1, and ENVISAT. Furthermore, good knowledge of the area is available, thanks also to a detailed ground truth, manually extracted from very high-resolution images of the area. Such ground truth delineates the impervious surfaces (red) against agricultural surrounding fields (green) and the Ticino river (blue) flowing south of the town.

In this work, several Landsat scenes of the city were collected to perform a joint classification based on the MRF model. The considered optical dataset is made up by two of these images, acquired on 7 April 1994 and 8 October 2000, which were added to the SAR dataset, made by two single-polarization Precision Mode ASAR images, recorded on 25 November 2002 and on 8 December 2002, and one ERS-2 image acquired on 13 August 1992. The overall considered area covers an 8×8 km^2 square-shaped area around the town.

According to the fusion rule base, the original data were considered and no spatial analysis was performed. Both a single-source and a joint classification of radar and optical data were instead performed using the previously mentioned neuro-fuzzy classifier, to provide a first analysis of the classification accuracy available to use these classification maps as a priori information for the classifier based on the MRF model (see Section 8.3.2).

Table 8.1 compares the overall as well as the single-class accuracy values of the fuzzy ARTMAP NN (A) and MRF classifier (M) maps considering some combinations of the available SAR data.

Table 8.2, instead, shows the accuracy for any combination of the Landsat datasets using the same classifiers (ARTMAP and MRF). Both tables show that the overall accuracy values improve with the second approach.

Finally, Table 8.3 shows the classification results of the whole available dataset considered together again using both classifiers. In all of these classifications, the choices for the parameters in Equation 8.3 were as

TABLE 8.1

Comparison of User Class and Overall Accuracy Values (Percentages) for NN and MRF Classification of Radar Data

Images	Water	Vegetation	Urban Areas	OA
13 August 1993 (A)	75.25	88.13	52.6	83.2
13 August 1993 (M)	94.31	85.77	65.57	83.35
25 November 2002 (A)	56.48	71.37	56.33	69.06
25 November 2002 (M)	45.29	72.32	61.83	70.31
08 December 2002 (A)	29.73	72.96	40.59	67.72
08 December 2002 (M)	58.79	77.18	46.6	72.76
25 November + 08 December (A)	61.58	80.05	56.44	76.54
25 November + 08 December (M)	45.07	81.41	60.86	77.87
25 November + 08 December + 13 August (A)	62.9	91.48	61.41	86.89
25 November + 08 December + 13 August (M)	46.75	95.98	63.07	90.53

TABLE 8.2

Comparison of User Class and Overall Accuracy Values for NN
and MRF Classification of Optical Data

Images	Water	Vegetation	Urban Areas	OA
07 April 1994 (A)	82.88	60.29	91.26	64.85
07 April 1994 (M)	90.81	60.78	90.73	65.39
08 October 2000 (A)	91.13	65.41	89.12	69.11
08 October 2000 (M)	91.25	74.06	87.71	76.24
07 April + 08 October (A)	94.31	73.61	88.46	76.03
07 April + 08 October (M)	76.77	79.6	89.24	80.79

follows: α is the overall accuracy of the ARTMAP classification, $\beta = 1.5$, and a second-order neighborhood and a pixel threshold percentage of 5% were used. With these parameters the ICM algorithm reaches the convergence in two, three, or at most in four iterations.

Apart from accuracy values, a visual analysis of the maps in Figure 8.6 also shows that the MRF approach results in, sometimes, only slightly but in any case higher overall accuracy for the classification, and in particular the urban areas are better recognized. It is important to remark here that we also considered, as a priori information for the MRF model, the classification obtained with spatial fuzzy ARTMAP using a spatial window of 3×3 pixels around each pixel of the image [20]. The classification results of this approach for the classification of the three SAR images are shown in the last two lines of Table 8.1, while either of the two classification maps is shown in Figure 8.5f.

This approach, tested on the whole dataset, allows achieving the same classification results, but with the MRF procedure terminating after just one iteration. Thus, it results in a faster classification.

Although this is not surprising, still it is to be noted that in the Pavia case the joint classifications improve the overall accuracy for both classifiers. This is probably due to the usual "averaging" effect over several, noise-plagued images. Generally, the better detected class is vegetation, regardless of the number of images input; the improvement in the overall accuracy springs from various fluctuations in the class accuracy for "water" and "urban," although neither of those shows a clear trend toward improvement when

TABLE 8.3

Comparison of User Class and Overall Accuracy Values for NN and MRF
Classification of the Joint Radar and Optical Datasets

Images	Water	Vegetation	Urban Areas	OA
ERS + ASAR + LAND 08 October (A)	87.09	69.24	87.27	72.01
ERS + ASAR + LAND 08 October (M)	48.99	82.53	84.14	81.94
ERS + ASAR + LAND 21 June (A)	60.06	92.9	67.67	90.14
ERS + ASAR + LAND 21 June (M)	54.07	98.23	71.58	93.71

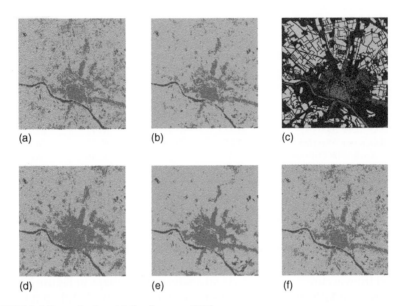

FIGURE 8.6 (See color insert following page 292.)
Classification maps of the three SAR available images obtained with ARTMAP (a) and with the MRF approach (b); the land-cover ground truth used to evaluate classification results (c); classification maps of the joint classification of the LANDSAT image of 21 June and the three SAR images with ARTMAP (d); and MRF (e); classification map of the three SAR available images obtained with spatial ARTMAP (f).

the number of jointly classified images increases. Notwithstanding the limited class accuracy, both the river and the urban area are well outlined, especially when MRF classification is concerned. This suggests that the spatial dependency need not be introduced before classification, and a spatial postprocessing, possibly heavier than the current one (e.g., morphological closing), could serve the purpose of increasing the overall accuracy. As a general comment to the results, the joint classification of LANDSAT and SAR data provides a better accuracy in the characterization of urban areas with respect to the use of SAR data alone. These accuracy values are comparable with the ones obtained from the classification of Landsat data alone; but in the joint classification of the two sensors we obtained a better discrimination of all the classes. This reflects the original idea that the use of more than one sensor in land-cover mapping and the use of an MRF model may improve the spectral discrimination between impervious and natural surfaces, using spatial analysis.

8.5 Conclusions

This chapter deals with the joint use of radar and optical data for identification of human settlement. While the topic is not completely new, a complete framework for a more complex task than simple urban area detection has been proposed. Spatial analysis, for instance, has been widely used for improving SAR interpretation in urban areas, but a methodology to adapt the spatial analysis to the impervious surface legends has not been provided in technical literature so far. Similarly, the use of texture in both SAR and optical data for formal vs. informal settlement discrimination is a new research area, still to be further investigated.

The main achievements of this work are therefore listed as follows.

1. This work provides the framework for a semiautomatic procedure for human settlement mapping, able to deal with spectral and spatial features and apt to the goal of providing maps at different geographical scales of interest and corresponding land-use legends.

2. The proposed spatial analysis, initially developed for SAR data, was validated using also optical data and shows therefore to be equally valid for data at different wavelengths. The spatial features of human settlement can also be used to discriminate between formal and informal settlements and achieve a better accuracy in land-use mapping of the anthropogenic landscape. However, as noted earlier, more research on this topic is needed.

3. From an operational point of view, the evaluated experiments show that the recently proposed MRF classifier provides better accuracy values than the originally used neuro-fuzzy classifier and similar capability to cope with multiple inputs, each one with possibly different statistics.

The possibility of "rapid mapping" following the procedure of this work is therefore justified on the one hand by the use of the semiautomatic methodology based on spatial and spectral features, which are very easy and fast to compute and to select. On the other hand, the results, while interesting, do not match the requirements of precise and ready-to-market mapping techniques, which are basically based on manual extraction/correction of the maps. Therefore, applications that may benefit from the proposed procedure are mainly "rapid mapping" ones, like for instance change detection or land-use mapping for disaster or relief management.

Acknowledgments

The authors wish to thank the Joint Research Center in Ispra (Italy) for sharing the ASAR and SPOT images of Al Fashir. Many thanks also to M. Stasolla for performing the classifications for this area and to G. Trianni for the analyses of the Pavia dataset.

References

1. F. Dell'Acqua, P. Gamba, G. Trianni: Fully vs dual polarization satellite sensors for urban area analysis, in *Proceedings of POLinSAR 2005*, January 2005, Unformatted CD-ROM.
2. F. Caltagirone, G. Manoni, P. Spera, R. Vigliotti: SkyMed/COSMO mission overview, in *Proceedings of IGARSS98*, June 1998, pp. 610–621.
3. Available online at: http://www.caf.dlr.de/tsx/start_en.htm, accessed on May 22, 2006.
4. R.J. Dekker: Texture analysis and classification of SAR images of urban areas, in *Proceedings of 2nd GRSS/ISPRS Joint Workshop on Remote Sensing and Data Fusion over Urban Areas*, Berlin, May 2003, pp. 258–262.
5. F. Dell'Acqua, P. Gamba: Texture-based characterization of urban environments on satellite SAR images, *IEEE Transactions on Geoscience and Remote Sensing*, 41(1), 153–159, 2003.
6. F. Dell'Acqua, P. Gamba: A preliminary statistical analysis of textural features in RADARSAT-1 images of an urban area, in *Proceedings of URSI Commission F Symposium*, Ispra, April 2005, Unformatted CD-ROM.
7. F. Dell'Acqua, P. Gamba, A. Iodice, G. Lisini, D. Riccio, G. Ruello: Simulation and analysis of fine resolution SAR images in urban areas, in *Proceedings of the 2nd GRSS/ISPRS Joint Workshop on Remote Sensing and Data Fusion over Urban Areas* (URBAN2003), Berlin, Germany, 22–23 May 2003, pp. 132–135.
8. L. Alparone, S. Baronti, A. Garzelli, F. Nencini: Landsat ETM+ and SAR image fusion based on generalized intensity modulation, *IEEE Transactions on Geoscience and Remote Sensing*, 42(12), 2832–2839, 2004.
9. L. Fatone, P. Maponi, F. Zirilli: Fusion of SAR/optical images to detect urban areas, in *Proceedings of IEEE/ISPRS Joint Workshop on Remote Sensing and Data Fusion over Urban Areas*, Rome, 2001, pp. 217–221.
10. G. Trianni, P. Gamba: A novel MRF model for multisource data fusion in urban areas, in *Proceedings of URSI GA*, New Delhi (India), October 2005, Unformatted CD-ROM.
11. D. Hindson, J. McCarthy (Eds.): Defining and gauging the problem, in: *Here to Stay: Informal Settlements in KwaZulu Natal*, Indicator Press, Durban, 1994, pp. 1–28.
12. H. Ruther, H. Martine, E.G. Mtalo: Application of snakes and dynamic programming optimisation technique in modelling of buildings in informal settlement areas, *Journal of Photogrammetry and Remote Sensing*, 56, 269–282, July 2002.
13. Y. Baudot: Geographical analysis of the population of fast-growing cities in the third world, in: *Remote Sensing and Urban Analysis*, Taylor & Francis, London & New York, pp. 225–241, 2001.
14. J.-P. Donnay, M.J. Barnsley, P.A. Longley: Remote sensing and urban analysis, in: *Remote Sensing and Urban Analysis*, Taylor & Francis, London & New York, 2001, pp. 3–18.
15. P. Hofmann: Detecting informal settlements from IKONOS image data using methods of object oriented image analysis – an example from Cape Town, South Africa, in *Proceedings of the Conference on Remote Sensing of Urban Areas*, Regensburg, July 2001, pp. 107–118.
16. R.M. Haralick, K. Shanmugam, I. Dinstein: Texture features for image classification, *IEEE Transactions on Systems, Man, and Cybernetics*, 3, 610–621, November 1973.

17. D.G. Barber, E.F. LeDrew: SAR sea ice discrimination using texture statistics: a multivariate approach, *Photogrammetric Engineering and Remote Sensing*, 57(4), 385–395.
18. F. Dell'Acqua, P. Gamba: Discriminating urban environments using multi-scale texture and multiple SAR images, in *Proceedings of the Workshop on Advances in Techniques for Analysis of Remotely Sensed Data*, CD-ROM, Greenbelt, MD, October 2003.
19. M. Pesaresi: The Remotely Sensed City, Final report of the post-doctoral fellowship at the JRC Ispra, Italy, March 2000.
20. P. Gamba, F. Dell'Acqua: Improved multiband urban classification using a neuro-fuzzy classifier, *International Journal of Remote Sensing*, 24(4), 827–834, 2003.
21. G.A. Carpenter, M.N. Gjaja, S. Gopal, C.E. Woodcock: ART neural networks for remote sensing: vegetation classification from Landsat TM and terrain data, *IEEE Transactions on Geoscience and Remote Sensing*, 35, 308–325, 1997.
22. A.H. Schistad Solber, A.K. Jain, T. Taxt: Multisource classification of remotely sensed data: fusion of Landsat TM and SAR images, *IEEE Transactions on Geoscience and Remote Sensing*, 32(4), 768–778, 1994.
23. P. Gamba, F. Dell'Acqua, G. Trianni: Semi-automatic choice of scale-dependent features for satellite SAR image classification, *Pattern Recognition Letters*, 27(4), 244–251, 2006.
24. Available online at: http://beta.sedac.ciesin.columbia.edu/gpw/, accessed on 1 July 2006.
25. Available online at: http://www.africover.org, accessed on 1 July 2006.
26. P. Gamba: A collection of data for urban area characterization, in *Proceedings of IGARSS'04*, Anchorage, USA, September 2004, Vol. I, pp. 69–72.

Part III

Transport-Related Impervious Surfaces

9

Transportation Infrastructure Extraction Using Hyperspectral Remote Sensing

Ramanathan Sugumaran, James Gerjevic, and Matthew Voss

CONTENTS

9.1 Introduction

Accurate and up-to-date information regarding road location and condition is essential data for transportation infrastructure planning and management. This information is used for a variety of purposes including traffic safety, construction projects, traffic engineering studies, and evaluation of maintenance needs [1,2]. Additionally, spatially accurate and up-to-date transportation networks are vital in ambulance and rescue dispatch emergency situations [3,4]. However, current road infrastructure databases are often outdated due to the dynamic nature of road networks where roads deteriorate or are improved and reconstructed [4,5]. In other cases, infrastructure databases may not even exist at all. This is particularly true for areas experiencing rapid road network expansion [3].

Currently, one of the most common methods of gathering transportation infrastructure information is global positioning system (GPS) mapping [3]. This method, which is the primary method used by the U.S. Department of Transportation (DOT), has been supported by the increasing availability of quality GPS technology [1]. However, GPS methods, along with other data collection methods, which include manual methods, digitization of aerial photographs, and video or photo-log vans, are all methods that require every road within a DOT's jurisdiction to be field-visited to obtain accurate information. These methods are effective, but are also inefficient since significant amounts of time and resources are required to cover even a minor amount of roadway [1,3,6]. This is problematic since both state DOT and local areas are responsible for significant street network systems. Iowa, for example, has a street system consisting of ~110,000 linear miles [1].

As a result of these issues, the USDOT is interested in identifying alternative technologies, such as remote sensing, that could be used along with current methods to meet road infrastructure informational needs more effectively [3]. To promote research in this area the USDOT, Research and Special Programs Administration (RSPA), and the National Aeronautics and Space Administration (NASA) collaborated to establish the National Consortia on Remote Sensing in Transportation in early 2000 with funding under the Transportation Equity Act for the twenty-first century [6]. The goal of this consortium was to identify alternative technologies, such as remote sensing, that could be used along with current methods to meet road infrastructure informational needs more effectively [3]. Studies that specifically attempt to extract road information using multispectral data have been conducted [7–10]. Although multispectral imagery continues to be among the most widely used remote sensing data, hyperspectral imagery is maturing into one of the most powerful and fastest growing sources of remotely sensed information [11]. Studies have also shown good results from hyperspectral data for road feature extraction [11,12]. Further, studies have shown relationships between spectral response and road deterioration [13].

The goal of this study was to assess the potential of AVIRIS hyperspectral remote sensing data for use with transportation infrastructure with different classification methods. Hyperspectral remotely sensed data were chosen for this study because of the availability of high spectral and spatial resolution. This choice was supported by previous studies that conclude that hyperspectral data have the highest potential for mapping the complex urban environment due to the fact that man-made features are often too small for sensors with lower spatial resolution to detect [14,15]. Additionally, these features are often too compositionally similar in respect to other man-made materials for sensors with lower spectral resolution to discriminate between them [14,15].

9.2 Data

The particular data chosen for this study were hyperspectral imagery acquired in the spring of 1999 over Shelton, Nebraska (Figure 9.1) by the high-resolution Airborne Visible Infrared Imaging Spectrometer (AVIRIS) sensor available from the Jet Propulsion Laboratory at the California Institute of Technology. The Shelton Nebraska site was chosen for this study for three reasons: (1) the location represents one of the few areas in the Midwest in which data have been collected using a hyperspectral sensor with high spatial resolution; (2) imagery for this area is presently archived and readily available; and (3) there is a variety of road infrastructure present in the image (e.g., urban streets, rural roads, highways, and unpaved roads).

The AVIRIS instrument contains 224 different detectors, each with a wavelength-sensitive range (also known as spectral bandwidth) of ~10 nm, allowing it to cover the entire spectral range between 380 and 2500 nm. The pixel size and swath width of the AVIRIS data depend on the altitude from which the data are collected. In this study, each ground pixel is 4 m^2, and the swath is 2 km wide. Although there are multiple hyperspectral sensors currently in use, the AVIRIS spectrometer was determined to be the best source of data for this research for four reasons: (1) the high-resolution data (4 m resolution) were archived and immediately available through the Jet Propulsion Laboratory; (2) the data's study area (Shelton, Nebraska) was

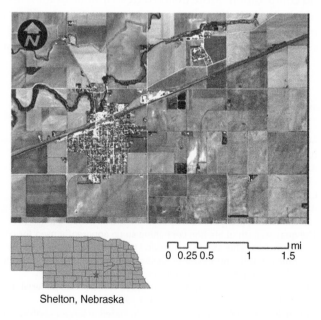

Shelton, Nebraska

FIGURE 9.1
Location of the study area.

easily accessible for ground truthing; (3) the data contained a variety of transportation features (i.e., various types of road surfaces); and (4) the data were available at no cost for graduate research.

Along with the AVIRIS hyperspectral image, other data used in this study included GPS data collected during the spring of 2003 for transportation features such as roads and parking lots, as well as for features and surfaces that could be easily confused with transportation features and surfaces (e.g., fence lines, baseball field infield, parking lots). This GPS information was used to help accurately define regions of interest. Lastly, a corresponding high-resolution digital orthophotographic quadrant (DOQ) with 1 m spatial resolution was used as reference data for classification accuracy assessment. DOQ imagery was acquired 2 months before the AVIRIS data, and thus, there should have been little or no change in the study area's transportation infrastructure.

9.3   Methodology

This research required the use of various image-processing techniques. The initial image processing involved the procedures to mosaic, georegister, and compensate for sensor detector and environmental attenuation error in the AVIRIS hyperspectral data. Classes based on materials were selected following a preliminary survey of the area. The classes were selected on the basis of materials and composition. More detailed descriptions of the classes may be found in Table 9.1. Next, the minimum noise fraction transformation and Classification and Regression Tree (CART) approaches were used to reduce the dimensionality of the AVIRIS data and simplify further processing.

TABLE 9.1

Transportation Infrastructure Class Descriptions

Class	Description
Brick	Brick road surfaces
City	Asphalt mixed with stone road surfaces located within the city of Shelton
Concrete	Concrete road surfaces
Gravel	Gravel road surfaces
Highway	Asphalt road surface north of Shelton running east–west and one highway west of Shelton running north–south
Quarry	The quarry south of Shelton containing stone apparently used for city street construction. This area was included to test for confusion with those roads
Railroad	The railroad at the north end of Shelton was included to test for confusion with roads due to proximity and the similar linear features
Roof	Often composed of asphalt, a common road construction material. Rooftops were included to test for confusion with roads
Track	Shelton high school's asphalt track was included to test for confusion with asphalt roads

Then, image classification was performed using the selected classification algorithms. Finally, the classification algorithms were statistically evaluated to determine which produced the most accurate results.

9.3.1 Preprocessing

After mosaicking and georeferencing the image, an image-based flat-field calibration algorithm was used. The flat-field correction method assumes that there is an area in the scene that is spectrally neutral (no variation in reflectance with wavelength) [16]. The radiance spectrum from this area is assumed to be composed primarily of atmospheric effects and the solar spectrum. Thus, the average radiance of this "flat field" is used as an estimate of atmospheric and solar attenuation, which can then be used as a correction factor to convert the radiance in each pixel to reflectance. The appeal of this image-derived approach is its ability to atmospherically correct remote sensing data without the dependence on external information or measurements, which were unavailable in this study. However, as pointed out by Clark et al. [17], it should be noted that while this method is independent of external information or measurements, it also relies on assumptions about the surface materials in the scene, which are rarely encountered and, thus, result in apparent reflectance, which shows deviations from spectra of comparable materials measured in the field or laboratory. Nevertheless, the flat-field calibration method was the best available approach for use in this study and was therefore used.

9.3.2 Data Reduction

An MNF transform was used to reduce the number of spectral dimensions to be analyzed. The MNF transformation is a linear transformation related to principal components, which orders the data according to signal-to-noise ratio [18]. It can be used to determine the inherent dimensionality of the data, to segregate noise in the data, and to reduce the computational requirements for subsequent processing [18,19]. The MNF transformation can be used to partition the data space into two parts: one associated with large eigenvalues and coherent eigenimages and a second with near-unity eigenvalues and noise-dominated images. By using only the coherent portions in subsequent processing, the noise is separated from the data, thus improving spectral-processing results. The higher numbered MNF bands contain progressively lower signal-to-noise ratio [20]. Using this method, the 224 band image was reduced to 30 bands that contained the best signal-to-noise ratio and therefore the highest quality information.

In addition to the MNF transform, a CART was also tested for its ability to select the best available bands from the hyperspectral dataset. The CART method, developed by Breiman et al. [21], is a popular decision tree tool that automatically sifts large, complex databases, searching for and isolating significant patterns and relationships. In previous remote sensing studies, the CART method has been used to analyze reflectance data extracted from each band for rule-based land-cover mapping [22]. In this study, the CART method was used to reduce the previously MNF-transformed AVIRIS dataset even further, essentially reducing the high-quality, noise-free bands of the MNF dataset to those that only contain most of the information about transportation features. This was accomplished by importing the results of the MNF transform into S-Plus statistical software where the CART operation could be performed. The goal of determining a list of best bands was an attempt to improve the current data reduction techniques, which are standard for hyperspectral data. To test this, the classification results generated using the CART process were compared with the classification results generated using the entire MNF transform subset.

9.3.3   Image Classification

Hsu and Tseng [23] call attention to the fact that using traditional multispectral classification methods with hyperspectral data usually results in disappointing efficiency, needing a large amount of training data, and hard improvement of classification accuracy. As a result, this study utilized whole pixel classification methods that have been approved for hyperspectral imagery such as spectral angle mapper (SAM) as well as subpixel classification methods such as mixture tuned matched filtering (MTMF), which is capable of targeting multiple materials per pixel in hyperspectral data [11]. Object-oriented classification that uses groups of pixels was also examined. The specific classification methods used in this research are discussed in greater detail in the following sections.

9.3.3.1   *Spectral Angle Mapper*

The SAM classification is an automated method that compares and maps the spectral similarity of image spectra to reference spectra [24]. The reference spectra used with SAM can be derived from either laboratory or field work or extracted from the image itself. Additionally, the SAM classification assumes that the data have been reduced to apparent reflectance [24]. The SAM algorithm determines the similarity between two spectra by treating them as vectors in space with dimensionality equal to the number of bands and calculating the "spectral angle" between them [20,24]. It should be noted that poorly illuminated pixels fall closer to the origin (the dark point) than pixels with the same spectral signature but greater illumination; however, the angle between the vectors is the same regardless of their length [24]. In other words, the SAM classification has the advantage of being totally insensitive to changes in illumination throughout an image. This is because the SAM method uses only the vector direction of the spectra and does not consider the vector length [20]. As a result, laboratory, field, or image spectra can be directly compared with the remotely sensed apparent reflectance spectra with the result being a classification image showing the best SAM match at each pixel. Additionally, gray-scale rule images can also be calculated, which show the actual angular distance in radians between each spectrum in the image and the reference spectrum. In this rule image, darker pixels represent smaller spectral angles, thus spectra are more similar to the reference spectrum [20].

9.3.3.2 Mixture Tuned Matched Filtering

MTMF is a hybrid method based on the combination of well-known signal-processing methodologies and linear mixture theory [25]. This method combines the strength of the matched filter method (no requirement to know all the endmembers) with physical constraints imposed by mixing theory (the signature at any given pixel is a linear combination of the individual components contained in that pixel). MTMF consists of two separate processes. The first process, matched filtering, is used to find the abundances of user-defined endmembers using a partial unmixing. Technically, matched filtering only partially unmixes pixels rather than trying to fully unmix a pixel by identifying every material present. However, as pointed out by Boardman et al. [26], a complete spectral unmixing of an image may not be possible or even desired. On the other hand, partial unmixing provides a method of solving only the small percentage of the data inversion problem that directly and specifically relates to the goals of the investigation. The matched filtering technique works by maximizing the response of the known endmember and suppressing the response of the composite unknown background, thus "matching" the known signature. Matched filter (MF) results are presented for each class or training set with a matched filter score, also known as an MF image. The matched filter image is simply a gray-scale image with values from 0 to 1.0, which provide a means of estimating relative degree of match to the reference spectrum (where 1.0 is a perfect match). Overall, it provides a rapid means of detecting specific materials based on matches to library or image endmember spectra while at same time it does not require knowledge of all the endmembers within an image scene. The partial unmixing employed by matched filtering is an ideal method for transportation feature identification in that it does not require all endmembers in the image to be known. This is significant since the only materials of interest in this research are transportation materials, and everything else can simply be considered background.

However, because the matched-filtering technique alone is susceptible to finding false positives, the MF images are often combined with a mixture-tuning technique to improve accuracy. Mixture-tuning works by using linear spectral mixing theory to constrain the MF result to feasible mixtures and reduce false alarms [25]. MTMF results are presented as two sets of images, the MF score (matched filter image), presented as gray-scale images with values from 0 to 1.0, which provides a means of estimating the relative degree of match to the reference spectrum (where 1.0 is a perfect match) and the infeasibility image, where highly infeasible numbers indicate that mixing between the composite background and the target is not feasible. The best match to a target is obtained when the MF score is high (near 1) and the infeasibility score is low (near 0).

9.3.3.3 Object-Oriented Nearest Neighbor

Object-oriented classification is a more recent development in image classification. The strength in this classification system lies in its ability to group pixels into segments, which are then classified based on form, textures, and spectral information [27]. These segments are generated based on similarity of pixel values as well as user-defined parameters such as size, shape, and compactness. The goal of the segmentation process is to create objects that correspond to the features they represent on the ground. A benefit of this method is the ability to classify segments based on their physical characteristics and spatial relationships with other pixels as well as with pixels from different layers of segmentation. Marangoz et al. [27], Coe et al. [28], Zhang and Couloigner [29], and Zhu and Scarpace [30] have used object-oriented classification to extract road and infrastructure features from aerial imagery.

The process of object-oriented classification involves first creating segments based on the user-defined criteria. Then a class hierarchy is created, which may have parent and child classes for advanced classification. For example, a parent class of vegetation may be created, which could be classified based on a vegetation index. Then a grass child class may be classified based on greenness. Coe et al. [28] used a similar classification hierarchy in their study. Then the method of classification is defined for all classes or it may be defined on a class-by-class basis. There are two primary methods of classification available to the eCognition user—nearest neighbor and user-defined membership functions. Nearest neighbor classification plots samples as vectors in n-dimensional space and classifies each segment based on the distance from its n-dimensional vector to the sample vector. Alternatively, a membership functions-based classification scheme may be developed based on knowledge of the spectral signatures of each material. eCognition also allows the combination of nearest neighbor classification and membership functions for each sample. A benefit of eCognition is that it uses a soft or fuzzy classification method [31]. Fuzzy classification is preferred because it allows for variations in spectral profiles within a class. In this case, membership to each class is not a matter of yes or no. There are degrees of membership to each class. The conversion from fuzzy classification to the hard classification necessary to create an output image is based on the highest fuzzy classification value.

9.3.4 Accuracy Assessment

Accuracy assessment was performed using stratified random sampling for the SAM, CART-MTMF, and MTMF classifications. In this case, there were

50 random samples for each class. For the object-oriented classification, a manually classified layer was created that contained all of the known, well-defined objects for each class. This was used as a reference for accuracy assessment.

9.4 Results and Discussion

Four classification methods were tested in this study, SAM, object-oriented nearest neighbor, MTMF, and MTMF combined with CART-selected bands. Figure 9.2 shows the classified outputs from the MTMF, SAM, MTMF combined with CART (MTMF-CART), and MTMF object-oriented. The overall accuracies for the four classifications were 88.92%, 81.89%, 84.32%, and 95.27%, respectively (Table 9.2). The overall measures of accuracy provided by the kappa statistic were 87.61%, 79.87%, 82.46%, and 94.33%, respectively (Table 9.2).

In terms of reliability, the best overall accuracy and highest kappa coefficient value were produced by the MTMF object-oriented classification, whereas the SAM classifier resulted in the lowest overall accuracy and kappa. The primary reason why the SAM method results were worse than the MTMF, MTMF-CART, and MTMF object-oriented methods is its greater difficultly in distinguishing the gravel roads class from the background class. In this case, the SAM classifier overrepresented the gravel roads class compared with the other two classifiers by incorrectly identifying a number of areas in the image as gravel roads that may have been gravel but technically were not roads (e.g., gravel driveways, gravel parking lots). Thus, despite the fact that these pixels may correctly identify gravel material, they do not correctly identify gravel roads. As a result, these pixels were recorded as incorrect in the accuracy assessment. Nevertheless, it should be pointed out that the SAM classifier was the quickest and easiest method to perform, making it an excellent tool for creating initial classification images. Further, examination of the error matrix demonstrates that in terms

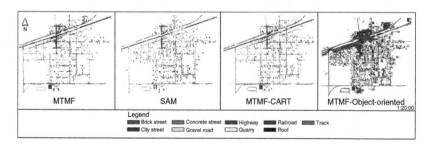

FIGURE 9.2 (See color insert following page 292.)
Comparison of classifications.

TABLE 9.2

Accuracy Assessment

Land-Cover Class	SAM			MTMF			MTMF-CART			MTMF Object-Oriented		
	Producer's Accuracy (%)	User's Accuracy (%)	Kappa	Producer's Accuracy (%)	User's Accuracy (%)	Kappa	Producer's Accuracy (%)	User's Accuracy (%)	Kappa	Producer's Accuracy (%)	User's Accuracy (%)	Kappa
City street	92.86	61.90	0.5703	100.00	71.43	0.6797	78.72	60.66	0.5493	100.00	95.48	1.000
Quarry	82.76	97.96	0.9758	100.00	100.00	1.0000	100.00	100.00	1.0000	—	—	—
Gravel road	100.00	63.83	0.6064	100.00	100.00	1.0000	100.00	100.00	1.0000	100.00	100.00	1.0000
Highway	97.73	97.73	0.9742	88.89	100.00	1.0000	87.50	84.00	0.8161	100.00	92.85	1.000
Railroad	100.00	91.67	0.9085	96.88	86.11	0.8480	97.06	86.84	0.8551	100.00	94.52	1.000
Concrete road	100.00	54.29	0.5181	100.00	55.26	0.5257	100.00	60.00	0.5759	72.88	92.09	0.7145
Roof	90.91	88.24	0.8708	93.62	93.62	0.9269	88.57	86.11	0.8466	88.19	93.15	0.86
Brick road	96.88	96.88	0.9658	96.88	100.00	1.0000	93.75	100.00	1.0000	83.99	100.00	0.8304
Track	100.00	100.00	1.0000	100.00	100.00	1.0000	100.00	93.33	0.9279	100.00	100.00	1.000
	Overall accuracy 81.89%		Overall Kappa 79.87%	Overall accuracy 88.92%		Overall Kappa 87.61%	Overall accuracy 84.32%		Overall Kappa 82.46%	Overall accuracy 95.27%		Overall Kappa 94.33%

of producer's accuracy the concrete, gravel, railroad, track, and to a lesser degree brick classes were classified consistently well when using all four classification methods. Quarry material was also classified well with the exception of the SAM classification. On the other hand, no class consistently underperformed for all four classification methods.

In terms of user's accuracy, only brick and quarry performed exceptionally well in all four classification methods. On the downside, city streets and concrete both performed consistently poorly in terms of user's accuracy. However, it should be noted that, like the previously described SAM results for the gravel class, the majority of error in the user's accuracy for the concrete and city street classes was caused by confusion with the background class. In other words, pixels were classified as either concrete or city street when in fact they should have been classified as background. Interestingly, these errors were only partial errors. In nearly all cases, this error was a result of a concrete driveway classified as concrete street, or a parking lot made from the same material as the city streets classified as a city street. Since driveways and parking lots are technically not streets, there was no other choice but to record them as misclassifications in the accuracy assessment. This indicates a limitation of remote sensing in general. While remote sensing classification may allow for materials to be identified, there is still no way to determine the use of that material. This limitation proved to be particularly significant in this study.

Other possible explanations for the poor results of the city streets class using the SAM classifier can be attributed to spectral mixing that is likely to occur in image pixels representing city streets. This spectral mixing is caused by overhanging trees, cars parked on the street, and close proximity to similar materials such as concrete driveways and rooftops. This might explain why the MTMF-based methods, which possess the ability to detect and extract specific materials even in mixed pixels, resulted in its better performance compared with the SAM method. This then raises the question about the performance of the MTMF-CART method and why its overall accuracy was only slightly greater than that of the SAM classifier despite the fact that it was based on the MTMF method. The likely reason that the MTMF-CART classification method performed worse is twofold. First, the CART method used to reduce the data dimensionality and select the best bands for transportation feature extraction is heavily dependent on the size of training sets. In other words, the CART method required larger training information to make the CART process more accurate. However, this contradicts with the MTMF classifier and hyperspectral remote sensing in general where a smaller number of pixels which are spectrally pure are preferred over larger training sets that likely contain impure or errant pixels. This idea is reiterated by the work of Shrestha et al. [31] who point out that endmember selection is critically important for hyperspectral classification since choosing a wrong one can make a significant difference in the classification result. In the case of this study, the larger regions of interest most likely resulted in inaccurate endmember collection and classification. A second possibility is that by attempting to choose only the best bands the CART process simply reduced the data dimensionality too much. In other words, it is very possible that the CART method may have eliminated bands that actually contained useful information about transportation features.

The object-oriented results were quite promising. By using very few eCognition's tools, high accuracies were obtained, particularly city street and roof classes. The increase in overall accuracy could likely be attributed to fuzzy classification as well as the nearest neighbor classification scheme. The track and gravel classes each had 100% in each measure of accuracy. These two classes were significantly different in spectral signature than any of the other classes, thus they yielded higher accuracy. The concrete class, which was problematic in the other classifications, provided the worst class accuracy here as well. The greatest sources of confusion with this class were city streets and rooftops. Possible reasons for this have been discussed earlier; however with object-oriented classification, there is potential for further errors. If the image objects are not correctly defined, then the training classes are not truly representative of the class. Due to spectral similarities or physical proximities to differing classes, these may be incorporated into the image segment. If this segment is selected for training data, it will then affect the results of the classification. Further rules based on size, shape, or proximity to neighboring classes could be developed and might enhance class accuracies.

9.5   Conclusion and Future Directions

The object-oriented classifier outperformed the SAM, MTMF-CART, and MTMF methods. Overall, all four classification methods were able to identify the classes of quarry, highway, brick road, railroad, roof top, and fitness track with a relatively high degree of success. In contrast, the city street and concrete street classes consistently recorded low accuracies. As mentioned previously, this is likely due to confusion between parking lots and driveways created from the same materials as street classes. Additionally, all four classifiers were able to successfully discriminate between road surfaces and roofing materials. This is in contrast to the results of similar studies, such as the study conducted by Gardner et al. [3], where confusion between roofs and roads was reported to be problematic.

In general, this study was considered to be successful, but at the same time there is still room for improvement beginning as far back as the date of data acquisition. Typically when a hyperspectral scene is flown by an airborne sensor such as AVIRIS, a ground crew is present the very same day to record spectrometer readings from various materials on the ground. Ground data availability could potentially improve overall results in two major ways. First, the atmospheric correction process could be greatly improved with ground information collected the same day as the scene was flown. Temperature, humidity, time of day, and spectral readings gathered in the field can all be used to remove the attenuation caused by atmospheric conditions. Second, correcting the image in such a manner essentially standardizes the image and the spectral responses it contains. This makes it possible to make use of the spectral signatures collected from ground data and from existing spectral libraries to help improve endmember development and classification accuracy. However, the AVIRIS data obtained for this study did not include any ground reference information so there was no option but to extract the information from the image. While this proved to be sufficient for this study, future research could benefit from the advantages of ground calibration data.

Additionally, higher-resolution imagery would potentially help eliminate the complications caused by mixed pixels. This would be particularly useful in the urban areas where many objects and materials are often crowded together. More importantly, higher-resolution imagery might also allow a method to be created to determine road condition. This method would greatly increase the utility of remote sensing methods compared with the current methods for gathering transportation data (e.g., GPS vehicles, photo-log vehicles, and even digitizing). However, current remote sensing imagery is simply unable to detect the small cracks and bumps that are typically associated with poor road condition. Imagery with resolutions of 1 in. or even less may be required before the extraction of road condition can be made reliable.

Whether it is pixel-based or object-based, a final important aspect to be addressed in the future is vectorizing the classified images. Vectorizing the image classifications is necessary to make the extracted data compatible with common GIS software packages. This is a crucial step since current transportation managers rely heavily on GIS for transportation data storage and analysis. Current software packages available to perform the task of vectorization include those such as SoftSoft's WinTopo Professional and Able Software's R2V. Overall, this study has shown that transportation features can successfully be extracted from hyperspectral imagery. The research also concluded that the object-oriented MTMF classification method produced the best overall results, making it superior to the SAM classifier and MTMF-CART methods. On the other hand, there are still many difficulties that must be overcome. In particular, mixed pixels, the complex urban environment, and the inability of sensors to detect material use opposed to material type create significant hurdles, which need to be addressed in the future.

Acknowledgments

This work was supported by the Midwest Transportation Consortium (MTC) and Iowa DOT. We also wish to thank NASA-JPL for providing the AVIRIS dataset for this research.

References

1. Hallmark, S., Mantravadi, K., Veneziano, D., and Souleyrette, R. *Evaluating Remotely for in Inventorying Roadway Infrastructure Features.* Ames: Center for Transportation Research and Education, Iowa State University, 2001.
2. Khattak, A. and Gopalakrishna, M. *Remote Sensing (LIDAR) for Management of Highway Assets for Safety.* Ames: Midwest Transportation Consortium, Center for Transportation Research and Education, Iowa State University, 2003.
3. Gardner, M., Roberts, D., Funk, C., and Noronha, V. Road extraction from AVIRIS using spectral mixture and Q-tree filter techniques (Technical Report, May, 2001). University of California, Santa Barbara, National Consortium on Remote Sensing and Transportation: Infrastructure, 2001.
4. Church, R. and Sexton, R. *Modeling Small Area Evacuation: Can Existing Transportation Infrastructure Impede Public Safety?* Santa Barbara, California: Vehicle Intelligence & Transportation Analysis Laboratory and Department of Geography, University of California at Santa Barbara, 2002.
5. Fletcher, D. and Kunda, R. *Development and Automation of High Resolution Image Extraction Methodologies for Transportation Features.* Albuquerque: Affiliated Research Center, University of New Mexico, 1999.
6. National Consortia on Remote Sensing in Transportation. Remote sensing and spatial information technologies in transportation: Synthesis report 2001. Washington, D.C.: U.S. Department of Transportation; Stennis Space Center, MS, National Aeronautics and Space Administration.
7. Harvey, W., McGlone, J., Mckeown, D., and Irvine, J. User-centric evaluation of semi-automated road network extraction. *Photogrammetric Engineering & Remote Sensing*, 70(12), 1353, 2004.
8. Hu, X., Zhang, Z., and Tao, C. A robust method for semi-automatic extraction of road centerlines using a piecewise parabolic model and least square template matching. *Photogrammetric Engineering & Remote Sensing*, 70(12), 1393, 2004.
9. Kim, T., Park, S., Kim, M., Jeong, S., and Kim, K. Tracking road centerlines from high resolution remote sensing images by least squares correlation matching. *Photogrammetric Engineering & Remote Sensing*, 70(12), 1417, 2004.
10. Song, M. and Civco, D. Road extraction using SVM and image segmentation. *Photogrammetric Engineering & Remote Sensing*, 70(12), 1365–1371, 2004.
11. Shippert, P. Introduction to hyperspectral image analysis. Retrieved April 7, 2003, from http://satjournal.tcom.ohiou.edu/pdf/shippert.pdf
12. Herold, M., Gardener, M., Hadley, B., and Roberts, D. The spectral dimension in urban land cover mapping from high resolution optical remote sensing data. In: *Proceedings of the 3rd Symposium on Remote Sensing of Urban Areas*, Istanbul, Turkey, 2002.
13. Herold, M. and Roberts, D. Spectral characteristics of asphalt road aging and deterioration: Implications for remote sensing applications. *Applied Optics*, 44(20), 4327–4334, 2005.
14. Bhaskara, S. and Datt, B. Applications of hyperspectral remote sensing in urban regions. Poster session presented at the Asian Conference on Remote Sensing, Taipei, Taiwan, 2000.
15. NCRST (National Consortia on Remote Sensing in Transportation). Achievements of the DOT-NASA joint program on remote sensing and spatial information technologies application to multimodal transportation. Washington, D.C.:

U.S. Department of Transportation; Stennis Space Center, MS, National Aeronautics and Space Administration, 2002.

16. Roberts, D.A., Yamaguchi, Y., and Lyon, R.J.P. Comparison of various techniques for calibration of AIS data. In: *Proceedings of the 2nd Airborne Imaging Spectrometer Data Analysis Workshop*, JPL Publication 86-35, p. 21, 1986.

17. Clark, R., Swayze, G., Livo, K., Kokaly, R., King, T., Dalton, J., et al. Surface reflectance calibration of terrestrial imaging spectroscopy data: A tutorial using AVIRIS. United States Geological Survey Spectroscopy Lab. Retrieved December 8, 2002, from http://speclab.cr.usgs.gov/PAPERS.calibration.tutorial/calibntA.html

18. Green, R., Eastwood, M., Sarture, C., Chrien, T., Aronsson, M., and Chippendale, B. Imaging spectroscopy and the airborne visible/infrared imaging spectrometer (AVIRIS). *Remote Sensing of Environment*, 65, 227, 1998.

19. Boardman, J. and Kruse, F. Automated spectral analysis: A geological example using AVIRIS data, North Grapevine Mountains, Nevada. Paper presented at the meeting of the ERIM Tenth Thematic Conference on Geologic Remote Sensing, Environmental Research Institute of Michigan, Ann Arbor, MI, 1994.

20. Kruse, F., Richardson, L., and Ambrosia, V. Techniques developed for geologic analysis of hyperspectral data applied to near-shore hyperspectral ocean data. Paper presented at the Fourth International Conference on Remote Sensing for Marine and Coastal Environments, Orlando, FL, March 1997.

21. Breiman, L., Jerome, F., Richard, O., and Charles, S. *Classification and Regression Trees*. Belmont, CA: Wadsworth Int. Group, 1984.

22. Sugumaran, R., Pavuluri, M., and Zerr, D. The role of high-resolution imageries for identification of urban climax forest using traditional and rule-based classification approach. *IEEE Transactions on Geoscience and Remote Sensing*, 41(9), 1933, 2003.

23. Hsu, P. and Tseng, Y. Feature extraction for hyperspectral image. Paper presented at the Asian Conference on Remote Sensing, Hong Kong, China, November 1999.

24. Kruse, F., Lefkoff, A., Boardman, J., Heidebrecht, K., Shapiro, A., and Barloon, P. The spectral image processing system (SIPS)—interactive visualization and analysis of imaging spectrometer data. *Remote Sensing of Environment*, 44, 145, 1993.

25. Boardman, J. Leveraging the high dimensionality of AVIRIS data for improved sub-pixel target unmixing and rejection of false positives: Mixture tuned matched filtering. *Summaries of the Seventh Annual JPL Airborne Geoscience Workshop*, Pasadena, CA, p. 55, 1998.

26. Boardman, J., Kruse, F., and Green, R. Mapping target signatures via partial unmixing of AVIRIS data. Paper presented at the Fifth JPL Airborne Earth Science Workshop, Pasadena, CA, 1995.

27. Marangoz, A.M., Oruc, M., and Buyuksalih, G. Object-oriented image analysis and semantic network for extracting the roads and buildings from IKONOS pan-sharpened images. In: *Proceedings of the ISRPS 2004 Annual Conference*, Istanbul, Turkey, July 19–23, 2004.

28. Coe, S., Alberti, M., Hepinstall, J.A., and Coburn, R. A hybrid approach to detecting impervious surface at multiple scales. In: *Proceedings of the ISPRS WG VII/1 "Human Settlements and Impact Analysis" 3rd International Symposium Remote Sensing and Data Fusion Over Urban Areas (URBAN 2005) and 5th International Symposium Remote Sensing of Urban Areas (URS 2005)*, Tempe, AZ, March 14–16, 2005.

29. Zhang, Q. and Couloigner, I. Automated road network extraction from high resolution multi-spectral imagery. In: *Proceedings of ASPRS 2006 Annual Conference*, Reno, Nevada, May 1–5, 2006.

30. Zhu, H. and Scarpace, F.L. Aerial image matching incorporating object recognition. In: *Proceedings of ASPRS 2006 Annual Conference*, Reno, Nevada, May 1–5, 2006.

31. Shrestha, D., Margate, D., Anh, H., and Van Der Meer, F. Spectral unmixing versus spectral angle mapper for land degradation assessment: A case study in southern Spain. Paper presented at the 17th World Congress of Soil Science, Bangkok, Thailand, 2002.

10

Road Extraction from SAR Imagery

Uwe Stilla, Stefan Hinz, Karin Hedman, and Birgit Wessel

CONTENTS

10.1   Introduction

Road extraction from remote sensing data has been of considerable interest in recent years due to the rapid progress of geographic information systems (GIS) and the increasing importance of roads in our daily life. Detailed and up-to-date road information is an important issue for numerous applications. Logistics, tourism, car navigation systems are just a few fields of interest. Yet, to accommodate for the needs of these applications, digital road information requires frequent updates, whereby the main source for road data collection is digital aerial and satellite imagery. Despite numerous technological advances, the process of data acquisition still needs a lot of manual interaction of a human operator, which is of course both time-consuming and expensive. Consequently, much effort has been put into automatic road extraction approaches in recent years (Stilla et al., 2005).

Besides well-known drawbacks due to the specific viewing geometry and coherent imaging, synthetic aperture radar (SAR) holds some prominent advantages over optical images (Stilla and Soergel, 2006). For instance, SAR is an active system, which can operate during day and night. It is also nearly weather-independent and, moreover, during bad weather conditions, SAR is the only operational system available today. Road extraction from SAR images therefore offers a suitable complement or alternative to road extraction from optical images.

The remainder of this chapter is organized as follows: Section 10.2 sketches typical approaches for automatic road extraction from remote sensing data with their underlying models and extraction strategies before outlining an extraction system developed at Technische Universitaet Muenchen (TUM) in more detail. Section 10.3 focuses then on the adaptation of this (generic) approach to the particular challenges induced by the utilization of SAR images. Finally, Section 10.4 presents novel concepts for the extension of road extraction from SAR data concerning the employed models and strategies and illustrates the benefits of these extensions by examples.

10.2   Road Extraction from Remote Sensing Data

10.2.1   Modeling of Roads

The extraction of topographic objects from images usually relies on an object model, which represents an abstraction of the corresponding real-world object class (roads, buildings, etc.). Since humans tend to organize objects

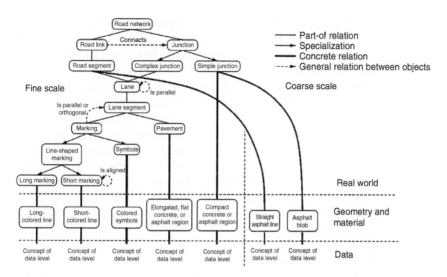

FIGURE 10.1
Example for road model as semantic network.

and object classes hierarchically using different levels of abstraction, it is reasonable to map this hierarchical structure to the model, especially when each level of abstraction can be linked with a particular scale in image space. Furthermore, the model should be described in a formalized way so that inconsistencies in the description are avoided. We will exemplify this type of modeling by semantic networks in the following sections, since semantic networks allow representing the knowledge about the objects in a very transparent fashion (see Figure 10.1). However, other modeling schemes like frames or production nets (Stilla, 1995) could be used as well.

The model comprises explicit knowledge about geometry (road width, parallelism of roadsides, etc.), radiometry (reflectance properties), topology (network structure), and context (relations with other objects, e.g., buildings or trees). The model described later consists of two parts: the first part describes characteristic properties of roads in the real world and in the used data, and represents a road model derived from these properties. The second part defines different local contexts and assigns those to the global contexts. In this way, the complex model for the object road is split into submodels that are adapted to specific contexts.

A description of roads in the real world can be derived from their function for human beings: roads are defined as a place, where one may ride, that is, an open or public passage for vehicles, persons, or animals. They are important for communication and transportation between different places. Therefore, roads are organized as a network. The denser an area is inhabited and the more intensively it is used, the denser the road network is. With respect to their importance, network components are classified into a hierarchy of different categories with different attributes. According to the different categories, roads differ with respect to minimum curvature radius and maximum allowed slope. Some important attributes for parts of the road network are the type and state of the road surface material, existence of road markings, sidewalks, and cycle tracks, or legal instructions, such as traffic regulations. The appearance of roads strongly depends on the sensor's spectral sensitivity, its viewing geometry, and its resolution in object space. In images with low resolution, that is, more than 2 m pixel, roads mainly appear as lines that form a more or less dense network. Contrary to this, in images with a higher resolution, that is, <0.5 m, roads are projected as elongated homogeneous regions with almost constant width. Here the attainable geometric accuracy is better, but background objects like trees or buildings disturb the road extraction more severely. On the other hand, in a smoothed image—which corresponds to a reduced resolution—lines representing road centerlines can be extracted in a stable manner even in the presence of these background objects. The smoothing reduces the influence of background objects or even eliminates them at all.

This scale-space behavior can be interpreted as abstraction, that is, the object road is simplified and its fundamental characteristics are emphasized, as shown in Mayer and Steger (1998). It follows that, when high-resolution data are available, not only the highest resolution should be considered but also lower resolutions, since the fusion of multiple scales contributes to improve the reliability and robustness of road extraction.

The road model shown in Figure 10.1 describes objects by means of "concepts," and is split into three levels defining different points of view. The real-world level comprises the objects to be extracted and their relations. On this level, a road network consists of junctions and road links that connect junctions. Road links are constructed from road segments. In fine scale, road segments are aggregated by lanes, which consist of pavement and markings. For markings there are two specializations: symbols and line-shaped markings. The concepts of the real world are connected to the concepts of the geometry and material level via concrete relations (Toenjes et al., 1999), which connect concepts representing the same object on different levels. The geometry and material level is an intermediate level which represents the 3D shape of an object as well as its material (Clément et al., 1993). The idea behind this level is that in contrast to the image/data level it describes objects independently from sensor characteristics and viewpoint.

Road segments, for instance, are linked to the "straight asphalt lines" of the geometry and material level in coarse scale. In contrast to this, the pavement as a part of a road segment in fine scale is linked to the "elongated, flat asphalt region" on this level. In case of markings that are painted on the road, they are modeled as colored thin lines or symbols.

While real-world level and geometry and material level describe the object independent of the sensor, the concepts at data level are of course strongly dependent on the sensor used for data acquisition. The pavement concept, for instance, differs significantly for aerial images (see, e.g., Baumgartner et al., 1999) and SAR images. While roads appear as bright

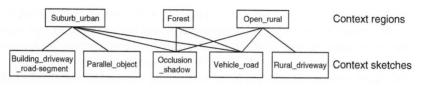

FIGURE 10.2
Context regions and context relations for modeling the influence of background objects. (Adapted from Baumgartner, A., Eckstein, W., Mayer, H., Heipke, C., and Ebner, H., *Automatic Extraction of Man-Made Objects from Aerial and Space Images* (*II*), Birkhauser Verlag, Basel, Switzerland, 1997.)

region in aerial images, they are usually darker than their surroundings in SAR images. Section 10.3 will elaborate this in more detail.

The road model presented earlier (Figure 10.1) comprises knowledge about radiometric (i.e., data-related), geometric, and topological character-istics of roads. This model is extended by knowledge about context: so-called context objects, that is, background objects like buildings, trees, or vehicles, can support road extraction, but they can also interfere. In addition, external GIS data can be regarded as context object. Experience has shown that modeling this interaction between road objects and context objects on a local level as well as a global level is an aid for guiding the extraction since the interpretation problem is split into smaller subpro-blems, which can be solved more efficiently by using specific models and extraction strategies.

In order to capture the varying appearance of roads globally, the so-called context regions "urban," "forest," and "rural" are distinguished (cf. Baumgartner et al., 1997, 1999; Hinz and Baumgartner, 2003; Wessel, 2006). Furthermore, the local context is modeled with so-called context rela-tions, that is, certain relations between a small number of road and context objects, which describe the influence of neighboring objects on the appear-ance of roads in a certain context region. Figure 10.2 shows typical context relations and their dependence on context regions, while two typical instances of a context relation "occlusion_shadow" are depicted in Figure 10.3.

Note, however, that the use of knowledge about local context and the verification of specific relations between local objects will in most cases be possible in high-resolution imagery only, because the image features which

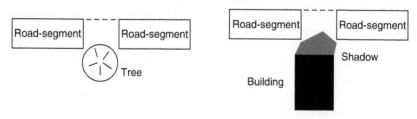

FIGURE 10.3
Sketches for context relation "occlusion_shadow" (see Figure 10.2).

contribute to the local context are usually not very prominent. Therefore, the local context is more tightly connected with the high resolution, whereas information about global context usually can be derived from images of lower resolution and is useful to guide the road extraction in both scales.

Not every context sketch has to be taken into account everywhere. The relevance of objects and relations depends on the global context. Roads in urban or suburban areas look quite different and have other relations compared with roads in rural or forest areas. In contrast to buildings in rural areas, buildings in downtown areas are very close to and highly parallel to roads; sidewalks and cycle tracks are more likely to appear in urban areas; in rural areas single trees and single buildings might hinder extraction, whereas in forest regions mainly shadows and occlusions pose problems. Therefore, it is wise to use different context sketches not only at multiple resolution levels but also within context regions.

10.2.2   Related Work on Road Extraction

In this section, we briefly outline some earlier fundamental work on road extraction in general. The ideas incorporated in these approaches are mainly independent of the sensor used, and many of them have been used in later developments.

The existing approaches cover a wide variety of strategies to extract roads automatically from digital aerial or satellite imagery, or at least to automate parts of the manual extraction process. As GIS-driven approaches for road extraction (Stilla, 1995; Bordes et al., 1997; deGunst and Vosselman, 1997; Zhang et al., 2001; Zhang 2004; Gerke et al., 2004) are more useful for verification than for extraction of new roads, we focus in the discussion of previous approaches on those that also aim at extraction of previously unknown roads. In semiautomatic approaches an operator provides, for example, starting points and starting directions on the road for a road following algorithm (McKeown and Denlinger, 1988; Vosselman and de Knecht, 1995). If an operator measures more than one point on the road, an algorithm like F-algorithm can be applied to find an optimal path, that is, the road between these points (Fischler et al., 1981; Merlet and Zerubia, 1996). If multiple views are used, this can also be done in 3D (Gruen and Li, 1997). The advantage of the approaches with multiple points is that the path of the road is more constrained, which results in a more reliable handling of critical areas. A similar approach based on so-called "ziplock" snakes is presented in Neuenschwander et al. (1995). By automatic detection of the seed points, semiautomatic schemes can be extended to fully automatic ones.

An automatic approach is described in Barzohar et al. (1997). The selection of starting points is based on gray-value histograms. Further assumptions about geometry and radiometry are described in a Markov random field. Road extraction is then performed by dynamic programming. Another fully automatic approach for the extraction of road networks from digital aerial imagery has also been proposed by Ruskone (1996): hypotheses for connections between automatically detected seed points are checked using geometrical constraints. The approach of Baumgartner et al. (1997, 1999) employs line and edge extraction and grouping at two different resolutions to find reliable initial road hypotheses. These hypotheses are verified and connected using global and local context models as described earlier. While these approaches mainly exploit the local road characteristics like parallel roadsides and homogeneous interior, the work of Steger et al. (1997), Wiedemann and Hinz (1999), and Wiedemann and Ebner (2000) focuses on the function of roads connecting various places in a scene. Thus, the exploitation of road network characteristics by optimal path calculations through a graph of road hypotheses plays a major role. A combination of the approaches of Baumgartner et al. (1999) and Wiedemann and Hinz (1999) including a detailed evaluation is presented in Hinz et al. (2001). It shows that both approaches complement each other and achieve superior results when combined. Finally, the approaches of Price (2000) and Hinz and Baumgartner (2003) extend the previous work for dense, complex urban areas. Major focus is put on the incorporation of detailed road elements like markings and lanes as well as their context relations (see Hinz and Baumgartner, 2003). To handle the huge complexity of such scenes, a self-diagnosis scheme is employed in the extraction system.

In the following section, we describe the approach "TUM-LOREX" of Steger et al. (1997), Wiedemann and Hinz (1999), and Wiedemann and Ebner (2002) in more detail. This road extraction system is especially designed to exploit the network characteristics of roads and relies on a generic architecture that—once primary linear features including their attributes have been extracted from the data—is independent of the spectral characteristics of the underlying remote sensing data.

10.2.3 TUM-LOREX

10.2.3.1 *Model*

The underlying road model of this approach consists of the top part of the road model displayed in Figure 10.1. The relevant part of the real-world level is depicted in Figure 10.4.

The system expects linear features (road segments) and their attributes as input. These may stem from line extraction, for instance using the approach of Steger (1998), or from other approaches, which are able to deliver initial road hypotheses as exemplified in Baumgartner and Hinz (2000). Based on

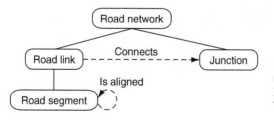

FIGURE 10.4
Real-world level of road model of TUM-LOREX.

these primitives so-called regional characteristics of roads are introduced. These incorporate the assumption that mostly roads are composed of long and straight segments having constant width and reflectance. Finally, roads are described globally in terms of functionality and topology: the intrinsic function of roads is to connect different—even far distant—places. They thus form a network wherein all road segments are topologically linked to each other. Though, since the image covers only a part of the whole road network, different subnetworks may occur in the image, which are not necessarily connected.

10.2.3.2 Extraction Strategy

The extraction strategy is derived from the road model and is composed of different steps (see Figure 10.5). After extracting, evaluating and introducing road segments as linear primitives, a weighted graph is constructed from the primitives and the gaps, that is, connection hypothesis, between the end points of different primitives. Road network generation is carried out by calculation of "best paths" among various pairs of points, which are assumed to lie on the road network with high probability. Finally, the extracted road network is topologically completed and refined. In the following text, a detailed description of each step is given.

Line primitive processing has three different tasks:

1. Increase the probability that line primitives either completely correspond to roads or to linear features not being roads
2. Fuse linear primitives extracted from different sources (channels, sensors, aspects, etc.)
3. Prepare the primitives for the generation of junctions

Ad 1: Linear primitives will not be split during the final extraction step, road network generation, to avoid costly iterative calculations. That means, they will either be completely added to the road network or they will not be added at all. Therefore, it is necessary to ensure that primitives completely correspond to roads or to linear structures not being roads, that is, linear primitives have to be split at points where they potentially cross a roadside. The most significant feature for a change in the semantics of a primitive

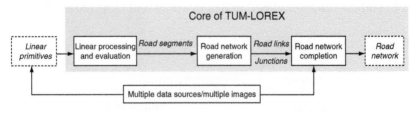

FIGURE 10.5
System architecture of TUM-LOREX.

(road/not road) is high local curvature. Hence, they are split at points where the curvature exceeds a given threshold. A potential overpartitioning of linear primitives that result from this threshold is acceptable since, during the following network generation, the individual parts of partitioned line can be merged again.

Ad 2: To make use of complementary and supplementary properties of multiple input sources, the linear primitives of different sources are fused by a union operation, whereby redundantly extracted primitives are eliminated. "Redundancy" is defined as follows:

1. Linear primitives of different channels overlap each other within a buffer of a certain width.
2. The direction difference of—even partly—overlapping lines does not exceed a given threshold.
3. Primitives extracted with high redundancy are weighted higher than primitives extracted in only one source.

Ad 3: Junctions are an essential part of the road network, but they are rarely detected during primitive extraction. To prepare the generation of missing junctions, lines are split at points which are a priori candidates for junctions. These are points close to other line ends, as, for example, point P in Figure 10.6, which lies on line l_1 closest to the end of line l_2.

The results of these three preprocessing steps are road segments, which are input to the global-grouping algorithm in the next step.

For road network generation, a weighted graph is constructed from the road segments to introduce regional characteristics into the extraction strategy. The nodes of the graph correspond to the end points of the road segments and the edges are the road segments themselves. Piecewise linear fuzzy functions are used to transform the following properties of lines into fuzzy values (Zadeh, 1989):

1. Length
2. Straightness, that is, standard deviation of local orientation
3. Mean width of a line and width constancy, that is, standard deviation of the width along the line

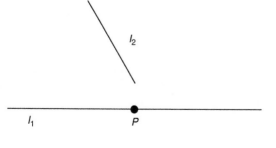

FIGURE 10.6
Candidate for a junction.

4. Mean gray value and gray-value homogeneity along a line
5. Degree of overlap with lines extracted (redundantly) from other channels

These fuzzy values are derived from the set of primitives, which results from the second preprocessing step (fusion of multiple data sources). Calculating the fuzzy values after the third preprocessing step (splitting linear primitives due to junction candidates) would lead to an incorrect evaluation of the road segments in many cases, because splitting affects the primitives' properties, for example, the length, and a decision if a junction candidate truly represents a road junction or if it is only caused by blunder is not possible at this state of processing.

An overall fuzzy value for each road segment is derived by aggregation of the individual fuzzy values using the fuzzy-"and"-operator (Zadeh, 1989). A final weight for each line is eventually calculated by dividing the length of the line by its overall fuzzy value. The final weights thus correspond to costs which are assigned to the respective edges of the graph. Optionally, a weighting of the different input sources can be introduced, for example, while using different geometric or radiometric resolutions. This means that the final weight of a road segment is scaled by a factor depending on the confidence one may concede the specific source, which a particular road segment's underlying primitive originally stems from.

Since the road segments are in general not connected to each other, especially, if they originate from different input sources, such connections have to be made possible. Each pair of end points of different road segments therefore defines a gap (see, e.g., Figure 10.7), and, as in the case of road segment weighting, fuzzy values are derived from:

1. The absolute gap length
2. The relative gap length (compared with the adjacent road segments)
3. The direction differences between the gap and the adjacent road segments, whereby collinearity (within a road) and orthogonality (e.g., at junctions) are preferred
4. An additional clipping threshold, which ensures that the weight of a gap cannot become higher than that of the adjacent road segments

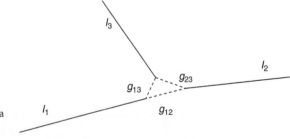

FIGURE 10.7
Multiple connections at a junction.

As before, an overall fuzzy value for each gap is calculated by aggregating the individual fuzzy values. For each gap yielding an overall fuzzy value greater than zero, a new edge is inserted into the weighted graph which receives as final weight (cost) the length of the gap divided by its overall fuzzy value. Thus, bridging of gaps is made possible. Gap weighting is purely based on regional criteria, but the decision which gaps are finally to be added to the road network should also consider the global network characteristics of roads. These are introduced in the following step of road network generation.

Various seed points, that is, road segments with high weights and thus high probability of being truly a road are selected and connected pairwise by calculating the optimal path through the weighted graph using the Dijkstra algorithm (Sedgewick, 1992). However, the optimal path is only calculated if the distance between two seed points exceeds a certain threshold, for example, 1 km. In doing so, the function of roads connecting places far away from each other is emphasized, whereas it is still possible to detect isolated parts of the road network as long as they are large enough. The combination of all resulting paths forms the road network.

There are still topological deficiencies in the extracted road network, especially in the vicinity of junctions where connections might be missing. In addition, the result might contain some separated subnetworks that can be possibly connected. To accommodate such situations, a final topological completion and refinement step is employed based on the calculation of network-specific parameters, especially a so-called detour factor (Wiedemann and Ebner, 2000). Figure 10.8 shows a part of a sample network, which could be the result of the previous steps, consisting of four nodes (A, B, C, D) and three edges (AB, BC, CD). Basically, between each pair of points a link hypothesis is generated, for which two types of distances are calculated:

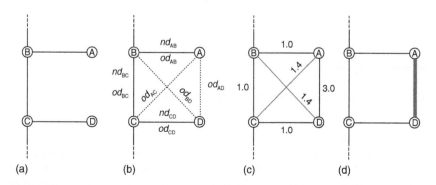

FIGURE 10.8
(a) Network, (b) network and optimal distances, (c) detour factors, and (d) selected link hypothesis.

1. The network distance nd along the shortest path within the existing network where, for example, nd_{BD} is the sum of nd_{BC} and nd_{CD} (see Figure 10.8b). It is computed for all possible pairs of points lying on the network and thus depends on the actual length and classes of the roads along which the shortest path has been found.

2. The optimal distance od along a hypothetical optimal path. It is intended to represent the requirements for fast and cheap transports and depends—besides the Euclidian distance between the two points—on factors like topography, land use, and environmental conservation.

For each point pair, the "detour factor" is calculated from the two distances defining detour_factor $= nd/od$. Figure 10.8c illustrates the detour factors for all point pairs. Here, both the network distance and the optimal distance are set to the Euclidean distance between the respective points for simplicity reasons.

The—naturally very large—number of link hypotheses is reduced assuming that only those links which show a locally maximal detour factor are of any interest for deeper investigation. Hence, a nonmaximum suppression (NMS) is performed on the set of all link hypotheses. In the earlier example, only the link hypothesis AD passes the NMS (see Figure 10.8d) yet, in general, more than one link hypothesis will be kept per point.

All remaining link hypotheses are sorted according to their detour factor and processed beginning with the best one. This connection hypothesis is sent to a module sought to verify the link hypothesis based on image data. For instance, the module designed for road primitive extraction might be used, but now called with relaxed parameter settings. Once a link hypothesis is accepted, it is inserted into the road network. Then the procedure of generating link hypotheses has to be started anew since the topology of the road network has changed. This process iterates until no further link hypothesis is generated or one can expect that the verification module will reject any further hypothesis.

10.3 Road Extraction from SAR Images

Compared with road extraction from optical images, rather few approaches have concentrated on road extraction from SAR images. When dealing with SAR imagery, one has to cope with the typical drawbacks of SAR-specific phenomena such as speckle-affected images, foreshortening, layover, and shadow. These phenomena arise from the side-looking scene illumination of the SAR sensor and make an object extraction in general difficult (Stilla et al., 2003, 2004).

As extraction approaches designed for optical images, work on road extraction from SAR images can be separated in the extraction of linear primitives and the following network generation step. Some approaches developed special line detectors for SAR images (Tupin et al., 1998; Chanussot et al., 1999) or use a combination of a previous classification as input for the line extraction (Dell'Acqua et al., 2003). In addition, a dedicated preprocessing is performed, if well-sophisticated line detectors for optical images are applied (Jeon et al., 2002; Wessel, 2006). In contrast to the line detection, techniques for grouping and network generation algorithms are similar to those used for optical images, as this step is relatively data-independent.

In the following, we describe the adaption of the abovementioned TUM-LOREX road model for the specific characteristics of SAR images.

10.3.1 Model for Roads in SAR Images

As described in Section 10.2.1, a road model comprises knowledge about radiometric (i.e., data-related), geometric, topological, and context-related characteristics of roads. In SAR images with a resolution of 1–2 m, roads appear in general as dark lines with low curvature and constant width. Geocoded SAR images do not show significant geometric deformations. Therefore, it can be stated that the *geometrical* properties of the real-world model of roads can be kept despite the side-looking geometry of the SAR sensor.

The *radiometry* of SAR images relies on the physical parameters of the surface. In case of impervious roads, the most important parameter is the surface roughness. Since the surface of roads is smooth compared with the wavelength of the imaging radar, specular reflection of the incident energy is the most prominent scattering effect at roads and thus only a small amount of energy is returned to the sensor. Beyond the physical properties of the object, the so-called speckle effect—a consequence of the coherent imaging—has an impact on the radiometry of the image. The speckle effect adds multiplicative noise to the images, that is, the noise level in a homogeneous area is proportional to its intensity. However, as roads have low intensity in SAR images, the noise level is also relatively low compared with the road's surroundings. All this leads to a low and relatively homogeneous dark appearance of roads in SAR images.

Whenever the complexity of the scene increases, interactions between roads and other objects appear frequently. Especially when interactions with high-elevated objects occur, layover and shadow effects arise. Layover occurs whenever the emitted radar signal reaches the top of a target before reaching the ground, that is, when the top is closer to the sensor as the ground. As a result, the top is displaced toward the sensor and its returning echo is superimposed with the echoes of the scatterers at the same distance to the sensor, which ultimately results in a very bright or even oversaturated image area. As has been empirically investigated by

Stilla et al. (2004), a road can be partly or totally covered by layover and shadow, especially in urban areas. In addition, bright double bounces or trihedral scattering influence the appearance of roads. For such situations, more information about local and global context has to be introduced into the road extraction.

10.3.2 Example: TUM-LOREX Applied to SAR Images

In Section 10.2.3, the TUM-LOREX approach for automatic extraction of roads from remote sensing data was described. The core of TUM-LOREX consists of a generic approach that is relatively independent of the data source. In this section, we apply the TUM-LOREX approach to SAR images. Therefore, mainly the first part, the extraction of linear primitives, has to be adopted for SAR data (Figure 10.9). For the extraction of linear primitives a specific strategy was developed, which is described later.

The extraction of linear features from SAR data is a critical point since the multiplicative noise of SAR data complicates this task drastically. Multiplicative noise leads for common line operators (developed for optical images) to an increase of extracted line segments as the intensity level increases. As these operators are gradient-based approaches and designed for optical images, they usually assume a Gaussian noise distribution, which is not valid for SAR. On the other hand, approaches especially designed for SAR data take correctly into account the multiplicative noise. However, they are often less flexible because they use template-based methods that are less sensitive to different line widths and orientations (Bovik, 1988; Tupin et al., 1998; Borghys et al., 2002). For roads, this task can be facilitated by using some road characteristics for the extraction of lines, like for example, Gamba et al. (2006).

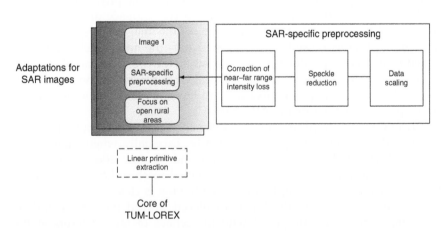

FIGURE 10.9
Strategy for road extraction from SAR data with TUM-LOREX.

For our extraction system, we chose a combination of a sophisticated gray-value curvature-based approach for line extraction—Steger's differential geometric approach (Steger, 1998)—and an elimination of line segments extracted at higher intensity levels. Steger's line detector is independent of the line direction, robust, subpixel precise, and well capable of extracting lines of several widths. Thus a maximum of candidates for roads is ensured before selecting line segments possibly being false alarms, that is, those line segments that are not consistent with the road model are eliminated during the evaluation process within the core of TUM-LOREX (Section 10.2.3). The rest is introduced as road segments for road network generation.

Before applying the line detector, we perform a number of preprocessing steps in particular regarding the near–far range intensity loss, the speckle effect, and data scaling. Furthermore, the line detector is only applied to open rural areas (Wessel, 2006). Those preprocessing steps are described in the following subsections.

10.3.2.1 Correction of the Near–Far Range Intensity Loss

The backscattering is subject to a near–far range intensity decrease. Objects in far range have a lower backscattering as the same objects in near range due to the greater incidence angle. This relation is also true for roads. In Figure 10.10a, the dependency between the backscattering coefficient of roads in X-band data and the incidence angle is shown. A notable near–far range loss can be observed, which makes the elimination of incorrect road candidates based on global thresholds more difficult. A correction of the near–far range intensity loss is necessary to achieve an equal backscattering intensity of roads for all incidence angles (Figure 10.10b). Another option would be to consider the near–far range effect directly within the extraction.

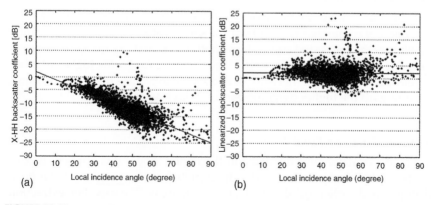

FIGURE 10.10
Dependency of backscatter coefficient of roads in X-band data on the incidence angle (backscatter in dB vs. local incidence angle in degree). (a) Standard calibration and (b) corrected calibration.

In general, a standard calibration (Equation 10.1) delivers the correction yielding:

$$\sigma_0 = (10 \log DN^2 + K) \sin \theta_{loc},\qquad(10.1)$$

where
  DN is the reflection values
  K is a calibration constant
  θ_{loc} is the local incidence angle

However, for the object class "roads" this correction performs an under-correction. As can be seen in Figure 10.10a, the standard correction even forces the intensity loss in far range. A possible alternative is a correction according to an—empirically found—incidence angle dependency of roads (Figure 10.10b). In studies, we found that calibrated data in airborne SAR images have an inclination of about −0.25° compared with well equalized gray values.

10.3.2.2 Speckle Reduction

Speckle reduction is an important step as we apply a gradient-based approach that would extract a lot of false alarms in single-look intensity images. First of all, we use amplitude images that are less infected by speckle. Additionally, we tested different speckle filters as the Frost-filter (Frost et al., 1982), refined Lee-filter (Lee, 1981), and multilooking. It has been shown that the line extraction on multilook images delivers less falsely extracted lines than on other filtered single-look data. This can be explained by the noise distribution of multilook data (four looks), which can be approximated by a Gaussian distribution (Ulaby et al., 1986).

10.3.2.3 Data Scaling

Suitable data scaling can bring out contrasts more clearly; this facilitates the extraction of some objects or at least the setting of parameters. Data scaling transforms the measured values into another interval. As the gray values of roads are located at the lower part of the dynamic range, the higher part has no relevance for roads. Therefore, it is possible to reduce the dynamic range by cutting off the higher dynamic part without loss of information for roads.

10.3.2.4 Focus on Rural Areas

The road extraction also depends on the region where it is applied, that is, on the global context. For roads, three global context regions have been distinguished earlier: rural, urban, and forest areas. The used line extraction is most suitable to extract lines in rural areas, because in those areas the line model fits best to the appearance of roads. In order to avoid false road extractions in other context regions, the search area is limited to rural areas.

Hence, a classification of the global context areas "forest," "urban," and "rural" is carried out. The classification is designed for X- and L-band data. The advantage in using these two frequency bands relates to the complementary radiometric characteristics of the global context areas in these bands, which eventually facilitates the discrimination. Forest has a high backscattering in L-band because of its volume scattering; urban areas show high contributions in both bands due to double-bounce returns. Open rural areas have low to middle backscattering. A result of this classification is shown in Figure 10.11. Areas of low intensity are segmented as further reduction of the search area for roads.

10.3.2.5 Line Extraction with Steger's Algorithm

After performing the preprocessing steps and the classification of rural areas, the line extraction algorithm of Steger (Steger, 1998) is applied. This operator is based on differential geometry and captures the local radiometric road characteristics. It assumes a parabolic profile of roads and can extract bright or dark lines. To initialize the procedure, a few semantically meaningful parameters have to be determined: The maximum width of the lines to be extracted can be chosen according to the road width scaled to the image. The two threshold values, which control the process of linking individual line pixels into pixel chains, can be derived from the gray value contrast between roads and their surroundings. Additionally, local line

(a)

FIGURE 10.11
Classification of context areas. (a) Section of a scene.

(continued)

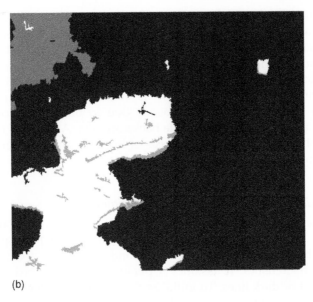

(b)

FIGURE 10.11 (continued)
(b) classification result: rural areas (black), forest (white) urban areas (dark gray), shadows (bright gray).

attributes like width, direction, and contrast are obtained. The result of the line extraction is a set of pixel chains and junction points for each image in subpixel precision. Due to the exclusive use of local road characteristics, the result is not complete and contains false alarms, that is, some roads are not extracted and some extracted lines are not roads. The linear primitives are introduced into the core of TUM-LOREX, whose next step is an evaluation.

10.3.2.6   Evaluation of Linear Primitives

In the evaluation step of TUM-LOREX, all extracted linear primitives are evaluated according to the road model. The fuzzy-logic weights are derived for each linear primitive from the evaluation measures. Especially the evaluation according to the maximum acceptable mean gray value is an important measure for SAR images. Because of the low backscatter of roads, this measure acts effectively for an elimination of incorrect hypotheses. The evaluation of linear primitives guarantees in cooperation with the line extraction of Steger a good basis for the network generation.

The resulting set of linear features after the preprocessing steps, line extraction and line evaluation, is fed into the TUM-LOREX system for road network generation (see Section 10.2.3.2 and Figure 10.5). A final result of road extraction from SAR data applying TUM-LOREX with the previously described specific strategy is shown in Figure 10.12.

FIGURE 10.12
Road extraction result with SAR-specific strategy (white lines: extracted roads; black lines: missing roads; dashed white lines: urban contours).

10.4 Extended Concepts for Road Extraction from SAR

The results of the TUM-LOREX approach show that the automatic extraction results are in general quite, complete, and reliable. But if the road is disturbed, for example, with layover of adjacent high objects like trees or buildings the extraction is interrupted. These are SAR-inherent difficulties because of the side-looking geometry of the SAR sensor. These effects make road extraction complicated, primarily—yet not purely—in urban areas. In case of adjacent high buildings, roads might only be partly visible in the radar image (Stilla et al., 2004). Furthermore in urban areas, the complexity arises through dominant scattering caused by building structures, traffic signs, and metallic objects in cities. In order to compensate for possible gaps, additional information needs to be considered.

In the following, we will outline three different concepts to include additional information:

1. The introduction of context information: Adjacent objects to roads like metallic bridges, road signs, or vehicles have a direct contextual relationship to roads and can be incorporated into the road extraction procedure. The same is true for rows of buildings, which cause layover and shadow areas (Wessel and Hinz, 2004; Amberg et al., 2005). We will show the potential of using local and global context for the extraction (Wessel, 2004).

2. The incorporation of specific detailed models: A road may have different geometry and appearance in SAR images depending on its respective road class. Primary roads, for instance, are usually straight and wider than secondary roads; and highways have usually two carriageways separated by crash barriers. We will exemplify the road-class-specific modeling by an additional road model for the road class "highway."

3. The use of multiview imagery: That means, multiple SAR images of the same scene are used, but with varying illumination from different directions. The illumination from different directions employs both supplementary and complementary information so that the general detectability of roads is increased. A numerical analysis for the potential improvement can be found, for example, in Stilla et al. (2004). It is obvious that, in urban areas, best results for the detectability of roads are obtained whenever the illumination direction coincides with the local orientation of the road axis. Preliminary work has shown that the utilization of multi-aspect SAR images also improves automatic road extraction. This has been tested both for real and simulated SAR scenes (Tupin et al., 2002; Dell'Acqua et al., 2003). We present an approach based on Bayes' theory for a combination of different views.

10.4.1 Integration of Context

Context can be divided into local and global contexts. Local context refers to context objects that only appear in connection with roads, for example, vehicles, bridges, or larger traffic signs. Global context is a design driver for the road model, respectively the used extraction strategy. The potential of the use and integration of context to support road extraction will be discussed in the following subsections.

10.4.1.1 *Local Context for Road Extraction from SAR Images*

Neighboring or contextually related objects to roads are called context objects, like trees aligned to the road or traffic signs. They have a special contextual relation to roads and can interrupt and disturb line extraction. Situations in which background objects in SAR images make road extraction difficult are mainly caused by the following local context objects.

1. Layover and shadow regions caused by buildings and trees
2. Blurred bright stripes caused by vehicles moving in along-track direction
3. Bright reflections caused by metallic objects like traffic signs or bridges

By modeling the influence and incorporating these objects into the road extraction process, one can take advantage of the fact that these objects are strong evidence of roads. Further objects such as a row of trees might as well be situated nearby a road and can be represented as a minor evidence of roads.

The relations of objects like buildings, vehicles, or vegetation with a road are given by (1) closeness, (2) parallelism, or (3) coverage. Many of the context objects are characterized by a high backscattering caused by metallic structures or by multiple bounces.

Studies for an automatic extraction of the mentioned context objects from SAR images are still topics of research (Kirscht, 1998; Lee et al., 2006; Stilla and Soergel, 2006). In the following examples, we assume that the context objects have been extracted by an external module and need to be integrated into the road extraction approach. For this, it is important to (1) estimate the evidence each context object provides for roads and (2) choose an appropriate representation form for each context object: High evidence for roads is provided by context objects that almost exclusively appear in conjunction with roads and rarely elsewhere. Therefore, vehicles blurred in azimuth direction, and also bridges, get high weights. Their representation form is a line. Other objects provide less evidence for roads. For example, rows of trees appear nearby roads but also elsewhere. They are henceforth represented as lines attached with low weights. Large traffic signs only appear together with roads. However, their correct (automatic) interpretation is assumed to be quite hard, so that they are added to the graph as medium-weighted short straight lines.

10.4.1.2 *Global Context for Road Extraction from SAR Images*

The knowledge about global context in a scene is a relevant design driver for the extraction; on the one hand for choosing the most suited extraction strategy, on the other hand for supporting exclusion/inclusion zones or triggering the extraction.

As mentioned earlier, three global context regions distinguished for roads are: rural, urban, and forest areas. In rural areas, the local context objects may appear differently and may have different influence on roads as in forest and urban areas where trees and buildings, respectively, are more frequent than in rural areas. Forest areas are excluded from the extraction since roads are rarely visible in SAR images of forest regions due to the oblique viewing geometry. By exclusion of regions, we cannot extract any longer a real network. This emphasizes the need for good seed information—a general difficulty of every automatic extraction algorithm. Urban areas, on the other side, are a good indicator for roads. Usually roads start or end at urban areas and cities are connected by roads. Hence, they are used to define reliable seed information for the road network completion. For this task, the information about the border of the cities is introduced with a weight into the network generation.

Figure 10.13 illustrates the benefits for road extraction when including local and global context information. Some gaps can be closed by using local context and some new roads are found by the introduction of cities as seed information. But the result also shows that still some gaps and false extractions are remaining.

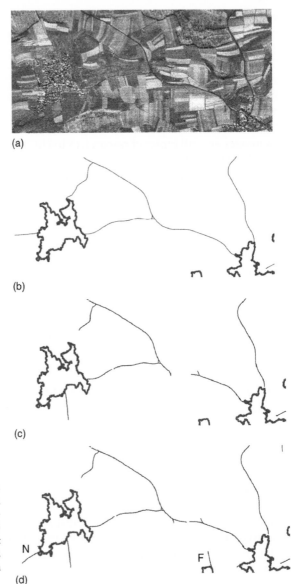

(a)

(b)

(c)

FIGURE 10.13
Road extraction in rural areas
without and with integration
of context information. (a) Section of the SAR image, (b)
ground truth data, (c) result
without exploitation of context,
and (d) result with exploitation
of context.

(d)

10.4.2 Integration of Road-Class-Specific Modeling: Example "Highways"

The extraction of highways sometimes fails due to the more complicated
structure of highways compared with roads. Especially the central crash
barrier separating the two driving directions is a salient feature, and the
approach for rural roads is not able to cope with it.

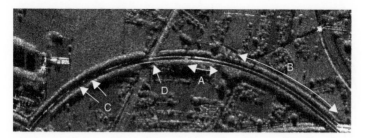

FIGURE 10.14
Orientation-dependent effects of the appearance of highways in SAR images.

10.4.2.1 Model for Highways

In contrast to city streets that are often completely occluded for side-looking radar, highways are mostly wide enough to be imaged with typical incidence angles. The model describes a highway comprising two (anti)parallel roads that are bordered by crash barriers. To extract such types of objects, the use of a multiscale model has proven to be very important (Hinz and Baumgartner, 2003). In the highest resolution, a highway is characterized by two parallel dark lines separated by a thin bright line, the central crash barrier. This crash barrier is a reliable feature, because it consists of differently arranged metallic components that act like small corner reflectors in almost every direction— though with different strengths. Figure 10.14 illustrates this and some more typical effects depending on the viewing angle: (A) Crash barriers in azimuth direction act like a corner reflector and appear very bright. (B) In other orientations, the direct metallic reflection dominates. (C) There are also some areas without any reflection. These are either caused by mirror reflections or by radar shadows in case of very high objects nearby. (D) Context objects like bridges or tunnels, vehicles or traffic signs complicate the extraction (see Section 10.4.1.1). In the same image with a reduced resolution (about 6 m), the fundamental structure of a highway is emphasized. It appears as a dark, smooth-curved line, and the crash barriers are no more visible.

10.4.2.2 Extraction of Highways

The extraction strategy for highways consists of four different steps: (1) hypotheses formation in low resolution, (2) hypotheses formation in high resolution, (3) fusion of both resolutions, and (4) network generation.

1. To create highway hypotheses in low resolution, dark and wide lines are extracted (Steger, 1998). The resulting lines are weighted with respect to highway construction parameters (width, length, curvature).

2. In the high resolution, dark lines and thin bright lines are extracted, that is, candidates for the individual roads and the crash barrier in

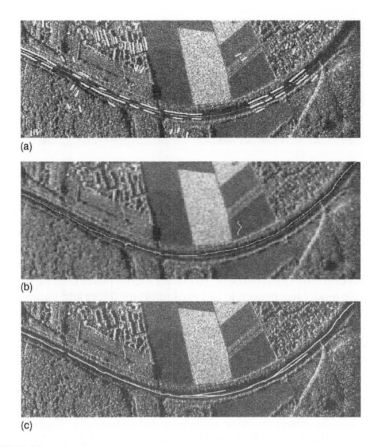

(a)

(b)

(c)

FIGURE 10.15
Extraction in fine scale (a), in coarse scale (b), and final result (c).

between. To get initial highway hypotheses, parallel dark lines enclosing a bright line are selected. These line aggregations are rated according to highway construction constraints and, in addition, according to the gray-value difference of the parallel dark lines.

3. All hypotheses are fused now using a "best-first" strategy. Thereby, hypotheses extracted in both resolutions get the highest weights.

4. Finally, the network is extracted by the graph-based grouping algorithm.

A typical result of this extended road extraction approach is depicted in Figure 10.15.

10.4.3 Multiaspect Fusion of SAR Images

Multiaspect SAR imagery illuminates the scene from different directions and has the advantage of delivering complementary information. If the line

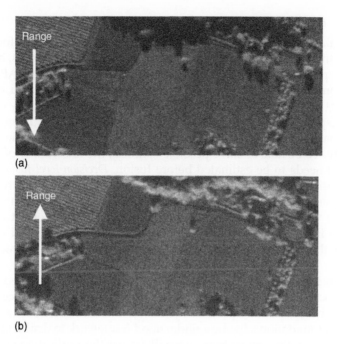

(a)

(b)

FIGURE 10.16
Antiparallel SAR views. Sensor: MEMPHIS (FGAN-FHR), 35 GHz, resolution of 1 m, $\theta \approx 60°$. (a) Illuminated from the left and (b) illuminated from the right.

extraction is not able to detect the line in one scene due to occlusions, it might be able to succeed in a second exposure. But critics of multiaspect images mean that multiaspect SAR images are associated with much effort during the flight campaign. However, a second exposure from an antiparallel view can be taken with less effort during the return flight. Figure 10.16 shows two antiparallel SAR views of a rural area close to Ravensburg, Germany. The secondary road, which ranges from the left to the right of the scene, is partly covered in one image, visible in the other image, and vice versa.

Multiaspect SAR images contain not only redundant and complementary information, but also contradicting information. This requires a careful selection within the fusion process. A correct fusion step has the ability to combine information from different sensors, which in the end is more accurate and better than the information acquired from one sensor alone. In the following chapter, we will focus on a fusion approach applied to multiaspect SAR images. In the first part, we will discuss different fusion strategies, followed by a more detailed discussion of the underlying theory of a probabilistic fusion strategy. In the end, we will present results of the fusion of line segments extracted from two suburban multiaspect SAR images.

10.4.3.1 Bayesian Fusion Approach

In general, better geometric accuracy is obtained by fusing information closer to the source. But contrary to multispectral optical images, a fusion of multiaspect SAR data on *pixel level* makes hardly any sense. SAR data is far too complex. Besides the speckle, the appearance of elevated object is dependent on the sensor viewing geometry and may appear totally different in multiaspect SAR images. The line extraction from SAR images often delivers partly fragmented and erroneous results. Oversegmentation occurs frequently. To be able to solve possible conflicts, the uncertainty of the line primitives needs to be estimated before fusion. *Decision-level fusion* means that an estimate (decision) is made based on the information from each sensor alone and these estimates are subsequently combined in a fusion process. If we put this into practice, the first step would consist of a line extraction in each image, followed by attribute extraction. Based on these attributes the uncertainty of each line primitive is estimated, followed by a fusion.

Many methods, both numerical and symbolic, can be applied for the fusion process. Some frameworks worth mentioning are evidence theory, fuzzy-set theory, and the probability theory. The last one is, regarding its theoretical foundations, the best understood framework to deal with uncertainties. In the following sections, we will discuss the application of a fusion process, which accommodates for these aspects. The chain of a decision-level fusion based on Bayes' statistical theory is depicted in Figure 10.17.

The underlying theory of the approach originates from Bayesian probability theory and can be drawn from the well-known Bayes' theorem

$$p(Y|X,I) = \frac{p(X|Y,I) \times p(Y|I)}{p(X|I)}. \tag{10.2}$$

The strength of Bayes' theorem is that it relates the probability that the hypothesis Y is true given the data X to the probability that we have

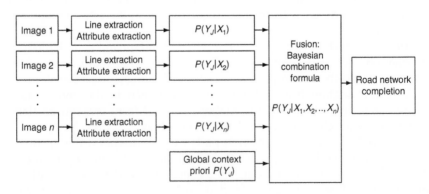

FIGURE 10.17
Fusion concept.

observed the measured data X if the hypothesis Y is true. The latter term is much easier to estimate. All probabilities are conditional on I, which is made to denote the relevant background information at hand. Since we are mostly interested in the solution, which yields the greatest value for the probability of the measured data, usually referred to as the maximum likelihood estimate, we can write Bayes' theorem in a compact form:

$$p(Y|X,I) \propto p(X|Y,I) \cdot p(Y|I). \tag{10.3}$$

The main feature involved in the road extraction process is the line primitive L, which can either be identified as a ROAD or as a FALSE_ALARM, which represent our two classes Y_1 and Y_2. If relevant, these hypotheses can be extended with more classes $Y_3, \ldots, Y_n$ (e.g., river, shadows). The identification is done by means of our measured data X, which in our case corresponds to the geometric and radiometric attributes of the line primitive. If two or more images are available, we shall combine data from multiple sources. Then the earlier assumptions will be extended to the assumptions whether a ROAD truly exists in the scene or not. We need to add a third term to our measured data X; the fact that a line has been extracted (L) or not extracted ($\bar{L}$) from one or more images. Hence, we deal with the following hypotheses:

$Y_1 = $ A ROAD exists in the scene

$Y_2 = $ A FALSE_ALARM exists in the scene

The images can be regarded as independent observations. The probability that an object Y_j exists given the measurements $X_1, \ldots, X_n, L_1, \ldots, L_n$ can by means of Bayes' theorem and by means of the product rule be expressed as

$$p(Y_j|X,L,I) \propto \prod_{i=1}^{N} (p(X_i|L_i,Y_j,I) \cdot p(L_i|Y_j,I) \cdot p(Y_j|I)) \tag{10.4}$$

where
$p(L|Y_1, I)$ is the posterior probability that a line is extracted if a ROAD truly exists
$p(X|L,Y_1, I)$ is the posterior probability that the data X is measured if a ROAD exists AND a line has been extracted
$p(Y_1|I)$ is the prior or subjective probability that a road exists in the image

The last one represents a subjective probability and can be defined by the user. Global context can here be especially useful. The frequency of roads is proportionately low in some context regions, for instance in forestry regions. In these regions, we normally have problems of frequently occurring false alarms. The second posterior terms can be estimated out of training data.

The selection of attributes of the line primitives is based on the knowledge about roads and is similar to the selection of attributes (fuzzy evaluation) in

Chapters 2 and 3. Radiometric attributes such as mean and constant intensity, and contrast of a line as well as geometrical attributes like length and straightness are all good examples. Since we deal with several attributes, an attribute vector is created. It should be pointed out that more attributes do not necessarily mean better results, instead also the opposite may occur. Hence, a careful selection including a few, but significant attributes is recommended. If there is no correlation between the attributes, the likelihood $p(X|Y_i)$ can be assumed equal to the product of the separate likelihoods for each attribute.

$$p(X|Y_j) = p(x_1, x_2, \ldots, x_n|Y_j) = p(x_2|Y_j)p(x_1|Y_j)\ldots p(x_n|Y_j). \qquad (10.5)$$

Each separate likelihood $p(x_i|Y_j)$ can be approximated by a probability density function learned from training data. Learning from training data means that the extracted line primitives are sorted manually into two groups, ROADS and FALSE_ALARMS. A fitting carried out in a histogram with one dimension is relatively uncomplicated, but as soon as the dimensions increase the task of fitting becomes more complicated. Figure 10.18a and b show the histogram of the attribute length and its fitted lognormal distributed curve.

The estimated probability density functions should represent a degree of belief rather than a frequency of the behavior of the training data. The obtained probability assessment shall correspond to our knowledge about roads. At a first glance, the histograms in Figure 10.18a and b seem to overlap. However, Figure 10.18c exemplifies for the attribute length that the discriminant function

$$g(x_i) = \ln\left(p(x_i|Y_1)\right) - \ln\left(p(x_i|Y_2)\right), \qquad (10.6)$$

increases as the length of the line primitive increases. The behavior of the discriminant function shall correspond to the belief of a human interpreter. Preferably the discriminant function shall be tested for each attribute.

After the separate likelihood is estimated for each line primitive, all line primitives are sorted according to their discriminant value. The line primitive with the highest discriminant value is chosen first and its neighboring line primitives are searched for. Redundant line primitives are removed according to the same buffer width and collinearity criteria as described in Section 10.2.3, but the weight is now estimated by means of Equation 10.4. As a result, the line primitives obtain an uncertainty assessment instead of a fuzzy value.

10.4.3.2   Examples

The fusion approach was tested on two multiaspect SAR images (X-band, multilooked, ground range SAR data) of a suburban scene near the airport of DLR in Oberpfaffenhofen, southern Germany. One image was illuminated from the south (Figure 10.19a) and one from the south-east (i.e., with roughly 45° difference). Manually extracted global context was incorporated into the process and was used as prior information (see Table 10.1 and

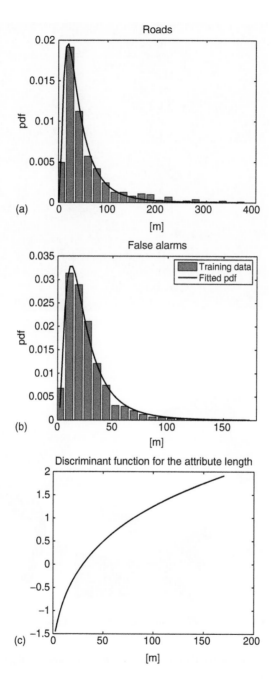

FIGURE 10.18
Approximation of the histogram of the attribute, length in *m* from (a) ROADS, (b) FALSE_
ALARMS by a lognormal distribution, and (c) discriminant function.

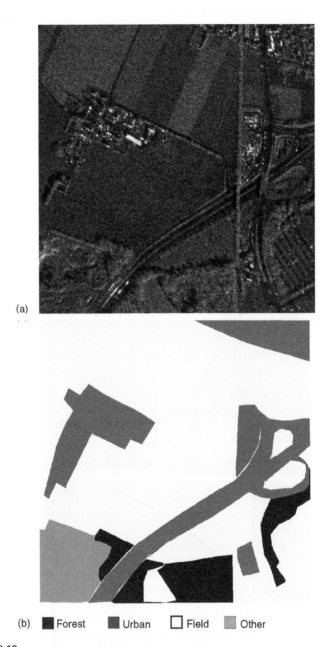

(a)

(b)   ■ Forest   ■ Urban   □ Field   ■ Other

FIGURE 10.19
(a) One of the SAR scenes analyzed in this work, (b) manual extraction of global context from previous SAR scene, and

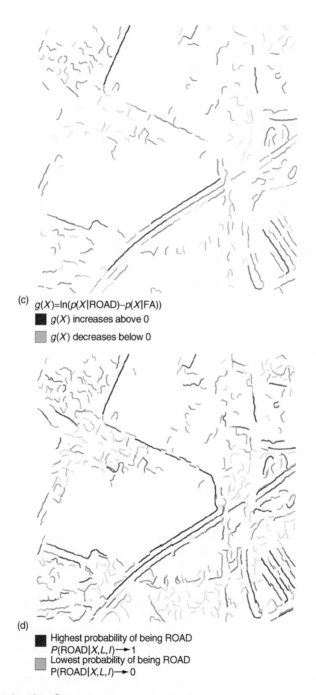

(c) $g(X)=\ln(p(X|\text{ROAD})-p(X|\text{FA}))$

■ $g(X)$ increases above 0

▨ $g(X)$ decreases below 0

(d)

■ Highest probability of being ROAD
$P(\text{ROAD}|X,L,I)\rightarrow 1$
▨ Lowest probability of being ROAD
$P(\text{ROAD}|X,L,I)\rightarrow 0$

FIGURE 10.19 (continued)
(c) results of line extraction from previous SAR scene. The computed discriminant value $g(X)$ of each line primitive is coded in gray, that is, the darker the line the higher is the probability that the line primitive belongs to a ROAD. (d) Results after fusion. This time, the lines are coded in their finally computed probability of being ROADS.

TABLE 10.1

Prior Terms for Different Global Context Areas

Global Context	p(ROADS)	p(FALSE_ALARMS)
Fields	0.2	0.8
Urban areas	0.3	0.7
Forest	0.05	0.95
Other areas	0.2	0.8

Figure 10.19b). The posterior probabilities, $p(L|Y_1,I)$ and $p(L|Y_2,I)$ were set to 0.6 and 0.4, respectively. The fusion was tested on a line extraction carried out in a scene taken by the same sensor as the training data but now performed with different parameter settings. In order to test the derived likelihood functions in terms of sensitivity and ability to discern roads from false alarms, we allowed a significant oversegmentation (Figure 10.19c). Most line primitives that correspond to roads got a good evaluation. As can be seen from Figure 10.19c and d, the detection rate is higher for the fusion of two images compared with the line extraction in one view.

10.5 Discussion and Conclusion

In this chapter, we have given the reader an introduction to road extraction from SAR data. Starting with a general description of modeling roads independent of the sensor and describing the generic TUM-LOREX system, we then have put emphasis on the adaption of the automatic road extraction approach toward the particular challenges of SAR data. The proposed approach aims at open rural areas and is suitable for imagery with a pixel size of ~2 m or less on ground.

In general, automatic road extraction from SAR data is challenging, simply on the basis of the effects of the side-looking geometry of the SAR sensor. Therefore, today, hardly any automatic road extraction procedures dealing with SAR data alone is able to deliver sufficiently correct results for an operational use. TUM-LOREX was originally designed for optical data and it was recently validated that TUM-LOREX still stays highly competitive compared with other approaches when applied to optical imagery (Mayer et al., 2006). But still, TUM-LOREX is far from getting transferred to a fully automatic operational mode even for optical data, which is considered as less complicated as SAR data. An operational mode requires results with high correctness, which is hard to obtain by a fully automatic system. One has to keep in mind that the results presented in this work are obtained by an experienced user. A potential user may need some training. One advantage of SAR images is the fact that the setting of parameters can be reduced to 2–3 relevant parameters to be tuned for an average result, respectively to 6–8 for a refined result.

Considering the transition into operational use, any automatic road extraction system needs to be supplemented by an interactive editing tool so that a human operator may check and edit the automatically obtained result. The utilization of confidence values attached to each extracted road is of highest importance for reducing the interactive efforts, as has been demonstrated in Hinz and Wiedemann (2004). In this context, it is desirable to base the whole extraction process onto a statistical formulation instead of fuzzy functions (Section 10.4.3). This will reduce the amount of sensitive parameters to be set by the user. In this way, the automatic part of the road extraction process would increase and involve a certain time-saving for the user. As soon as one has passed the critical step of parameter setting, the computational time is reasonable even for relatively large SAR data. If the learned likelihood functions are robust enough to be applied to different images—of course under the condition that the image characteristics do not differ too heavily—still remains to be seen. One should also keep in mind that the Bayesian fusion still has to be connected with the following step, the network-based grouping. First after the fusion has been fully implemented in TUM-LOREX, the performance of the statistical approach can be analyzed in depth.

The work presented in this chapter, showed that the demand on further research on automatic road extraction from SAR data still remains. For a transition to more complicated areas, such as suburban or forestry areas, neighbored information (bridges, moving vehicles, and traffic signs) needs to be incorporated to a higher extent in the road modeling. In this work, these features were extracted manually. It remains still to be seen, to what extent a robust extractor of several local features is feasible or not. Consequently, also these features involved in local context relations should be attached with confidence values. Preferably, when available, the incorporating local context information shall be complemented with multiview data to increase the detection rates of the roads. Besides, a multiview approach has the ability to involve a reasoning step, which is based on the sensor geometry and its influence on the relations between the extracted features (local context and possible road candidates). One of the advantages of TUM-LOREX is its flexible architecture, which allows a future development of implementation of both context information and multiview image data.

Acknowledgment

The authors would like to thank the Microwaves and Radar Institute, German Aerospace Center (DLR) as well as FGAN-FHR for providing SAR data.

References

Amberg, V., Coulon, M., Marthon, P., Spigai, M., 2005. Improvement of road extraction in high resolution SAR data by a context-based approach. In: *Geoscience and Remote Sensing Symposium, 2005.* IGARSS '05, Vol. 1, pp. 490–493.

Barzohar, M., Cohen, M., Ziskind, I., Cooper, D., 1997. Fast robust tracking of curvy partially occluded roads in clutter in aerial images. In: Gruen, A., Baltsavias, E., Henricsson, O. (Hrsg.) (eds.), *Automatic Extraction of Man-Made Objects from Aerial and Space Images (II)*, Birkhauser Verlag, Basel, Switzerland, pp. 277–286.

Baumgartner, A., Eckstein, W., Mayer, H., Heipke, C., Ebner, H., 1997. Context-supported road extraction. In: Gruen, A., Baltsavias, E., Henricsson, O. (Hrsg.) (eds.), *Automatic Extraction of Man-Made Objects from Aerial and Space Images (II)*, Birkhauser Verlag, Basel, Switzerland, pp. 299–308.

Baumgartner, A., Hinz, S., 2000. Multi-scale road extraction using local and global grouping criteria. *International Archives of Photogrammetry and Remote Sensing* 33(B3/1): 58–65.

Baumgartner, A., Steger, C., Mayer, H., Eckstein, W., Ebner, H., 1999. Automatic road extraction based on multi-scale, grouping, and context. *Photogrammetric Engineering and Remote Sensing* 65(7): 777–785.

Bordes, G., Giraudon, G., Jamet, O., 1997. Road modeling based on a cartographic database for aerial image interpretation. In: Gruen, A., Baltsavias, E., Henricsson, O. (Hrsg.) (eds.), *Automatic Extraction of Man-Made Objects from Aerial and Space Images (II)*, Birkhauser Verlag, Basel, Switzerland, pp. 123–139.

Borghys, D., Lacroix, V., Perneel, N., 2002. Edge and line detection in polarimetric SAR images. In: *International Conference on Pattern Recognition*, Band 2, Quebec, pp. 921–924.

Bovik, A.C., 1988. On detecting edges in speckle imagery. *IEEE Transactions on Acoustics, Speech, and Signal Processing* 36(10): 1618–1627.

Chanussot, J., Mauris, G., Lambert, P., 1999. Fuzzy fusion techniques for linear features detection in multitemporal SAR images. *IEEE Transactions on Geoscience and Remote Sensing* 37(3): 1292–1305.

Clément, G., Giraudon, G., Houzelle, S., Sandakly, F., 1993. Interpretation of remotely sensed images in a context of multisensor fusion using a multispecialist architecture, *IEEE Transactions on Geoscience and Remote Sensing* 31(4): 779–791.

Dell'Acqua, F., Gamba, P., Lisini, G., 2003. Improvements to urban area characterization using multitemporal and multiangle SAR images, *IEEE Transactions on Geoscience and Remote Sensing* 41(9): 1996–2004.

Fischler, M.A., Tenenbaum, J.M., Wolf, H.C., 1981. Detection of roads and linear structures in low-resolution aerial imagery using a multisource knowledge integration technique. *Computer Graphics and Image Processing* 15: 201–223.

Frost, V., Stiles, J., Shanmugan, K., Holtzman, J., 1982. A model for radar images and its application to adaptive digital filtering of multiplicative noise. *IEEE Transactions on Pattern Analysis and Machine Intelligence* 4(2): 157–166.

Gamba, P., Dell'Acqua, F., Lisini, G., 2006. Improving urban road extraction in high-resolution images exploiting directional filtering, perceptual grouping, and simple topological concepts, *IEEE Geoscience and Remote Sensing Letters* 3(3): 387–391.

Gerke, M., Butenuth, M., Heipke, C., Willrich, F., 2004. Graph supported verification of road databases. *ISPRS Journal of Photogrammetry and Remote Sensing* 58(3/4): 152–156.

Gruen, A., Li, H., 1997. Linear feature extraction with LSB-snakes. In: Gruen, A., Baltsavias, E., Henricsson, O. (Hrsg.) (eds.), *Automatic Extraction of Man-Made Objects from Aerial and Space Images (II)*, Birkhauser Verlag, Basel, Switzerland, pp. 287–298.

de Gunst, M., Vosselman, G., 1997. A semantic road model for aerial image interpretation. In: Foerstner, W., Pluemer, L. (eds.), *Semantic Modeling for the Acquisition of Topographic Information from Images and Maps*, Birkhauser Verlag, Basel, Switzerland, pp. 107–122.

Hinz, S., Baumgartner, A., 2003. Automatic extraction of urban road nets from multi-view aerial imagery. *ISPRS Journal of Photogrammetry and Remote Sensing* 58(1–2): 83–98.

Hinz, S., Baumgartner, A., Ebner, H., 2001. Modelling contextual knowledge for controlling road extraction in urban areas. In: *IEEE/ISPRS Joint Workshop on Remote Sensing and Data Fusion over Urban Areas*.

Hinz, S., Wiedemann, C., 2004. Increasing efficiency of road extraction by self-diagnosis. *Photogrammetric Engineering and Remote Sensing* 70(12): 1457–1466.

Jeon, B., Jang, J., Hong, K., 2002. Road detection in spaceborne SAR images using a generic algorithm. *IEEE Transactions on Geoscience and Remote Sensing* 40(1): 22–29.

Kirscht, M., 1998. Detection, velocity estimation, and imaging of moving targets with single-channel SAR. In: *Proceedings of European Conference on Synthetic Aperture Radar*, EUSAR '98, pp. 587–590.

Lee, J.S., 1981. Refined filtering of image noise using local statistics. *Computer Graphics and Image Processing* 15: 380–389.

Lee, J.-S., Krogager, E., Ainsworth, T.L., Boerner, W.-M., 2006. Polarimetric analysis of radar signature of manmade structure. *IEEE Geoscience and Remote Sensing Letters* 3(4): 555–559.

Mayer, H., Hinz, S., Bacher, U., Baltsavias, E., 2006. A test of automatic road extraction approaches. *International Archives of Photogrammetry, Remote Sensing, and Spatial Information Sciences* 36(3): 55–60.

Mayer, H., Steger, C., 1998. Scale-space events and their link to abstraction for road extraction. *ISPRS Journal of Photogrammetry and Remote Sensing* 53(2): 62–75.

McKeown, D., Denlinger, J., 1988. Cooperative methods for road tracking in aerial imagery. *Computer Vision and Pattern Recognition* 662–672.

Merlet, N., Zerubia, J., 1996. New prospects in line detection by dynamic programming. *IEEE Transactions on Pattern Analysis and Machine Intelligence* 18(4): 426–431.

Neuenschwander, W., Fua, P., Szekely, G., Kübler, O., 1995. From ziplock snakes to velcro surfaces. In: Gruen, A., Kuebler, O., Agouris, P., (eds.) *Automatic Extraction of Man-Made Objects from Aerial and Space Images*, Birkhauser Verlag, Basel, Switzerland, pp. 105–114.

Price, K., 2000. Urban street grid description and verification. In: *5th IEEE Workshop on Applications of Computer Vision*, pp. 148–154.

Ruskone, R., 1996. Road Network Automatic Extraction by Local Context Interpretation: Application to the Production of Cartographic Data. PhD thesis, Universit'e Marne-La-Valle'e.

Sedgewick, R., 1992. *Algorithms in C++*. Addison-Wesley Publishing Company, Inc. Boston, MA.

Steger, C., 1998. An unbiased detector of curvilinear structures. *IEEE Transactions on Pattern Analysis and Machine Intelligence* 20(2): 113–125.

Steger, C., Mayer, H., Radig, B., 1997. The role of grouping for road extraction. In: Gruen, A., Baltsavias, E., Henricsson, O. (Hrsg.) (eds.), *Automatic Extraction of Man-Made Objects from Aerial and Space Images (II)*, Birkhauser Verlag, Basel, Switzerland, pp. 245–256.

Stilla, U., 1995. Map-aided structural analysis of aerial images. *ISPRS Journal of Photogrammetry and Remote Sensing* 50(4): 3–10.

Stilla, U., Michaelsen, E., Soergel, U., Hinz, S., Ender, H.J., 2004. Airborne monitoring of vehicle activity in urban areas. In: Altan, M.O. (ed.), *International Archives of Photogrammetry and Remote Sensing* 35(B3): 973–979.

Stilla, U., Rottensteiner, F., Hinz, S. (eds.), 2005. Object extraction for 3D City Models, road databases, and traffic monitoring—concepts, algorithms, and evaluation (CMRT05). *International Archives of Photogrammetry and Remote Sensing* 36, Part 3W24.

Stilla, U., Soergel, U., 2006. Reconstruction of buildings in SAR imagery of urban areas. In: Weng, Q., Quattrochi, D. (eds.), *Urban Remote Sensing*, Taylor & Francis, CRC Press, Boca Raton, FL, pp. 47–68.

Stilla, U., Soergel, U., Thoennessen, U., 2003. Potential and limits of InSAR data for building reconstruction in built up-areas. *ISPRS Journal of Photogrammetry and Remote Sensing* 58(1–2): 113–123.

Toenjes, R., Growe, S., Bueckner, J., Liedke, C.-E., 1999. Knowledge-based interpretation of remote sensing images using semantic nets. *Photogrammetric Engineering and Remote Sensing* 65(7): 811–821.

Tupin, F., Houshmand, B., Datcu, M., 2002. Road detection in dense urban areas using SAR imagery and the usefulness of multiple views. *IEEE Transactions on Geoscience and Remote Sensing* 40(11): 2405–2414.

Tupin, F., Maˆtre, H., Mangin, J.-F., Nicolas, J.-M., Pechersky, E., 1998. Detection of linear structures in SAR images: application to road network extraction. *IEEE Transactions on Geoscience and Remote Sensing* 36(2): 434–453.

Ulaby, F.T., Moore, R.K., Fung, A.K., 1986. *From Theory to Applications, Band III of Microwave Remote Sensing: Active and Passive*, Artech House, Dedham, MA.

Vosselman, G., de Knecht, J., 1995. Road tracing by profile matching and kalman filtering. In: Gruen, A., Kuebler, O., Agouris, P., (eds.), *Automatic Extraction of Man-Made Objects from Aerial and Space Images*, Birkhauser Verlag, Basel, Switzerland, pp. 265–274.

Wessel, B., 2004. Road network extraction from SAR imagery supported by context information. *International Archieves of Photogrammetry, Remote Sensing and Spatial Information Sciences* 35(3B): 360–365.

Wessel, B., 2006. Automatische Extraktion von Straßen aus SAR-Bilddaten, PhD Thesis, Deutsche Geodätische Kommission, Reihe C, Nr. 600.

Wessel, B., Hinz, S., 2004. Context-supported road extraction from SAR imagery: transition from rural to built-up areas. In: *Proceedings of EUSAR 2004*, pp. 399–402.

Wiedemann, C., Ebner, H., 2000. Automatic completion and evaluation of road networks. *International Archives of Photogrammetry and Remote Sensing* 33(B3/2): 979–986.

Wiedemann, C., Hinz, S., 1999. Automatic extraction and evaluation of road networks from satellite imagery. *International Archives of Photogrammetry and Remote Sensing* 32(3–2W5): 95–100.

Zadeh, L., 1989. Knowledge representation in fuzzy logic. *IEEE Transactions on Knowledge and Data Engineering* 1(1): 89–100.

Zhang, C., Baltsavias, E., Gruen, A., 2001. Updating of cartographic road databases by image analysis. In: Baltsavias, E., Gruen, A., van Gool, L. (Hrsg.) (eds.), *Automatic Extraction of Man-Made Objects from Aerial and Space Images (III)*, Balkema Publishers, Lisse, The Netherlands, pp. 243–253.

Zhang, C., 2004. Towards an operational system for automated updating of road databases by integration of imagery and Geodata. *ISPRS Journal of Photogrammetry and Remote Sensing*, 58(3/4): 166–186.

11

Road Networks Derived from High Spatial Resolution Satellite Remote Sensing Data

Renaud Péteri and Thierry Ranchin

CONTENTS

11.1   Introduction

There is a strong demand for accurate and up-to-date road network information. Road network knowledge is crucial for the creation and the update of maps, geographic information system (GIS) database, transportation, or land planning. For local authorities, cartography of the road network is needed for urban planning, dirty water collection through gutter network (most often located under roads), traffic flow analysis, or pollution mapping. Closely related applications are geomarketing, electricity and telecommunication networks, databases for car navigation, and so on. Currently, road network cartography is essentially done by human interpretations of high-resolution aerial images and additional in situ information. This is a long and tedious work that requires to be done again for each update of the road network.

High spatial resolution imagery is now available for civilian applications and reveals the very fine details of the imaged area. Examples of high-resolution satellites are SPOT 5, IKONOS, QuickBird, OrbView, or EROS. The term "high resolution" is relative and refers to satellites with spatial resolutions better than 5 m in the panchromatic channel (one can even talk about very high resolution when the image resolution is better than 1 m).

The current availability of high spatial resolution images represents an undeniable asset to Earth observation. The urban environment, which is the most difficult context because of its high complexity and information density, could benefit the most from high-resolution imagery (Puissant and Weber, 2002). In addition to the increased precision for road detection and location, high-resolution satellite imagery can be used in numerous cases where access to the studied area is difficult: administrative constraints, authorization to overfly the area, conflicts, wars, or natural catastrophes. Moreover, satellite means is significantly cheaper than aerial or in situ data acquisition campaigns.

As promising as it is, the use of high-resolution images for road extraction induces a change in the road representation and a significant increase in noise. Moreover, quantitative assessment of the results has to be redesigned when dealing with such images. In this chapter, a new method suitable for high-resolution images is proposed. Originally designed for urban areas, this method can naturally be applied to easier cases such as rural or semi-rural areas. The chapter is structured as follows.

The change in the road representation induced by high resolution imagery is first presented. A short survey on road extraction with the evolution from linear to surface models is proposed (Section 11.2).

A new method for extracting road networks from high spatial resolution images is then described. It models roads as a surface and is built on cooperation between linear and surface representation of roads. In order to overcome local artifacts, the method makes use of advanced image processing tools, such as active contours and the wavelet transform (Section 11.3).

An example of application of the method on a high-resolution image from the QuickBird satellite is proposed. The result is quantitatively assessed compared with human interpretation (Section 11.4).

This chapter concludes with a discussion on the principal benefits of the method and on future prospects (Section 11.5).

11.2 Automatic Extraction of Road Networks: From Linear to Surface Methods

11.2.1 Linear and Surface Representation of Roads

Automatic extraction of road networks from satellite images is not a recent problem. It has been the topic of numerous works in the field of remote sensing for more than 20 years (see Mena, 2003, for an overview). However, due to its complexity, it is still an active field of research. Before the availability of high resolution, the extraction methods published were dedicated to images with a spatial resolution of 10 m at best. At this resolution, roads are represented by lines of 1–3 pixels, leading to methods extracting linear road networks. Visible roads at this resolution are primary roads, freeways, highways, or boulevards in town. Secondary roads are not visible or have a fragmented aspect in the image.

With the availability of high spatial resolution images, road perception changes from linear to surface representation (>3 pixel wide). Surface representation means that the road is no longer represented as a line of a certain thickness, but as two parallel contours defining a surface. The radiometry of this surface is often inhomogeneous, especially in the urban context.

11.2.2 New Sensors, New Cartographic Scale

Road representation on a map is also highly related to the cartographic scale of interest (Figure 11.1). According to cartographic generalization principles, roads at a scale of 1:25,000 are represented by lines well located on the road centerline, but with a width thicker than in reality (Figure 11.1a).

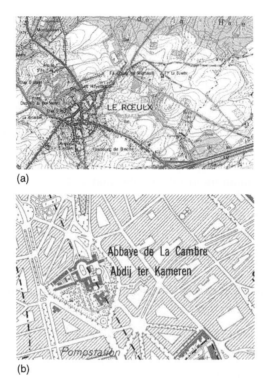

(a)

(b)

FIGURE 11.1

The cartographic representation of roads depends on map scale. IGN Belgique (a) map at scale of 1:25,000 and (b) map at scale of 1:10,000.

The Belgium geographic institute (IGN Belgique) has recently achieved the mapping of the whole country at a scale of 1:10,000. At this scale, the road is represented as a surface element, and its width on a map corresponds to its real width at the ground level (Figure 11.1b).

Road network extraction methods can be classified into two groups: linear approaches and surface approaches. These two approaches differ in the road representation, depending on the image resolution and the cartographic scale of interest. The special issue of *Photogrammetric Engineering & Remote Sensing* on linear feature extraction from remote sensing data for road network delineation and revision (PE&RS, 2004) gives a good overview of recent methods available. In the following sections, the two groups of approaches are presented.

11.2.3 Road Extraction: Linear Approaches

Before 1999, the best spatial resolution delivered by satellite sensors was 10 m. Moreover, satellite or even aerial images were mainly used for the creation of maps at scales smaller than 1:25,000. Hence, the first works about road extraction were linear approaches.

The first methods for line extraction were based on generic tools of image processing, such as linear filtering (Wang and Howarth, 1987) and mathematical morphology (Destival, 1987; Serendero, 1989). Later more advanced techniques were used, such as Markov fields (Merlet and Zérubia, 1996) or point process (Lacoste et al., 2005), neural networks (Bhattacharya and Parui, 1997; Doucette et al., 2001), dynamic programming (Gruen and Li, 1995), or multiscale analysis (Baumgartner et al., 1999). Other approaches were based on Kalman filtering (Véran, 1993; Vosselman and De Knech, 1995) cooperation of different algorithms (Mac Keown and Denlinger, 1988), multisource approaches (Rellier et al., 2000; Jin and Davis, 2005) or by using the third dimension (Zhang et al., 2000), combination of multispectral channels provided by satellite sensors (Xiaoying and Davis, 2003; Bacher and Mayer, 2005; Zhang and Couloigner, 2006), or expert systems (Garnesson et al., 1992; Eidenbenz et al., 2000).

A more complete review of techniques for extracting linear features from imagery can be found in Quackenbush (2004).

11.2.4 Road Extraction: Surface Approaches

11.2.4.1 Introduction

Methods for surface extraction were first applied to airborne images where roads appear as surface elements. The appearance of high spatial resolution images has rebosted this research. These methods are often based on a road model, that is, the radiometry along one road is relatively homogeneous and contrasted compared with its background. Moreover, the width of the road and its curvature are supposed to vary slowly. Active contours were used for surface road extraction by several authors (Fua and Leclerc, 1990; Péteri and Ranchin, 2004; Amo et al., 2006). Other approaches were based on light propagation simulation (Guigues and Viglino, 2000), the use of the third dimension (Wang and Trinder, 1998), and multiscale approaches (Couloigner and Ranchin, 2000; Laptev et al., 2000).

11.2.4.2 A Better Precision, New Kinds of Artifacts

Figure 11.2 illustrates the level of details reached by available civil sensors. This image of an urban scene comes from the QuickBird satellite and has a spatial resolution resampled at 70 cm in the panchromatic channel. At this resolution, roads have two visible sides and are localizable with high precision. One can notice the high level of visible details (cars, ground markings, projected shadows, etc.) that represent noise or artifacts in automatic extraction methods. Other kinds of artifacts that can be encountered are trees, tarred areas (parking, airport), or buildings with radiometry similar to roads and with an important contrast compared with their environment.

On the one hand, this significant increase in noise in the image complicates the task of road extraction algorithms. On the other hand, this increase in the spatial resolution potentially enables a more precise geographic

FIGURE 11.2
QuickBird image from the city of Strasbourg, France. Copyright DigitalGlobe.

location, a better identification of the different road types (highways, main streets, small lanes, etc.), and an estimation of the road network area.

Taking advantage of high-resolution imagery for automatic mapping implies defining a method able to take into account this change from linear to surface model for roads. The method should also be robust with respect to noise (often encountered in urban environment) and to deal with all the distinguishable types of roads.

In the following sections, a new method suitable for high-resolution satellite images and meeting the mentioned requirements is proposed.

11.3 A Collaborative Method for Surface Road Extraction from High Spatial Resolution Images

11.3.1 Description

The method proposed in the following section was originally developed for dealing with the new context of high-resolution images over urban areas (Péteri, 2003; Péteri and Ranchin, 2003). This method can naturally be applied to easier cases such as rural or semirural areas where roads are often more visible and artifacts less numerous.

In order to extract and characterize the road network from high-resolution images, a modular method has been developed (Figure 11.3).

Inputs of the algorithm, besides the high-resolution satellite image, are models of roads, such as local parallelism of road sides (see Couloigner and Ranchin, 2000) and properties of the road network (such as connexity).

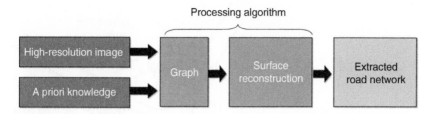

FIGURE 11.3
The methodology including topology management and road reconstruction.

Moreover, four classes of road width have been defined (according to established specifications). The algorithm is composed of two sequential modules: a graph module and a reconstruction module. They are working in a collaborative way to provide a surface extraction of the road network.

A topologically correct graph of the road network is first extracted. This step aims at giving correct spatial connections between roads as well as an approximation of their centerline location. The next step is the actual road reconstruction. Due to the high resolution of the images, a surface reconstruction has to be performed. This step uses the previous step of graph management as an initialization for the reconstruction.

In the following sections, the different modules of the process are described. For a more detailed description of the different steps of the method, one could refer to Péteri (2003).

11.3.2 Graph Management

This module intends to extract a topologically correct graph of the road network. It aims at giving correct spatial connections between roads as well as an approximation of their location.

The graph management module is composed of two steps.

11.3.2.1 Extraction of the Graph Polylines

The graph can come from a road database or be extracted automatically. In this second case, the graph extraction algorithm is based on the work of Airault and Jamet (1995). This algorithm is semiautomatic at the initialization step, where the user gives a seed point and an initial direction of propagation. The principle of this tracking algorithm is to generate a research tree of potential paths and to select the best path for the road to detect by minimizing a cost function. The cost function evaluates the homogeneity of the local radiometry variance for the potential directions of propagation. This homogeneity has a minimal variance in the direction corresponding to the road direction. The algorithm robustness is ensured by optimizing the directional homogeneity criterion on a long enough distance (the research tree depth).

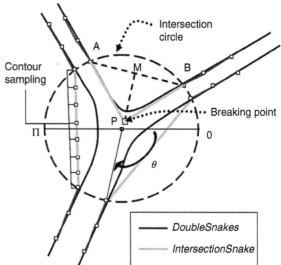

FIGURE 11.4
Initialization of *IntersectionSnakes*.

11.3.2.2 Getting the Complete Graph by Connecting the Extracted Polylines

A road is defined from an intersection to another and is represented in the network graph as a polyline. Roads are connected to each other at the intersections, which are the graph nodes.

Because of their difference in shape and topology, roads and intersections have to be processed separately, both for their detection and their extraction. The border between roads and intersections is defined by an *intersection circle* (Figure 11.4), which includes the whole intersection and is manually given by the user. It separates the two kinds of processing: extraction of road segments with parallel sides and extraction of intersections. Its center will be a graph node defining an intersection.

The graph network topology is reconstructed from the different unconnected polylines extracted from the tracking algorithm. These polylines are then linked to the centers of the intersection circles they are crossing. After this step, the extracted graph is topologically correct, but the different polylines are not necessarily well registered on the road centerline. Moreover, the road network has a linear representation.

From the extracted graph, polylines are then sampled and propagated along their normal direction to initialize the surface reconstruction module.

11.3.3 Reconstruction Module

11.3.3.1 Description

The goal of this module is to reconstruct roads as surface elements from the graph provided by the previous step. This module makes use of specific active contours (snakes) combined with a multiscale analysis. Active contours

(Kass et al., 1987) are deforming models that enable to introduce a priori information about the object to extract. Their evolution is controlled by an energy functional that should be minimized to fit to the shape to extract. Our snake implementation is based on the greedy algorithm described by Williams and Shah (1992). The joint use of the multiscale analysis with the wavelet transform enables to increase the algorithm robustness, by minimizing the problem of noise proper to high-resolution images (vehicles on the road, ground markings, etc.). Two sequential steps compose this reconstruction phase: the extraction of road segments with parallel sides and the extraction of road intersections. Indeed, these two objects present too many differences in both topology and shape to be processed in the same way. The intersection circle constitutes the frontier between these two steps (Figure 11.4).

11.3.3.2 Extraction of Parallel Road Sides

In order to extract segments of road with parallel sides, a new mathematical model has been defined: the *DoubleSnake*. It is composed of two discrete open active contours (called branches) with evolution constraints of simple active contours, and evolving jointly to maintain a local parallelism between them. A new energy term $E_{\parallel}$ in the *DoubleSnake* evolution has been introduced to take into account this notion of local parallelism between its two branches. Moreover, their extremity points are forced to minimize their energy while staying on the intersection circle (Figure 11.4).

The *DoubleSnake* energy functional controlling its evolution is defined as

$$E_j = \sum_i \left[\alpha^i E_{\text{cont}}^i + \beta^i E_{\text{curv}}^i + \gamma^i E_{2^j \text{image}}^i + \delta^i E_{\parallel}^i \right], \tag{11.1}$$

where
    i represents the point i of one of the branch
    $\alpha^i, \beta^i, \gamma^i$, and δ^i are weighting values for the different energies at point i
    E_{cont}^i and E_{curv}^i are internal energies that control the shape of the *DoubleSnake*. E_{cont}^i controls the space between snake's points
    E_{curv}^i controls the snake's curvature

Special attention is paid to the image energy term $E_{2^j \text{image}}^i$, as it is the one that attracts the *DoubleSnake* to the object to extract. The wavelet transform enables a multiscale representation of the contours in an image (Mallat and Zhong, 1992).

The image energy term is then computed at different spatial scales j, using the coefficients of the wavelet transform:

$$E_{2^j \text{image}}^i = -\sqrt{\left| W_{2^j}^1 f(i) \right|^2 + \left| W_{2^j}^2 f(i) \right|^2}, \tag{11.2}$$

where $W_{2^j}^{1,2} f(i)$ are the coordinates of the wavelet transform at scale j.

11.3.3.3 Extraction of Road Intersections

Once road segment extraction is finished, the intersection extraction starts. The active contour used at this step (the *IntersectionSnake*) consists of an open active contour. It is initialized by linking extremity points of the *DoubleSnakes* with segments and then sampling them. A "breaking" point is introduced by the algorithm if the angle θ between two extremities is less than $3\pi/5$ (see Figure 11.4).

11.3.3.4 Reconstruction Algorithm

Figure 11.5 illustrates the different steps of the reconstruction algorithm from the road network graph.

From the original image, a multiscale analysis is performed, decomposing the images hierarchically into a set of approximation images of coarser resolution and a set of wavelet coefficient images (Mallat, 1997). The *DoubleSnakes* are running on all the successive approximation images, from the coarsest resolution image given by the road class, till the original resolution image. For each approximation image, the image energy term is

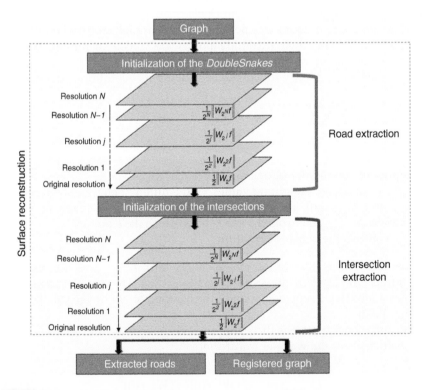

FIGURE 11.5
The surface reconstruction algorithm.

computed from the associated wavelet coefficient image. This energy term enables active contours to be attracted by details present at the resolution of the considered approximation image.

Once *DoubleSnakes* have all minimized their energy, the *IntersectionSnakes* are initialized and their extraction starts in the same multiscale process. Two constraints are added during their evolution: their extremities are fixed, and the active contour is constrained not to go out of the intersection circle. At the end of the different iterations, all the active contours are in an equilibrium state.

11.3.4 Properties of This Approach

This approach has the particularity to use a collaboration between the two kinds of road representation: the linear representation and the surface representation. The separation of the two processes enables the user to perform an intermediary check, and if needed a correction, between the extraction of the road network graph and the reconstruction step. The aim is a continuous check of the quality of the extracted road network.

We propose in this chapter the algorithms for the two modules, but one can also think of other methods for each step. For instance, one can think of the graph step of any algorithm enabling a linear extraction of the road network (see Quackenbush, 2004), possibly reducing the image resolution artificially. This method is therefore open to future evolutions.

11.3.5 Assessment of the Results

In an operational context, quantitative assessment is essential. It enables to characterize the results of an automatic algorithm, as well as to give a measure of reliability. A reference is needed to compare the obtained results and to get qualitative criteria.

Péteri and coworkers (Péteri and Ranchin, 2002; Péteri et al., 2004) have shown that a reference extracted from the digital image and based on only one human interpretation is not reliable for a very high spatial resolution scene with artifacts.

A method based on the acquisition of the reference by several image interpreters has then been proposed. It enables to reduce the human interpretation variability. From a "mean" interpretation that is taken as the reference, several criteria are used for comparing a road extracted by the algorithm to the reference: the Hausdorff distance, comparison between the lengths, and comparison between the areas. A tolerance zone has also been defined, representing the variability among the different human interpreters. The quantitative assessment is performed by computing the percentage of the extracted contour located in this tolerance zone. One could refer to Péteri et al. (2004) for a precise description of this reference and for the different comparison quantitative criteria that are used.

11.4 Application: QuickBird Image of the Area of Fredericton, Canada

11.4.1 Description of the Studied Scene

The following scene comes from the QuickBird satellite from the firm DigitalGlobe. This satellite acquires images in the panchromatic mode at a spatial resolution of 0.61 m. The presented image in Figure 11.6 has been resampled at 0.70 m. This image has been acquired on 31 August 2002 above the town of Fredericton, Canada (Figure 11.6). The town of Fredericton is typical of North American cities, where the street network is quadrilinear.

One can notice the several artifacts on the road surface, such as vehicles or ground marking (in particular roads 6 and 9). There is also a high number of partial occlusions of the roads, mainly due to trees and building shadows (roads 2, 3, 5, and 10). Finally, there are three types of "X" intersections and one type of "T" intersection (between roads 5, 7, and 8). The following sections detail the different steps of the extraction of this road network as well as its quantitative assessment.

11.4.2 Extraction of the Road Network Graph

Figure 11.7 shows the result of the road tracking algorithm.

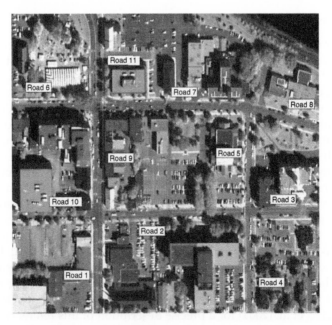

FIGURE 11.6
QuickBird image from the center of Fredericton, Canada DigitalGlobe.

FIGURE 11.7
After the extraction of polylines by the tracking algorithm.

Five seed points (black round dots in Figure 11.7) as well as the initial directions of propagation have been given by the user. The tracking results in five polylines, and some are crossing several intersections. During the whole tracking process, there were three cases when the user had to correct the trajectory (black lozenge-shaped dots). These mistakes are due to occlusions caused by shadows on roads 6 and 11, and due to the close radiometry of road 10 and a building next to it.

The next step consists in the extraction of the graph topology from this set of extracted polylines and the intersection circles. As described in Section 11.3.2, polylines where extreme points are located on the intersection circles are truncated and then linked to the center of this circle (Figure 11.8). When polylines are crossing several intersections, they are "divided" into several parts. The graph is then subsampled by keeping for each polyline one point out of three (Figure 11.9).

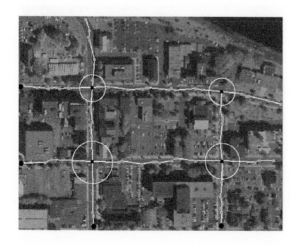

FIGURE 11.8
Getting the topological graph.

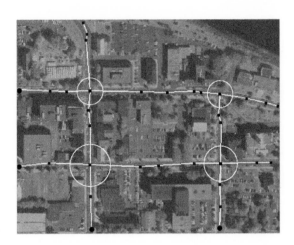

FIGURE 11.9
Topological graph subsampling.

At this stage, the graph network is complete and topologically correct, but the graph polylines are not well registered on the road centerline (for instance roads 2, 4, or 11). This graph enables to initialize the surface reconstruction module.

11.4.3 Reconstruction Module

From the obtained graph, polylines are then sampled and "propagated" at a distance defined by the road class. Here, roads are main streets of Fredericton and all are considered as class 3 (see Péteri, 2003). Coefficients controlling the evolution of active contours are defined by the user and are fixed on the whole image.

Figure 11.10 presents the result after extraction of road segments with parallel sides and initialization of intersections.

FIGURE 11.10
Intersection initialization after road segment extraction.

FIGURE 11.11 (See color insert following page 292.)
Final result after the extraction of intersections.

Finally, Figure 11.11 shows the final extraction result. One can note that the algorithm has succeeded in extracting badly initialized roads (as road 4). This is thanks to regularizations term and to the energy of parallelism. Thanks to the different wavelet coefficient images, the algorithm shows a good robustness with respect to noise on the road surface such as vehicles (case for roads 7 and 9) or ground marking (road 9). It has also managed to overcome several occlusions caused by trees (road 2) or shadows (roads 3, 6, and 7).

Some roads segmented have not been extracted so precisely: next to intersections between roads 1, 9, and 10, sides of road 2 are progressively moving apart. This is due to the presence of a parking lot and numerous occlusions. For road 11, the initialization far from the road centerline and the high number of projected shadows have caused a drift in the *DoubleSnake*, which has nevertheless kept a correct width.

In matter of intersections, some have been a bit cut, whereas others have been correctly and precisely extracted.

Figure 11.12 shows a zoom on a T-shaped intersection between roads 5, 7, and 8. This example illustrates the introduction of breaking points (refer to the intersection initialization in Figure 11.10). Visually in Figure 11.12, the intersections have been properly extracted; regularization terms of the active contours have enabled to overcome local occlusions and an area with poor contrast. Moreover, the introduction of breaking points at the intersection has made it possible to extract right angles properly.

11.4.4 Quantitative Assessment

According to the protocol defined in Péteri et al. (2004), eight image interpreters have been asked to acquire the road network of this scene. The result

FIGURE 11.12
Extraction result: zoom at intersection between roads 5, 7, and 8.

is shown in Figure 11.13. The complexity of the scene has generated interpretation variability among the interpreters, particularly for certain road segments (case for roads 4, 6, 8, and 10) and for almost all the intersections.

In Figure 11.14, a zoom on the intersection between roads 6, 7, 9, and 11 is presented. The tolerance zone (gray areas) and the reference (dotted line) are extracted from these interpretations (Section 11.3.5). The tolerance zone width traduces the important difference of judgment among the different image interpreters.

The contour extracted by the algorithm is drawn as a plain white line. This example is representative of results that can be obtained: because of projected shadows and a too distant initialization, road 11 is shifted compared with the reference and is entirely outside the tolerance zone. Conversely, roads 6 and 7 have been precisely extracted, are close to the reference, and completely inside the tolerance zone. Road 9 combines the two cases: the right side has not been correctly extracted due to a too long projected shadow that has generated a bad initialization of the intersection. Inversely, the left side has been very precisely extracted, which has enabled a good initialization of the intersection. One can notice in Figure 11.14

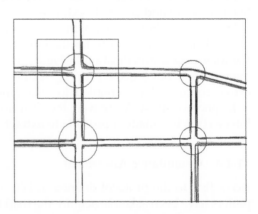

FIGURE 11.13
Superposition of eight interpretations of the scene.

FIGURE 11.14
Extracted road (white plain line), reference (dotted line), and tolerance zone (in gray).

that some intersections have been truncated a bit, whereas others have been correctly extracted.

Visual analyses are confirmed by numerical values deriving from the reference and the tolerance zone. Geometric criteria, reported in Table 11.1, also show a global match between the extracted network and the reference. Lengths of the extracted network and the referenced one are similar (about 4% of the relative difference). Difference between surfaces is small for road segments (2.2%). It is 16.9% for the intersections that have been more difficult to extract.

The Hausdorff distance, traducing the maximum discrepancy with the reference, is an average of 2.8 m for road segments, and of 4.9 m for the intersections.

Concerning the tolerance zone (Table 11.2), more than 70% of road sides are inside it, which is a good result considering the scene complexity and the noise in it. The rate of 52.5% for the intersections traduces the dependence of the *IntersectionSnakes* concerning the initialization provided by *DoubleSnakes*, and even before, by the tracking algorithm.

Indeed, a shift in the position of *DoubleSnakes* points located on the circle (see Figure 11.14) will generate a shift in the final result of the intersection extraction.

TABLE 11.1

Geometric Criteria on Lengths and Surfaces

	Length		Surface	
	Difference (m)	Relative Difference (%)	Difference (m²)	Relative Difference (%)
Road segment	36	1.9	294	2.2
Intersections	69	11	1156	16.9
Total	105	4.1	1410	7.0

TABLE 11.2

Percentage of Road Inside the Tolerance Zone

	Extracted Road
Road segments	71.3%
Intersections	52.5%

11.4.4.1   Gain in Time

Different interpreters have been timed during the scene acquisition process. For some highly occulted roads, they needed several successive adjustments. Including the possible rectification duration, the algorithm enables a gain of time that can be estimated by a factor of 10 compared with manual acquisition. This gain in time is potentially very promising for practical applications.

11.5   Conclusion and Prospects

This chapter deals with the use of high-resolution satellite images for road network extraction. A state-of-the-art on road extraction method has been proposed, describing two kinds of approaches for road extraction (linear and surface methods). These methods apply mainly according to the spatial resolution of the processed images and the scale of interest.

A global method for extraction of road networks from high-resolution satellite images has been then proposed. It aims at meeting the strong demand for automatic creation and update of maps, especially over urban areas, which could benefit the most from high-resolution imagery.

This method is modular and takes advantage of the cooperation between a linear representation of the road (graph module) and a surface representation (reconstruction module). Its application and evaluation on a Quick-Bird image over an urban area enhance the good behavior of the algorithm, even in the presence of artifacts. The method has also been applied on several other high-resolution images, coming from the IKONOS, QuickBird, or SPOT 5 satellites. In Péteri and Ranchin (2003), the method is applied on a semirural context, less noisy than urban environment: the inclusion rate in the tolerance zone reaches 90% for the roads and 80% for the intersections. The limit case of application of the method is reached on a 25 cm urban scene of the French Geographical Institute airborne camera (Péteri, 2003): at this very high spatial resolution, the method of road model is indeed not valid anymore and needs additional information.

Even if the method should be tested on a high number of cases to study its limits and its sensitivity to parameters, the algorithm enables a significant gain in time for the human operator (from a factor 6 to a factor 10, depending on the scene complexity). The next step of the work is to test the algorithm in operational conditions to evaluate the benefits of such an approach for the production of surface road maps.

One advantage of the proposed method is the use of few information sources. However, its modularity enables the introduction of external data. For instance, the method could benefit from the integration of dedicated works on intersections: one could mention the work of Boichis (2000), which proposes a precise extraction from a database of different types of intersections, including roundabouts. The elevation information could also enable to solve some ambiguities. One could consider using it at the end of the extraction process to distinguish the roadsides from the top of the buildings. Using multispectral images could also help the extraction process. For instance, the vegetation index (NDVI) can enable to mask trees along the roads and to restrict the zones to be extracted. Thermal infrared images can bring indications of the presence of moving vehicles, traducing potential roads.

While introducing external data, one should however take care of not significantly increasing the complexity and the computational time of the algorithm. Moreover, the gain in precision and reliability should be consistent with operational needs.

Acknowledgments

This work was supported by a CNRS/DGA grant of the French Ministry of Defense. Thanks go to Sylvain Airault from the French National Geographic Institute for providing the source codes of his programs. The authors would also like to thank the PNTS program and the Canadian GEOIDE Network (Geomatics for Informed Decision) for the QuickBird image.

References

Airault S., Jamet O., Détection et restitution automatique du réseau routier sur des images aériennes, *Traitement du Signal*, 12(2), 1995, 189–200.

Amo M., Martinez F., Torre M., Road extraction from aerial images using a region competition algorithm, *IEEE Transactions on Image Processing*, 15(5), 2006, 1192–1201.

Bacher U., Mayer H., Automatic road extraction from multispectral high resolution satellite images, in Stilla U., Rottensteiner F., Hinz S. (Eds.), *CMRT05*. IAPRS, Vol. XXXVI, Part 3/W24, Vienna, Austria, August 29–30, 2005, pp. 29–34.

Baumgartner A., Steger C., Mayer H., Eckstein W., Heinrich E., Automatic road extraction based on multi-scale, grouping, and context, *Photogrammetric Engineering and Remote Sensing*, 65(7), 1999, 777–785.

Bhattacharya U., Parui S.K., An improved backpropagation neural network for detection of road-like features in satellite imagery, *International Journal of Remote Sensing*, 18(16), 1997, 3379–3394.

Boichis N., Extraction automatique des carrefours routiers par interprétation d'images aèrienne guidée par une base de données cartographiques, PhD thesis, University of Cergy-Pontoise, Cergy-Pontoise, France, 2000, 142 p.

Couloigner I., Ranchin T., Mapping of urban areas: A multiresolution modeling approach for semiautomatic extraction of streets, *Photogrammetric Engineering and Remote Sensing*, 66(7), 2000, 867–874.

Destival I., Recherche automatique de réseaux linéaires sur des images SPOT, *Bulletin de la Société Française de Photogrammétrie et de Télédétection*, 66, 1987, 5–16.

Doucette P., Agouris P., Stefanidis A., Musavi M., Self-organised clustering for road extraction in classified imagery, *ISPRS Journal of Photogrammetry and Remote Sensing*, 55(5–6), 2001, 347–358.

Eidenbenz C., Käser C., Baltsavias E.P., ATOMI—Automated reconstruction of topographic objects from aerial images using vectorised map information, in IAPRS (Ed.), *19th ISPRS Congress*, Vol. 33, Amsterdam, The Netherlands, 2000, pp. 462–471.

Fua P., Leclerc Y.G., Model driven edge detection, *Machine Vision and Applications*, 3, 1990, 45–56.

Garnesson P., Giraudon G., Montesinos P., MESSIE: un système multispécialiste en vision. Application à l'interprétation d'image aérienne, *Traitement du Signal*, 9(5), 1992, 403–419.

Gruen A., Li H., Road extraction from aerial and satellite images by dynamic programming, *ISPRS Journal of Photogrammetry and Remote Sensing*, 50(4), 1995, 11–20.

Guigues L., Viglino J.-M., Automatic road extraction through light propagation simulation, in *International Archive of Photogrammetry and Remote Sensing*, Vol. XXXIII, Amsterdam, Holland, 2000.

Jin X., Davis C.H., An integrated system for automatic road mapping from high-resolution multi-spectral satellite imagery by information fusion, *Information Fusion* 6(4), 2005, 257–273.

Kass M., Witkin A., Terzopoulos D., Snakes: Active contour models, in *Proceedings of IEEE Conference on Computer Vision*, London, England, 1987, pp. 259–268.

Lacoste C., Descombes X., Zerubia J., Point processes for unsupervised line network extraction in remote sensing, *IEEE Transactions on Pattern Analysis and Machine Intelligence*, 40(10), 2005, 1568–1579.

Laptev I., Mayer H., Lindeberg T., Eckstein W., Steger C., Baumgartner A., Automatic extraction of roads from aerial images based on scale-space and snakes, *Machine Vision and Applications*, 12, 2000, 23–31.

Mac Keown D., Denlinger J., Cooperative methods for road tracking in aerial imagery, in *Proceedings of IEEE Computer Society Conference—Computer Vision and Pattern Recognition*, Ann Harbor, MI, USA, 1988, pp. 662–672.

Mallat S., *A Wavelet Tour of Signal Processing*, AP Professional, London, UK, 1997, 577 p.

Mallat S., Zhong S., Characterization of signals from multiscale edges, *IEEE Transactions on Pattern Analysis and Machine Intelligence*, 40(7), 1992, 2464–2482.

Mena J., State of the art on automatic road extraction for GIS update: A novel classification, *Pattern Recognition Letters*, 24(16), 2003, 3037–3058.

Merlet N., Zérubia J., New prospects in line detection by dynamic programming. *IEEE Transactions on Pattern Analysis and Machine Intelligence*, 18(4), 1996, 426–430.

Péteri R., Extraction de réseaux de rues en milieu urbain à partir d'images satellites à très haute résolution spatiale, thèse de l'Ecole Nationale Supérieure des Mines de Paris, formation "*Informatique, Temps Réel, Robotique et Automatique*," décembre 2003, 152 p.

Péteri R., Couloigner I., Ranchin T., Quantitatively assessing roads extracted from high-resolution imagery, *Photogrammetric Engineering & Remote Sensing*, 70(12), 2004, 1449–1456.

Péteri R., Ranchin T., Assessment of object extraction methods from satellite images: Reflections and case study on the definition of a reference, in Benes T. (Ed.), *22nd EARSeL Annual Symposium "Geoinformation for European Wide Integration,"* Prague, Czech Republic, 4–6 juillet 2002, Mill Press, Rotterdam, The Netherlands, pp. 141–147.

Péteri R., Ranchin T., Urban street mapping using QuickBird and IKONOS images, *Proceedings of IEEE International Geoscience and Remote Sensing Symposium*, Vol. III, Toulouse, France, juillet 2003, pp. 1721–1723.

Péteri R., Ranchin T., Multiresolution snakes for urban road extraction from IKONOS and QuickBird images, in Rudi Goossens (Ed.), *Proceedings of the 23rd EARSeL Annual Symposium "Remote Sensing in Transition,"* 2–4 June 2003, Ghent, Belgium, Mill Press, Rotterdam, The Netherlands, 2004, pp. 69–76.

Photogrammetric Engineering & Remote Sensing (PE&RS), Special Issue: Linear feature extraction from remote sensing data for road network delineation and revision, *Photogrammetric Engineering & Remote Sensing*, 70(12), 2004.

Puissant A., Weber Ch., The utility of very high spatial resolution images to identify urban objects, *Geocarto International*, 17(1), 2002, 31–41.

Quackenbush L.J., A review of techniques for extracting linear features from imagery, *Photogrammetric Engineering & Remote Sensing*, 70(12), 2004, 1383–1392.

Rellier G., Descombes X., Zérubia J., Local registration and deformation of road cartographic database on a SPOT satellite image, Technical Report 3939, INRIA, Sophia Antipolis, France, 2000, 23 p.

Serendero M., Extraction d'informations symboliques en imagerie SPOT: réseaux de communications et agglomérations, PhD thesis, Nice-Sophia Antipolis University, Nice, France, 1989, 160 p.

Véran J., Suivi de routes dans une image aérienne par filtrage de Kalman, Technical Report, Ecole Nationale Supérieure des Télécommunications, Paris, France, 1993, 30 p.

Vosselman G., De Knech J., Road tracking by profile matching and Kalman filter, in Verlag B. (Ed.), *Proceedings of Automatic Extraction of Man-Made Objects from Aerial and Space Images*, 1995, pp. 265–274.

Wang J.F., Howarth P.J., Automated road network extraction from Landsat TM imagery, in *Proceedings of the Annual ASPRS/ACSM Convention 1*, Baltimore, MD, USA, Vol. 1, 1987, pp. 429–438.

Wang Y., Trinder J., Use of topology in automatic road extraction, in *Proceedings ISPRS*, Commission III, Working Group 4, Columbus, OH, USA, 1998.

Williams D.J., Shah M., A fast algorithm for active contours and curvature estimation, in *CVIP: Image Understanding*, Vol. 55, January 1992, pp. 14–26.

Xiaoying J., Davis C., Automatic road extraction from high-resolution multispectral IKONOS imagery, in *Proceedings of IEEE International Geoscience and Remote Sensing Symposium*, Vol. III, Toulouse, France, 2003, pp. 1730–1732.

Zhang C., Baltsavias E., Gruen A., Knowledge-based image analysis for 3D road reconstruction, in *Proceedings of the 21st Asian Conference on Remote Sensing*, Taipei, Taiwan, 2000, pp. 100–105.

Zhang Q., Couloigner I., Benefit of the angular texture signature for the separation of parking lots and roads on high resolution multi-spectral imagery, *Pattern Recognition Letters* 27, 2006, 937–946.

12

Spectral Characteristics of Asphalt Road Surfaces

Martin Herold

CONTENTS

12.1 Introduction

Recent advances in imaging spectroscopy have shown capabilities to derive physical and chemical material properties on a very detailed level (Clark, 1999) with the potential of hyperspectral remote sensing to study and survey transportation assets and road surfaces being mentioned on several occasions (Usher and Truax, 2001; Gomez, 2002; Herold et al., 2004). Consequently, one would raise the questions: What are common spectral characteristics of roads and how are specific road surface conditions reflected in the spectral characteristics of these surfaces? The Santa Barbara asphalt road spectra library was developed to provide generic understanding about the spectral properties of road surfaces in various conditions and with different distresses (Herold and Roberts, 2005). The following examples and interpretations represent a subset of this spectra library and should support remote sensing researchers, transportation scientists, and others in their study of road surface conditions.

In general, spectral libraries contain pure spectral samples of surfaces, including a wide range of materials over a continuous wavelength range with high spectral detail and additional information and documentation about surface characteristics and the quality of the spectra (i.e., metadata). In February 2004, a ground spectra acquisition campaign was conducted in the area of Santa Barbara/Goleta, California. Ground spectra were acquired with an analytical spectral devices (ASD) full range (FR) spectrometer (Analytical Spectral Devices, Boulder, Colorado, USA). The FR spectrometer samples a spectral range of 350–2400 nm. The instrument uses three detectors spanning the visible and near-infrared (VNIR) and shortwave infrared (SWIR1 and SWIR2), with a spectral sampling interval of 1 nm. FR field spectrometer data are widely used and considered to provide accurate and high-quality spectral measurements. All acquired targets are documented and integrated into a spectral library, which is made available here. A pavement condition index (PCI) and a structure index (SI) were derived parallel from roadware (http://www.roadware.com) in situ vehicle inspections in December 2002. PCI and SI are a single road performance indicator with a scale usually between 0 and 100. A road surface in perfect condition would receive a score of 100 for both and deduct values for each in situ measured type and intensity of distress are subtracted. The spectra were preprocessed to apparent surface reflectance. For more detail on spectra acquisition and processing, refer to Herold and Roberts (2005).

12.2   Spectral Properties of Asphalt Road Surfaces

Asphalt pavements consist of rocky components and asphalt mix (or hot mix or bitumen). The mineral constituents of the crushed stone rocky components can vary depending on the geological region but usual major components in the aggregate are dominated by SiO_2, CaO, and MgO (Robl et al., 1991). The asphalt mix consists of oil, asphaltenes, and resins. The oils add viscosity and fluidity; asphaltenes cause strength and stiffness; and resins are important for interfacial adhesion and ductility in the pavement. This makes bitumen a complex substance that can vary in composition depending on the source of the crude oil and on the refining process. The chemical nature essentially is a mix of hydrocarbons with 50–1000 carbon atoms plus enough hydrogen, oxygen, sulfur, and nitrogen substituents to give some of the molecules a polar character. More specifically, the chemical components of asphalt mix are carbon (80%–87%), hydrogen (9%–11%), oxygen (2%–8%), nitrogen (0%–1%), sulfur (0.5%–1%), and some trace metals.

Figure 12.1 presents three spectral samples of pure road asphalt with no obvious structural damages or cracks. The age of the pavement, the PCI, and the structure index are shown with image examples of the surface. Spectrum A reflects a recently paved road. The surface is completely sealed with asphalt mix. The spectral reflectance is generally very low and hydrocarbon constituents determine the absorption processes. The minimum reflectance is near 350 nm with a linear rise toward longer wavelengths. Hydrocarbon

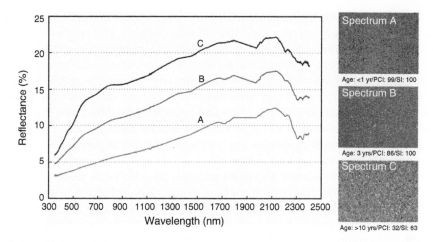

FIGURE 12.1
Spectral effects of asphalt aging and deterioration from the in situ spectral measurements (the major water vapor absorption bands are interpolated).

compounds exhibit electronic transitions arising from excitations of bonding electrons in the UV and visible region causing this strong absorption. The absorption is broad and there are no individual resolvable absorption bands in this spectral region due to the complex hydrocarbon nature of bitumen. Overlapping electronic processes with their absorption strength decreases toward longer wavelength causes these absorptions. This broad overall reflectance increase toward longer wavelengths is also seen in coals, oil shales, and chars (Cloutis, 1989; Hoerig et al., 2001).

For longer wavelengths, spectrum A in Figure 12.1 exhibits some obvious absorption features in the SWIR. However, there are a large number of fundamental organic absorption bands: aromatic C–H stretch, symmetric and asymmetric stretches and bands of CH_3 and CH_2 radicals, carbonyl/ carboxyl C–O stretch and the aromatic carbon stretch, and numerous combinations and overtones (Cloutis, 1989), the low overall reflectance suppresses most of the distinct features except the most prominent ones near 1700 nm and from 2200 to 2500 nm. Various C–H stretching overtones and combination bands dominate the feature in the 1700 nm region. If this feature is well developed, it is asymmetric and reflects a doublet with the strongest absorption at 1720 nm and a second less deep one at 1750 nm. The region between 2200 and 2500 nm is affected by numerous overlapping combination and overtone bands (Cloutis, 1989; Kuehn et al., 2004). This causes the strong reflectance decrease beyond 2200 nm. The absorption is strong in the 2300 nm region with a well-developed doublet at 2310 and 2350 nm with the 2310 nm feature, which is usually the stronger one.

Spectrum C in Figure 12.1 shows an old, deteriorated road surface. The image of the surfaces shows that the asphalt seal is widely eroded and the remaining asphalt mix has undergone an aging process. The natural aging of asphalt is caused by reaction with atmospheric oxygen, photochemical reactions with solar radiation, and the influence of heat, and results in three major processes (Bell, 1989): the loss of oily components by volatility or absorption, changes in composition by oxidation, and molecular structuring that influences the viscosity of the asphalt mix (steric hardening). The loss of oily components is a relatively short-term process; the other two are more long-term processes. With the erosion and aging of the asphalt mix, the road surface is less viscous and more prone to structural damages like cracking.

The spectral effects represent a mixture of both the exposition of rocky components and the asphalt aging. The vanishing of the complex hydrocarbon components causes a general increase in reflectance in all parts of the spectrum. This difference is highest in the NIR and SWIR with >10% reflectance. The electronic absorption processes in the visible region reflect the dominance of minerals and result in a concave shape with distinct iron oxide absorption features. They appear for 520, 670, and 870 nm. The typical SWIR hydrocarbon absorption features in 1700 and 2300 nm regions vanish for older road surfaces and are replaced by mineral absorptions. For example, there is a significant change in slope in the transition from hydrocarbon to mineral absorption. For older road surfaces, the slope increases between 2120 and 2200 nm as the 2200 nm silicate absorption gets more prominent. The slope is higher for new pavement materials 2250–2300 nm, which correlates with the intensity of the 2300 nm hydrocarbon feature.

Spectrum B in Figure 12.1 represents a road pavement of intermediate age and condition. The surface exhibits both the asphalt mix and exposed minerals. The spectral characteristics reflect this intermediate stage by showing absorption features from hydrocarbons and minerals. The intensity and characteristics of the features are less distinct than for "pure" very new and very old road surfaces. This shows that the aging and deterioration process is gradual and there is strong spectral evidence that this transition in surface material properties can be described in hyperspectral datasets. It should be noted that a road aging from 1 to 3 years, a change in PCI of 100 to 86, and a constant structure index of 100 have about the same spectral impact than from 3 to >10 years, a PCI decrease from 86 to 32, and a structure index decrease from 100 to 63. This suggests that the spectral signal is very sensitive to early stages of aging and deterioration and later, more severe road damages have a lower spectral impact.

Street paint also represents hydrocarbons with highly reflective properties (Figure 12.2). Spectra A and B of Figure 12.2 show reflectance values up to 55% in the VNIR region. The difference between spectra A and B in the visible region is due to color since spectrum B represents yellow street paint and blue wavelengths are absorbed. The street paint graphs highlight a typical asymmetric hydrocarbon doublet with the strongest absorption at 1720 nm and a second less deep one at 1750 nm (Cloutis, 1989). Numerous overlapping combinations and overtone bands cause the strong reflectance decrease beyond 2200 nm including a well-developed doublet at 2310 and

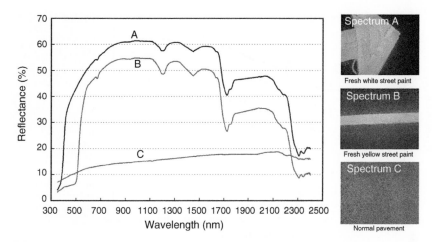

FIGURE 12.2
Spectral characteristics of street paint in different colors vs. normal asphalt pavement from the in situ spectral measurements (the major water vapor absorption bands are interpolated).

2350 nm with the 2310 nm feature, which is slightly stronger. These features nicely compare with characteristics of newly paved road surfaces (Figure 12.1), hence with different absorption intensity. From a remote sensing perspective, the presence of street paint will increase the brightness of a road surface especially in the VNIR region and emphasize the hydrocarbon absorptions in the SWIR.

12.3 Typical Asphalt Road Distresses

The most common road distress and indicator of pavement quality is cracking. Cracks, especially with Alligator pattern, indicate structural failure of the road surface due to traffic loads. Cracks allow moisture to infiltrate, increase road surface roughness, and may further deteriorate to potholes.

Figure 12.3 shows the spectral effects of structural damages or Alligator cracks with different severity on the spectral signal. The general road surface reflectance of the pavement is similar to spectrum C in Figure 12.1, with the spectrum dominated by mineral absorption processes. The main spectral impact of cracking is on object brightness in all parts of the spectrum. The increasing surface roughness and shadows cause reflectance differences of up to 7%–8% in the NIR and SWIR between the actual pavement and high-severity cracks. The concave shape in the VNIR region is more obvious for brighter, noncracked road pavements. There is also an indication that the cracked surfaces have more intense hydrocarbon absorption features in the 1700 and 2300 nm regions. The asphalt mix erosion and

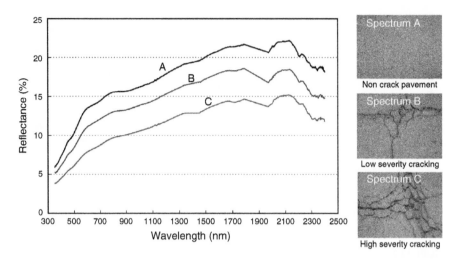

FIGURE 12.3
Spectral effects of severity of structural road damages from the in situ spectral measurements (the major water vapor absorption bands are interpolated).

oxidations happen on the road surface. Cracking exposes deeper layers of the pavement with higher contents of the original asphalt mix, which is then manifested in increased hydrocarbon absorption features. This fact highlights the contrary spectral signal between road deterioration of the pavement itself (Figure 12.1) and the severity of structural damages (Figure 12.3). An aging road surface gets brighter with decreasing hydrocarbon absorptions; structural distresses cause less reflectance with increasing hydrocarbon features. Although the reflectance difference and intensity of the hydrocarbon absorptions are less for cracks than for new asphalt surfaces, this fact indicates certain limitations in hyperspectral remote sensing of road conditions.

A second common road distress is raveling. The process of raveling describes the progressive dislodgement of pavement aggregate particles. This is mainly caused by increasing loss of bond between aggregate particles and the asphalt binder. Effects are the accumulation of loose aggregate debris on the road surface and less friction of vehicles, increasing surface roughness, and collecting water in the raveled locations causing vehicle hydroplaning.

The spectra A and B in Figure 12.4 compare a normal pavement with a raveled road surface. The raveling exhibits larger amounts of rocky components and raveling debris (gravel) on the surface. This generally increases the brightness of the surface due to increasing mineral reflectance and less prominent hydrocarbon absorptions. The raveling spectrum shows characteristics from both the normal pavement and spectrum C. Spectrum C reflects a gravel parking lot surface. In comparison with the pavements, this surface has higher reflectance in the visible and photographic

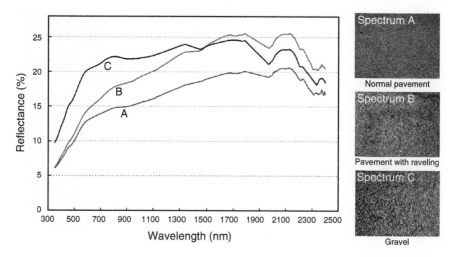

FIGURE 12.4

Spectral effects of raveling from the in situ spectral measurements (the major water vapor absorption bands are interpolated).

near-infrared due to the missing hydrocarbon absorptions. The mineral composition is reflected in more prominent features from iron oxide and other minerals like a calcite feature near 2320 nm.

12.4 Asphalt Road Surface Maintenance

Besides rehabilitation treatments, there are several maintenance methods to improve and maintain the quality of road surfaces. Their spectral characteristics are compared with a common asphalt road surface (Figure 12.5). Spectrum A shows a slurry crack seal that helps to prevent water or other noncompressible substances such as sand, dirt, rocks, or weeds entering the crack. Slurry seal crack fillings are mixtures of emulsified asphalt or rubberized asphalt spread with a machine onto the asphalt surface. This treatment material has a constant low reflectance on the order of 5% reflectance. Only very minor hydrocarbon absorption features are represented, which are similar to the ones found for parking lot surfaces (Herold et al., 2004).

Patches are used to treat an area of localized road distress. The material is similar to a usual pavement and spectrum B has similarity to a newly paved road (see Figure 12.1). Chip seal treatments include spraying an asphalt binder on the pavement, then immediately covering by a single layer of uniformly sized chips. The new surface treatment is then rolled to seat the aggregate, and broomed to remove any loose chips. The chip seal spectrum C has significantly higher reflectance than a usual asphalt road surface with more prominent mineral absorption features, similar to a raveled road surface (Figure 12.4).

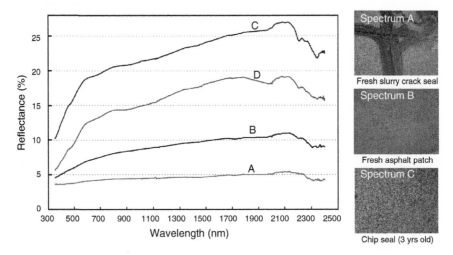

FIGURE 12.5

Spectral effects of different road surface treatments from the in situ ground spectral measurements. Spectrum D is an untreated road surface corresponding to spectrum C in Figure 12.2 (the major water vapor absorption bands are interpolated).

12.5   Other Surface Features

Road surfaces contain a variety of other surface features. Their spectral effects presented for asphalt road surfaces are exemplified in Figure 12.6.

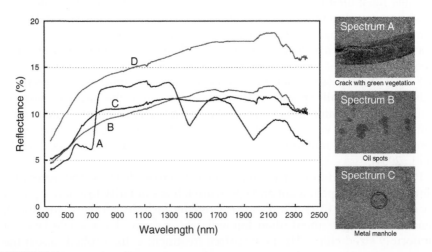

FIGURE 12.6

Spectral effects of different road surface features from the in situ ground spectral measurements. Spectrum D is an untreated road surface corresponding to spectrum C in Figure 12.2 (the major water vapor absorption bands are interpolated).

Spectrum A shows an older crack containing green vegetation. This has a strong impact on the spectral response exposing spectral features from chlorophyll, water content, and lignocellulose typical for vegetation (Herold et al., 2004). The accumulation oil on the pavement decreases the overall reflectance with more prominent hydrocarbon absorption characteristics (spectrum B). The metal manhole in spectrum C also indicates a decreasing reflectance compared with a usual pavement along with stronger iron oxide signal at the absorption bands at 520, 670, and 870 nm.

12.6 Summary and Remote Sensing Prospects

Asphalt road surfaces reflect distinct spectral responses from a number of absorption features. New asphalt pavements are dominated by hydrocarbon absorptions. Pavement aging and erosion of the asphalt mix result in a gradual transition from hydrocarbon to mineral absorption characteristics with a general increase in brightness and changes in distinct small-scale absorption features. Thus, there is spectral evidence for the aging and degradation processes of in situ asphalt pavements. Structural road damages (e.g., cracks) indicate a somewhat contrary spectral variation. Cracking decreases the brightness and emphasizes hydrocarbon absorption features. In addition, there are a number of additional features commonly found on road surfaces including gravel (i.e., raveling), metal, and vegetation. All show spectral impacts and add to the spectral complexity of asphalt road surfaces.

From a remote sensing mapping perspective, the general material characteristics and variability of road surfaces cause problems for their accurate spectral detection within urban areas. Other impervious surfaces (i.e., shingle and tar roofs) show similar material compositions (i.e., hydrocarbons) and are easily confused in the urban land cover and material classification process (Herold et al., 2003a,b). Additional information about three-dimensional urban structure (i.e., from LIDAR data) or from spatial image analysis algorithms is required to resolve such spectral similarity (Herold and Roberts, 2006). There have been positive examples of characterizing asphalt pavement deterioration using spectral information (Herold and Roberts, 2005). Other examples have successfully detected hydrocarbons from hyperspectral imagery (Hoerig et al., 2001; Kuehn et al., 2004). Understanding spectral characteristics and absorption processes (as provided here) is certainly essential for progress in this arena. In addition, fine spatial resolution on the order of 0.5 m ground instantaneous field of view (GIFOV) is required for clear identification of spectral road deterioration effects and to avoid false detections. Given the three-dimensional urban surface structure, roads are the "bottom layer" that can be covered or shadowed by surrounding surfaces such as trees, buildings, or cars.

In contrast to asphalt surfaces, concrete pavements (most commonly portland cement concrete) are composed of cement, mineral aggregates

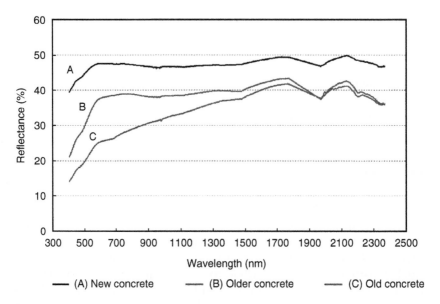

FIGURE 12.7
Spectral characteristics of concrete road surfaces from the in situ spectral measurements. The major water vapor absorption bands are interpolated.

including gravel or sand, water, and various other ingredients (admixtures and additions). Portland cement consists of a mixture of oxides of calcium, silicon, and aluminum and possibly some components of sulfate (i.e., gypsum). The material characteristics are reflected in distinct spectral signatures (Figure 12.7). Spectra of high reflectance dominated by the mineral absorption features are characteristic of concrete road surfaces. From a road aging point of view, Figure 12.7 highlights a general decrease in brightness for older concrete surfaces. This is contrary to asphalt roads that exhibit increasing brightness for older road surfaces. In situ oxidation, deterioration, and accumulation of dust and hydrocarbons may be responsible for such spectral effects. Certainly, the characteristics of concrete road surfaces should be studied in more detail and to the level asphalt roads have been investigated here. A generic spectral understanding is needed to further develop remote sensing applications for such transportation infrastructure surveying.

Acknowledgments

The ASD field spectrometer was kindly supplied by the Jet Propulsion Laboratory. The author would like to acknowledge the support of the U.S. Department of Transportation, Research and Special Programs Administration, OTA #DTRS-00-T-0002 (NCRST-Infrastructure). The authors thank CALTRANS, D. Roberts, V. Noronha, P. Dennison, M. Gardner, D. Prentiss,

J. Schuhrke at the University of California, Santa Barbara, and R. Souleyrette at the Center for Transportation Research and Education (CTRE) at Iowa State University, Ames, for their support to this study.

References

Bell, C.A. 1989. Summary Report on the Aging of Asphalt-Aggregate Systems, Strategic Highway Research Program (SHRP) Publications SHRP-A-305, 100 p., URL: http://gulliver.trb.org/publications/shrp/SHRP-A-305.pdf (accessed: March 2004).

Clark, R.N. 1999. Spectroscopy of rocks and minerals and principles of spectroscopy, in: A.N. Rencz (ed.), *Manual of Remote Sensing*, Chapter 1, John Wiley & Sons, New York, pp. 3–58.

Cloutis, A.E. 1989. Spectral reflectance properties of hydrocarbons: remote-sensing implications, *Science*, 4914, 165–168.

Gomez, R.B. 2002. Hyperspectral imaging: a useful technology for transportation analysis, *Optical Engineering*, 41, 9, 2137–2143.

Herold, M., Gardner, M., and Roberts, D.A. 2003a. Spectral resolution requirements for mapping urban areas, *IEEE Transactions on Geoscience and Remote Sensing*, 41(9), 1907–1919.

Herold, M., Gardner, M.E., Noronha, V., and Roberts, D.A. 2003b. Spectrometry and hyperspectral remote sensing of urban road infrastructure, Online *Journal of Space Communications*, 3, URL: http://satjournal.tcom.ohiou.edu/issue03/applications.html

Herold, M. and Roberts, D.A. 2005. Spectral characteristics of asphalt road aging and deterioration: Implications for remote sensing applications, *Applied Optics*, 44, 20, 4327–4334.

Herold, M. and Roberts, D.A. 2006. Multispectral satellites-imaging spectrometry-LIDAR: spatial-spectral tradeoffs in urban mapping, *International Journal of Geoinformatics*, 2, 1, 1–14.

Herold, M., Roberts, D., Gardner, M., and Dennison, P. 2004. Spectrometry for urban area remote sensing—development and analysis of a spectral library from 350 to 2400 nm, *Remote Sensing of Environment*, 91, 3–4, 304–319.

Hoerig, B., Kuehn, F., Oschuetz, F., and Lehmann, F. 2001. Hyperspectral remote sensing to detect hydrocarbons, *International Journal of Remote Sensing*, 22, 1413–1422.

Kuehn, F., Oppermann, K., and Hoerig, B. 2004. Hydrocarbon index—an algorithm for hyperspectral detection of hydrocarbons, *International Journal of Remote Sensing*, 25, 12, 2467–2473.

Robl, T.L., Milburn, D., Thomas, G., O'Hara, K., and Haak, A., 1991. The SHRP Materials Reference Library Aggregates: Chemical, Mineralogical, and Sorption Analyses, Strategic Highway Research Program (SHRP) Publications SHRP-A/UIR-91-509, 88 p., URL: http://gulliver.trb.org/publications/shrp/SHRP-91-509.pdf (accessed: September 2002).

Usher, J. and Truax, D. 2001. Exploration of Remote Sensing Applicability within Transportation. Remote Sensing Technologies Center final projects report, URL: http://www.rstc.msstate.edu/publications/99-01/rstcofr01-005b.pdf (accessed: March 2004).

Part IV

Roof-Related Impervious Surfaces

13

Urban 3D Building Model from LiDAR Data and Digital Aerial Images

Guoqing Zhou

CONTENTS

13.1 Introduction

The urban three-dimensional (3D) model has increasingly been needed for various applications such as town planning, microclimate investigation, and so on. Traditional photogrammetry is an important tool to acquire the 3D data. However, the photogrammetric method has encountered difficulties for complex scenes in dense urban areas due to the failures of image matching, which are primarily caused by, for example, occlusions, depth discontinuities, shadows, poor or repeated textures, and the lack of a model of man-made objects (Zhou et al., 1999). For this reason, the extraction of buildings and digital terrain model (DTM) are currently carried out using human-guided interactive operations, such as stereo compilation from screen. This process is both costly and time-consuming.

In recent years, light detection and ranging (LiDAR) data have been widely applied in urban 3D building extraction. A variety of methods have been proposed for this purpose. Yoon and Shan (2002) grouped the methods into two categories: the classification approach and the adjustment approach. The classification approach detects the ground points using certain operators designed on the basis of mathematical morphology (Lindenberger, 1993; Vosselman, 2001), terrain slope (Axelsson, 1999), or local elevation difference (Wang et al., 2001). The refined classification approach uses triangulated irregular network (TIN) data structure (Axelsson, 2000; Vosselman and Mass, 2001) and iterative calculation (Axelsson, 2000; Sithole, 2001) to consider the discontinuity in the LiDAR data or terrain surface. The adjustment approach uses a mathematical function to approximate the ground surface, which is determined in an iterative least adjustment process while outliers of nonground points are detected and eliminated (Kraus and Pfeifer, 1998, 2001). On the other hand, Baltsavias et al. (1995) discuss three different approaches: an edge operator, mathematical morphology, and height bins for detection of objects higher than the surrounding topographic surface. These approaches were used by authors like Haala (1995) and Eckstein and Munkelt (1995). They analyzed the compactness of height bins or used mathematical morphology (Eckstein and Munkelt, 1995; Hug, 1997). Other building extraction methods include extraction of planar patches, some of which use height, slope, and aspect images for segmentation (e.g., Morgan and Tempfli, 2000; Morgan and Habib, 2002).

Although plenty of efforts have been made in LiDAR data processing for urban 3D data extraction, all of the existing methods are not yet mature (Vosselman and Maas, 2001; Yoon and Shan, 2002). It has also been realized, also by many other photogrammetrists, that methods based on a single terrain characteristic or criterion can hardly obtain satisfactory results in all terrain types. For this reason, this chapter presents a combination of LiDAR data and orthoimage data for the urban 3D digital building model (DBM), digital surface models (DSM), and DTM generation.

13.2 Building Detection and Extraction

13.2.1 Edge Detection

As described earlier, the building extraction based on either single image data or single LiDAR data cannot reach a satisfying result. One of the main causes is the building's breaklines. It is thus very important to extract the breaklines before applying any interpolation technique; especially, the breaklines can be used to identify the sudden change in slope or elevation. Therefore, the detected breaklines will serve the purposes of both interpolation and building extraction. Almost all the breaklines represent parts of artificial objects in urban areas, while a breakline (edge) in digital image is a sharp discontinuity in gray-level profile. Thus, our implementation for building edge detection is to combine LiDAR data and orthoimage. LOG algorithm is first employed to extract the edges, and the LiDAR data are coregistered to orthoimage for extracting the building's edges using the principle that the buildings are higher than their surrounding topographic surfaces. After that, some postprocessing, such as merging line segment into line and deleting isolated point and line segment are carried out. Finally, a human–computer interactive operation is designated for extraction of complete edges of the objects. These extracted edges of objects, associated with the horizontal coordinates, are coded and saved in files in vector format for generating the DBM. The details include the following:

1. *Coregistration of LiDAR data and orthoimage.* Orthoimage is a geo-coded image, in which the geodetic coordinates for each pixel can be specified. The LiDAR data contain the geodetic coordinates. Thus, the coregistration of the two datasets is simple when their data are the same.

2. *Line extraction of building's roof.* Although many line feature extraction algorithms have been developed earlier, the line feature extraction in this chapter is conducted by (i) edge detection by LOG algorithm; (ii) edge vectorization by contour-tracing algorithm; (iii) initial straight line generation by applying splitting–merging algorithm to edge contour; (iv) definition of statistic values of lines to refine possible straight lines using a least-squares technique; and (v) establishment of image polygon. The details of the steps are

   (i) *Detecting edges from orthoimage.* The LOG algorithm is used to detect the edge of buildings, and then the contour-tracing algorithm is employed to link those discrete edges into a continuous curve. In these steps, a lot of edge contour, such as roads, parking lot, lake, and so on are extracted. This means that it is hard to disseminate a parking lot and house if only the extracted edge contour information is employed. For this reason, the LiDAR data will also be used for identifying the building.

   (ii) *Detecting edges from LiDAR data.* The extraction of buildings is based on a simple fact that buildings are higher than their surrounding topographic surface. The ability of the laser to penetrate vegetation, thus giving an echo from several heights, makes it possible to distinguish between the two classes: man-made objects and vegetation. The extraction procedures are based on an implementation of the minimum description length (MDL) criterion for robust estimation (Rissanen, 1983; Axelsson, 1992). A cost function (described later) is formulated for the two classes: buildings and vegetation based on the second derivatives of the elevation differences. The algorithm is as follows:

Assume that buildings consist of connected planar surfaces, for example, many neighboring TIN facets. Similarly, neighboring scan line points will lie on a straight line with the second derivatives of elevation differences zero. Buildings' roof ridges and other changes of direction of the buildings' roofs will cause a nonzero value of the second derivatives with respect to the elevation. The mathematical model is formulated along a scan line as

$$\frac{\partial^2 z}{\partial^2 x} = 0 \quad \text{point} \subseteq \text{straight line segment}$$

$$\frac{\partial^2 z}{\partial^2 x} \neq 0 \quad \text{point} \subseteq \text{breakpoint}$$

where
   x is the direction along the scan line
   z is the elevation

The cost function, or description length (DL), of the building model contains three parts. A detailed description can be found in Axelsson (1992).

(iii) *Forming an initial straight line*. Combing the results of the extracted edges from the LOG algorithm and LiDAR data, we can easily identify the building edge from other edges because the elevation of the building's edges is higher than the other edges. This chapter only considers the building, and thus other edges will not be taken into account. For the building edges detected from both orthoimage and LiDAR data, they are merged and processed using the splitting–merging algorithm to extract initial straight lines. In this method, a threshold, perpendicular distance of each edge point to a line no more than 2.0 pixels, is set up to detect the short straight line. On the other hand, this method results in some long straight lines, which may be split into several discontinuous short lines (line segments). For this reason, grouping and merging those short lines into a reasonable line is conducted using the colinear chain algorithm. After this method is carried out, many straight lines can be initially extracted.

(iv) *Refining the straight line*. The extracted straight lines in step (iii) deviate from their real edge positions due to noises.

For this reason, we employed the least-squares line-matching technique to remove the false lines and refine the straight lines. The basic idea is

Assuming that $L(x,y)$ and $L'(x',y')$ represent the extracted and real lines (i.e., at the real edge position), their relationship is expressed by an affine transformation, that is,

$$L(x,y) = L'(a_1 + a_2x + a_3y, b_1 + b_2x + b_3y), \tag{13.1}$$

where a_1, a_2, a_3, b_1, b_2, and b_3 are parameters of the affine transformation. Once they are determined, the straight line L' can be located. Assuming that $f(x,y)$ and $g(x,y)$ represent the intensity of the extracted edge and real edge, theoretically, we have $f(x,y) = g(x,y)$. However, with the existence of noise $e(x,y)$, that is, $f(x,y) - g(x,y) = e(x,y)$. Combining with Equation 13.1 and linearizing them, we have

$$f(x,y) - v(x,y) = g_0(x,y) + g_xda_1 + g_xxda_2 + g_xyda_3 \\ + g_ydb_1 + g_yxdb_2 + g_yydb_3. \tag{13.2}$$

For any pixel, one observation equation can be established using Equation 13.2. For a straight line, all observation equations are expressed by a vector form, that is,

$$-\mathbf{v} = AX - \mathbf{I}, \tag{13.3}$$

where

$\mathbf{I}$ is the vector of intensity error between $f(x,y)$ and $g(x,y)$
X denotes parameters of the affine transformation (i.e., a_1, a_2, a_3, b_1, b_2, and b_3)
A is the coefficient matrix
$\mathbf{v}$ is the residual vector

The parameters of the affine transformation can be solved via a least-squares estimation. The standard deviation (variance) σ_0 is calculated and used as the criterion to judge whether to remove a "misextracted" straight line. From statistics, if a pixel's variance, σ_i, which can be calculated from the variance matrix, is greater than $3\sigma_0$, this point will be removed; otherwise, the pixels are kept. The results from this method can better locate a building straight line.

(v) *Image polygon establishment.* Once the work is finished, a group of lines is selected to form building's polygon. To this end, a line segment is merged into a line and an isolated point is deleted. Moreover, a human–computer interactive operation is designated for final building extraction.

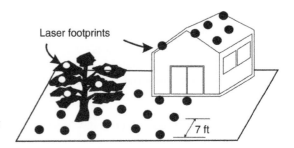

FIGURE 13.1
LiDAR footprints on building
roof, ground, and vegetation.

13.2.2 Building Polygon Extraction

After the complete edges of buildings have been detected, the algorithms for forming building polygon and extracting building geometrical parameters will be performed. The main steps are (1) linking the 2D complete image edges of the building with 3D LiDAR data using horizontal coordinates and (2) determining the three-dimensional building breaklines from image edges and exactly estimating the building boundary via integrating building edge results from both orthoimage and LiDAR data. Internal breaklines can be determined by intersecting the adjacent planar facades within the building. It is known that the LiDAR footprints do not exactly match building boundary. Therefore, one cannot determine the building boundary with only height data unless the density of LiDAR point cloud is like a gray image. Figure 13.1 shows a portion of a building near its boundary. Some laser footprints are located on the building roofs, whereas others are on the ground. The segments of LiDAR data are therefore from the image segments, which describe various buildings. Therefore, we have selected the georeferenced images whose 2D geodetic coordinates are known. We can directly use the horizontal coordinates of the boundary edges to obtain each 3D building model. The building boundary in addition to the internal facade parameters and the internal 3D breaklines will be the result of the building extraction process.

13.3 Creation of DBM and DTM

13.3.1 Establish Relationship between Building and LiDAR Data

After each building polygon is established, the next step is to establish the relationship between the building and LiDAR data for establishing 3D DBM. The orthoimages are stored as raster data, whereas the LiDAR point cloud is collected along the track. The linkage of the two datasets is implemented by the horizontal coordinates. Thus, we determine which LiDAR footprint is inside of the building boundary. We employ a filling algorithm, whose steps are (note that a rectangle is selected as a sample in Figure 13.2) as follows:

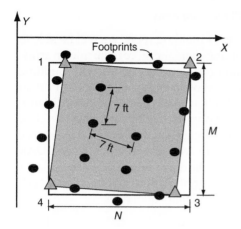

FIGURE 13.2
Determination of inside footprints in a building using filling algorithm.

1. *Create the polygon of building roof:* The edge of a roof surface that we extracted earlier is a set of point coordinates, for example $(x_1, y_1; x_2, y_2; \ldots \ldots; x_n, y_n)$. We can obtain the surface polygon by connecting the edge points orderly in this step (see Figure 13.2).

2. *Obtain the boundary of the polygon of the roof surface:* For a given roof surface in Figure 13.2, the coordinates of four corner points can be obtained by

$$\text{Corner 1: } (X_1, Y_1)$$
$$\text{Corner 2: } (X_2, Y_2)$$
$$\text{Corner 3: } (X_3, Y_3)$$
$$\text{Corner 4: } (X_4, Y_4)$$
$$\ldots \ldots$$
$$\text{Corner } n: (X_n, Y_n)$$

3. *Obtain the reduced LiDAR footprints within the boundary rectangular:* For speeding up the calculation, we reduce the LiDAR points via the test to see whether these points are in the roof surface or not. By simple comparison of the LiDAR point coordinates and the rectangular corners, we can obtain the reduced LiDAR points.

4. *Determine the LiDAR footprints in the reduced points:* The determination of the LiDAR footprints in the reduced points that are inside or outside is carried out by filling the algorithm. This algorithm was realized by Microsoft MFC function, that is, CRgn::PtInRegion in MS VC^{++}.

This procedure is then repeated for each building roof until all buildings are implemented.

13.3.2 Interpolation Algorithm via Planar Equation

After we obtained a complete extraction of building roofs, we will be able to obtain the LiDAR footprints within building roofs and store them into an array. Now, each building has its LiDAR footprint data, associated with building boundary information. We use the information to generate the DSM of urban areas. There are many interpolation methods available such as inverse distance weight (IDW), which calculate the unknown elevation by using the close known neighbors and give them different weight on the basis of the distance between them and the unknown points. Here, we suggest a method for LiDAR data interpolation. The basic principle is to fit the building roof using planar equation, which is solved by the LiDAR footprints within building roof boundary, which we already obtained earlier. The planar equation is

$$AX + BY + CZ = 1, \tag{13.4}$$

where
  A, B, and C are unknown parameters
  X, Y, and Z are coordinates of LiDAR data

At least three LiDAR footprints are requested to determine the planar equation (building surface). However, usually, more than three footprints are measured in each surface. The least-squares method is thus employed to calculate the parameters of the planar equation. The equation is

$$\begin{bmatrix} X_1, Y_1, Z_1 \\ X_2, Y_2, Z_2 \\ \cdots\cdots\cdots \\ X_m, Y_m, Z_m \end{bmatrix} \times \begin{Bmatrix} A \\ B \\ C \end{Bmatrix} = \begin{Bmatrix} 1 \\ 1 \\ \cdots \\ 1 \end{Bmatrix}. \tag{13.5}$$

where m is the number of LiDAR footprint in a building roof. This interpolation method for DSM generation via planar equation and the surface boundary can reach higher accuracy than the other method.

13.3.3 Establish Digital Building Model

In this chapter, an object-oriented data structure has been developed for the description of DBM. This model makes best use of the LiDAR datasets for better creating DBMs, for example, image boundary, which provides the boundary information of building roof and the LiDAR data, which provide height information of buildings. In this model, each building is an object of the building class, that is, an entity of the class. One building object consists of the attributes of the building ID, roof type ID, and the series of the roof surfaces. Each surface of a building object is also considered as an object. The surface class comprises the surface boundary, the LiDAR footprints within the building roof, and planar equation parameters describing the building roof. The boundary is composed of a set of points. One of the advantages of this model is its flexibility for future expansion, for example, adding other building attributes, such as wall surfaces, texture, and so on (see Figure 13.3). We implement this data structure in Section 13.3.4.

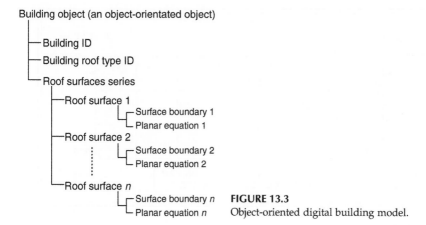

Building object (an object-orientated object)

— Building ID
— Building roof type ID
— Roof surfaces series
    — Roof surface 1
        — Surface boundary 1
        — Planar equation 1
    — Roof surface 2
        — Surface boundary 2
        — Planar equation 2
    — Roof surface *n*
        — Surface boundary *n*   **FIGURE 13.3**
        — Planar equation *n*   Object-oriented digital building model.

13.3.4 Create Digital Terrain Model (DTM)

LiDAR data present two aspects: ground and buildings. Thus, the data could be segmented into two types of regions corresponding, on one hand, to a surface linked to the ground, and, on the other hand, to a surface linked to surface objects. Therefore, DTM can be generated by separation of the surface objects from the DSM. DBM has been generated earlier, and DTM can be generated by removing the surface objects. The steps are as follows:

1. Based on the extracted boundary of the building in image processing, we can get the horizontal coordinates of these boundary points
2. Seeking for corresponding LiDAR footprints according to the planimetric coordinates
3. Removing those LiDAR footprints whose planimetric coordinates are the same as that of the other building boundary
4. Interpolating the DTM via IDW method

13.4 Experiments

13.4.1 Datasets

13.4.1.1 Control Field

The Virginia Department of Transportation (VDOT), contracting to Woolpert LLC at Richmond, Virginia, has established a high-accuracy test field in Wytheville, Virginia. This control field contains 19 targeted ground points for airborne LiDAR data accuracy evaluation. The target points are spaced at least several kilometers apart and distributed in a generally east-west direction. The field extends from the west side of Wytheville east ~11.4 mi with a north-south extent of ~4.5 mi centered on Wytheville, that is, from latitude 36°54′16″ to 36°59′54″ N, and from longitude 81°08′23″ to

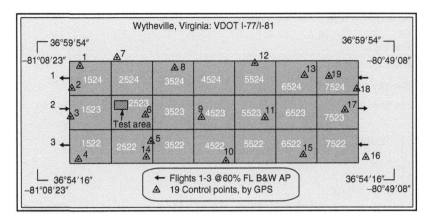

FIGURE 13.4
Geodetic control test field in Wytheville, Virginia.

81°49′08″ W (see Figure 13.4). The point accuracy can attain standard deviations better than 0.02, 0.02, and 0.01 m in X, Y, and Z, respectively. This level of accuracy is comparable with geodetic accuracy standard for Order C (1.0 cm plus 10 ppm).

13.4.1.2 Aerial Image Data

The analog black-and-white aerial photographs were also acquired along east-west flight lines over the test field on September 19, 2000 at a scale of 1:1000. The Woolpert camera at a focal length of 153.087 mm and Kodak 2405 film with a 525 nm filter were employed. A total of 96 exposures with 4 equal length flight lines (see Figure 13.4) were conducted. All the elevation data were referenced to NAVD88 datum; horizontal data were referenced to NAD83 Virginia State Plane Coordinate system. Aerial photos have a pixel resolution of 2.0 ft, and the orthoimage was produced using differential rectification techniques.

13.4.1.3 LiDAR Data

The LiDAR data were collected using the Optech 1210 LiDAR system in September 2000. The LiDAR data at a sampling density of 7.3 ft have an accuracy (on hard surfaces) of 2.0 ft at least. The specifications for LiDAR data collection are as follows:

- Aircraft speed: 202 ft/s
- Flying height: 4500 ft above ground level
- Scanner field of view (half angle): ±16°
- Scan frequency: 14 Hz

- Swath width: 2581 ft (1806 ft with a 30% side lap)
- Pulse repetition rate: 10 kHz
- Sampling density: average 7.3 ft

13.4.2   Experimental Results

We developed a system for semiautomation of urban 3D model generation from LiDAR data and orthoimage data using Microsoft Visual C^{++} under Microsoft 2000. The system consists of the following modules (Figure 13.5).

1. *Create new/open a project*: This menu opens an existing project or creates a new project.
2. *Data input (image and LiDAR)*: This module contains LiDAR data input, image display, and data format conversion (e.g., for raw image to bmp image, tiff image format, etc.).
3. *LiDAR data check*: This module checks the systematic error of LiDAR via various methods, such as overlay LiDAR data onto georeferenced image, ground control points checks, and so on.
4. *Image processing and interactive edit*: This module contains image filtering, enhancement, edge detection, line feature and area detection and description, image interpretation, interactive operation, and so on.
5. *DBM generation*: This module generates DBM using the object-oriented data structure described earlier.
6. *Urban DSM and DTM generation*: This module generates high-accuracy DSM by applying the method presented earlier.

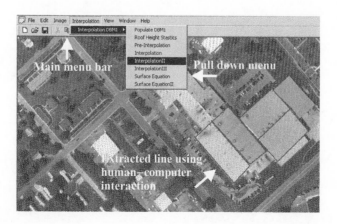

FIGURE 13.5
Semiautomatic urban 3D model generation (the green points are LiDAR point cloud; the gray images are aerial images; and the red lines are the detected edges of the building).

FIGURE 13.6
A patch of the original image (Image ID# 2523).

> Meanwhile, conventional interpolation methods, such as IDW, and bilinear interpolation methods are also available. The DTM is generated by removing surface objects.

We selected a patch of original photo (Photo ID: 2523) to test our method (Figure 13.6). This photo covers the city of Wytheville, Virginia. With this software, a group of experimental results are listed in Figures 13.5 through 13.13. Figure 13.7 is the result of automatic building edge detection, and Figure 13.8 depicts the detected buildings after human–computer

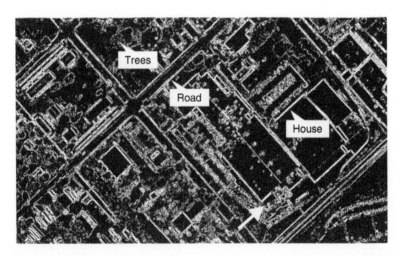

FIGURE 13.7
Automatically detected building edges.

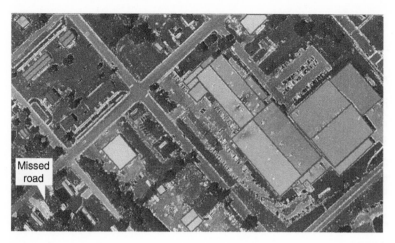

FIGURE 13.8
Building edges detected by human–computer interaction operation.

interactive operation. Figures 13.9 through 13.12 depict the DSMs, which are generated by IDW, Spline, and our algorithm to compare the interpolation accuracy between our method and other interpolation methods. As seen, both IDW and Spline interpretation methods cannot reach high accuracy. The building edges are not very clear. It appears that there are dim slopes to the ground. In addition, the roof surfaces are rough but most of the real roof surfaces are planar. Obviously, our interpolation result is much better than IDW and Spline methods. The edges and the roof surfaces are clearer. Figure 13.13 is a DBM. The most important fact is that each building is an

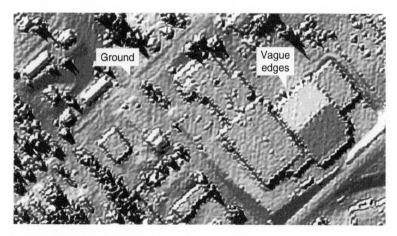

FIGURE 13.9
Result of LiDAR data interpolated by IDW.

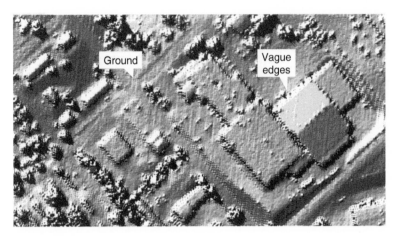

FIGURE 13.10
Result of LiDAR data interpolated by Spline (the Spline parameters are weight = 0.1, number of points = 12, type is regularized).

object in our program, which is convenient for future application, such as visualization.

13.4.3 Discussion

Table 13.1 lists statistic results of house recognition using the proposed method. As seen, the recognition rate achieves 88%. Those houses that cannot be recognized are probably too small because more than three LiDAR footprints are required in the proposed method. The accuracy of

FIGURE 13.11
Result of house information extraction by using our method.

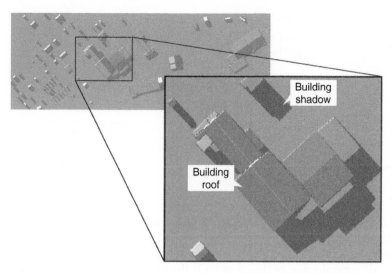

FIGURE 13.12
Digital building model (DBM) generated by our software system.

DBM is higher than other methods because this method suggests an integration of aerial image and LiDAR data processing, of which aerial image processing can recognize the boundary of each house, and LiDAR data provide elevation information, resulting in higher interpolation accuracy than other methods, such as IDW and Spline. In addition, this method requires at least three LiDAR footprints; this means that houses greater than 21 ft long can be recognized because the spacing size of the adopted LiDAR point cloud is 7.3 ft. Usually, this requirement can be reached. Thus, the proposed method is practicable.

FIGURE 13.13
Human–computer interaction for house interpretation, and the number of LiDAR footprints within a house.

TABLE 13.1

Successfully Interpreted Rates
of Houses and Roads

	Houses
Total	94
Interpreted number	83
Successful rate	88%

13.5 Conclusion

This chapter presented the generation of an urban 3D model, including 3D DSM and DBM via integrating image knowledge and LiDAR data. A human–computer interactive operation system has been developed for this purpose. The main contribution of this chapter is the development of a high-accuracy interpolation method for DBM/DTM/DSM generation and an object-oriented building model. In this model, we defined the roof types, roof boundary coordinates, planar equation parameters, and so on. Especially, for a roof surface, the model consisted of roof boundary and their planar equations, which are obtained from the combined processing of LiDAR and orthoimage data. For the planar equation of each roof surface, we first extract the LiDAR point data within the building roofs by their spatial relationship and calculate the planar equation parameters using LiDAR footprints. We use the planar equation to calculate the grid value within the roof boundary. The experimental results demonstrated that the developed method for DSM and DBM generation in urban areas is capable of reaching a high accuracy.

Acknowledgments

The experimental data were provided by the VDOT and Woolpert L.L.C. at Richmond, Virginia. We especially thank Frank Sokoloski and Qian Xiao very much for our discussion on the technology of LiDAR data processing and development of the system as well as for their kind help in LiDAR data check and delivery. All experiments were conducted by Ph.D. student Mr Changqing Song at the Civil Engineering Department of Old Dominion University.

References

Axelsson, P., 1992. Minimum description length as an estimator with robust properties. In Foerstner, W., Ruwiedel, S. (Eds.), *Robust Computer Vision*, Wichmann, Eerlag, Karlsruhe, pp. 137–150.

Axelsson, P., 1999. Processing of laser scanner data—algorithms and applications, *ISPRS Journal of Photogrammetry & Remote sensing*, 54, 138–147.

Axelsson, P., 2000. DEM generation from laser scanner data using adaptive TIN models, *International Archive of Photogrammetry and Remote Sensing*, XXXIII, Part B4, 110–117.

Baltsavias, E., et al., 1995. Use of DTMs/DSMs and orthoimages to support building extraction. In Gruen, A., Kubler, O., Agouris, P. (Eds.), *Automatic Extraction of Man-Made Objects from Aerial and Space Images*, Birkhauser, Basel, pp. 199–210.

Eckstein, W., Munkelt, O., 1995. Extracting objects from digital terrain models. In Schenk, T. (Ed.), *Remote Sensing and Reconstruction for Three-Dimensional Objects and Scenes. Proceedings of SPIE Symposium on Optical Science, Engineering, and Instrumentation*, Vol. 2572, San Diego.

Haala, N., 1995. 3D building reconstruction using linear edge segments. In Fitsch, D., Hobbie, D. (Eds.), *Photogrammetric Week*, Wichmann, Karlsruhe, pp. 19–28.

Hug, C., 1997. Extracting artificial objects from airborne laser scanner data. In Gruen, A., Baltsavias, E., Henricsson, O. (Eds.), *Automatic Extraction of Man-Made Objects from Aerial and Space Images (II)*, Birkhauser, Basel, pp. 203–212.

Kraus, K., Pfeifer, N., 1998. Determination of terrain models in wooded areas with airborne laser scanner data, *ISPRS Journal of Photogrammetry and Remote Sensing*, 53, 193–203.

Kraus, K., Pfeifer, N., 2001. Advanced DEM generation from LiDAR data. In Hofton, Michelle A. (Ed.), *Proceedings of the ISPRS workshop on Land Surface Mapping and Characterization Using Laser Altimetry, Annapolis, Maryland, International Archives of the Photogrammetry, Remote Sensing and Spatial Information Sciences*, Vol. XXXIV, Part 3/W4 Commission III.

Lindenberger, J., 1993. Laser-Profilmessungen zur topographischen Gelaedeaufnahme, Deutsche Geodaetische Kommission, Series C, No. 400, Munich.

Morgan, M., Habib, A., 2002. Interpolation of LiDAR data and automatic building extraction. In *ASPRS Annual Conference, CD-ROM*, Washington, D.C., April 19–25.

Morgan, M., Tempfli, K., 2000. Automatic building extraction from airborne laser scanning data. *International Archives of Photogrammetry and Remote Sensing*, Amsterdam, 33, B3, 616–623.

Rissanen, J., 1983. A universal prior for integers and estimation by minimum description length. *The Annals of Statistics*, 11, 2, 416–431.

Sithole, G., 2001. Filtering of laser altimetry data using a slope adaptive filter. In Hofton, Michelle A. (Ed.), *Proceedings of the ISPRS Workshop on Land Surface Mapping and Characterization Using Laser Altimetry, Annapolis, Maryland, The International Archives of the Photogrammetry, Remote Sensing and Spatial Information Sciences*, Vol. XXXIV, Part 3/W4 Commission III, pp. 203–210.

Vosselman, G., Mass, H., 2001. Adjustment and filtering of raw laser altimetry data. In *Proceedings of the OEEPE Workshop on Airborne Laser Scanning and Interferometric SAR for Detailed Digital Elevation Models*, March 1–3, Stockholm.

Wang, Y., Mercer, B., Tao, C., Sharma, J., Crawford, S., 2001. Automatic generation of bald earth digital elevation models from digital surface models created using airborne IFSAR. In *CD-ROM Proceedings of ASPRS Conference*.

Yoon, J.-S., Shan, J., 2002. Urban DEM generation from raw airborne LiDAR data. In *ASPRS Annual Conference, CD-ROM*, Washington, D.C., April 19–25.

Zhou, G., Albertz, J., and Gwinner, K., 1999. Extracting 3D information using temporal-spatial analysis of aerial image sequences, *Photogrammetry, Engineering & Remote Sensing*, 65, 7, 823–832.

14

Building Extraction from Aerial Imagery

Armin Gruen

CONTENTS

14.1 Introduction

Site recording and modeling have been important topics in photogram-metry from its very beginning in the middle of the nineteenth century. Since then technologies have changed several times fundamentally. Today the issue of full automation of all processes involved has led to widespread research activities in both the photogrammetry and the computer vision communities. However, progress is slow and the pressing need to produce precise, reliable, and complete datasets within reasonable time has had scientists and developers turn toward semiautomated approaches. While the tasks may differ in terms of required resolution (level of detail) of models, type of product (vector model, hybrid model, including mapped texture, attributed model with integrated thematic information), size of dataset, sensor platform (satellite, aerial, terrestrial), and sensor and data type (images in various forms, laser scans, scanned maps, etc.), one common problem remains in all cases and that is the automated extraction of objects from images. A typical example is the task of automated building detection and reconstruction, which is difficult for many reasons.

The most common sources of data are 2D images that lack direct 3D information. Aerial images may differ from each other with respect to scale, spectral range of recording, sensor geometry, image quality, imaging conditions (weather, lighting), and so on. Objects like buildings can be rather complex structures with many architectural details. They may be surrounded by other disturbing man-made and natural objects. Occlusion of parts is common and the geometrical resolution may be limited. Therefore, the corresponding images are of very complex content and highly unstructured. Solving the problem of building detection and reconstruction under these conditions is not only of great practical importance but also provides an excellent test bed for developing image analysis and image-understanding techniques.

The basic problem in object extraction stems from the fact that automated image understanding is still operating at a very rudimentary level. This applies both to close-range and aerial/space applications. However, there is a remarkable relation between image scale and success rate in extraction. At smaller image scales, the level of geometric modeling becomes lower and the image context is easier to grasp since the relationships between objects are less distorted by artifacts. Thus, the extraction of digital terrain models (DTMs) and objects like buildings, roads, rivers, land-use elements, and so on, becomes less complicated.

Over the last 15 years, photogrammetric approaches to building extraction and modeling have evolved. What started out as a pure research issue has now found firm grounds in professional practice. After the first phase of efforts to extract buildings fully automatically, the tight specifications of users have led to the development of efficient manual and semiautomated procedures. Actually, the need to extend modeling from simple to much more complex buildings and full ensembles and to even generate complete city models (including DTM, roads, bridges, parking lots, pedestrian walkways, traffic elements, waterways, vegetation objects, etc.) puts fully automated methods even further back in the waiting line of technologies for practical use. In a sense, the user requirements have outpaced the capabilities and performance of automated methods. However, to make it clear, automation in object extraction from images is and will continue to be a key research topic.

There are many fully automated approaches to building extraction, but only very few that were designed as semiautomated ones from the very beginning. Very often, procedures are declared as automatic but require so much postediting that their status as automatic methods becomes questionable.

In this chapter, we restrict ourselves to purely image-based approaches. Images inhibit a wealth of information, which is yet unmatched by other sensor products. Map scanning, laser scanning, radar, and other more rare techniques are not covered. More specifically, we focus on building extraction from aerial images, because this has recently found most prominent attention. A wider spectrum of our work, including city modeling from satellite images, has been presented in Baltsavias and Gruen [1].

There are a number of useful reviews and paper collections on building extraction techniques available [2–9].

14.2 Applications of City Models

Currently, the major users in Europe are in city planning (Figure 14.1a), facility mapping (especially chemical plants and car manufacturers, see Figure 14.1b), telecommunication, construction of sports facilities, and other infrastructure buildings. Others include environmental studies and simulations, location-based services (LBS), risk transports and analysis, car navigation, simulated training (airplanes, trains, trams, etc.), energy providers (placement of solar panels), real-estate business, virtual tourism, and microclimate studies. Interesting markets are expected in the entertainment and infotainment industries, for example, for video games, movies for TV and cinema, news broadcasting, sports events, animations for traffic and crowd behavior, and many more.

Diverse applications require different levels of detail in modeling, a great variety of different objects to be extracted, and the handling of different data types and manipulation functions. Therefore, when designing an efficient method for object extraction and modeling the following requirements should be observed:

1. Extract not only buildings, but other objects as well
2. Generate truly 3D geometry and, if a GIS platform is used, topology as well
3. Integrate natural image texture (for DTM, roofs, facades, and special objects)
4. Allow for object attribution
5. Keep level of detail flexible

(a)

(b)

FIGURE 14.1
(a) 3D model of ETH Zurich campus Hoenggerberg. Used for the purpose of city planning
(Science City Hoenggerberg). (b) Modeling of a chemical plant (combination of vector and
raster image data). (Courtesy CyberCity AG, Urdorf, Switzerland.)

6. Allow for a wide spectrum of accuracy levels in centimeter and decimeter ranges

7. Produce structured data, compatible with major CAD and visualization software

8. Provide for internal quality-control procedures, leading to absolutely reliable results

14.3 Techniques for 3D City Modeling

In this section, we discuss three major techniques, which are used in city model generation (there also exist combinations):

1. *Digitization of maps.* This gives only 2D information. The height of objects has to be approximated or derived with great additional efforts. It does not provide for detailed modeling of the roof landscape. This roof landscape is usually very important to the user, because city models are mostly shown from an aerial perspective. In addition, map data are often outdated.

2. *Extraction from aerial laser scans.* Laser scans produce regular sampling patterns over the terrain. Most objects in city models are best described by their edges, which are not easily accessible in laser scans and often cannot be derived unambiguously. Some objects of interest do not distinguish themselves through height differences from their neighborhood, and thus cannot be found in laser data. Finally, the resolution of current laser scan data is not sufficient for detailed models.

3. *Photogrammetric generation.* Aerial and terrestrial images are very appropriate data sources for the generation of city models. They allow to derive both the geometrical and the texture models from one unique dataset. The photogrammetric technique is highly scalable; it can adopt to required changes in resolution and accuracy in a flexible way. The processing of new images guarantees an up-to-date model. Images are a multipurpose data source and can be used for many other purposes as well.

The introduction of large format digital aerial cameras has further increased the efficiency for city modeling. Since ultra-high-resolution satellite imagers with stereo capabilities became available, this also became a viable tool for 3D city modeling, at reduced resolution [10]. However, the geometrical resolution of optical satellite imagers is further increasing. For example, Digital Globe has announced for 2007 a new satellite sensor (WorldView I) at 50 cm footprint.

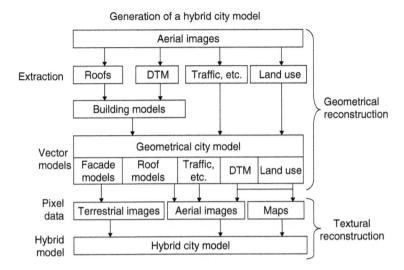

FIGURE 14.2
Image-based reconstruction of a hybrid city model.

As a result of this brief analysis, the photogrammetric approach must be considered the most relevant technique, because it can respond best to the various requirements. A scheme for image-based reconstruction of a hybrid city model is shown in Figure 14.2. Hybrid refers to the fact that both vector and raster data can be represented by the model. According to this scheme roof landscapes, DTM, transportation elements, land-use information, and so on can be extracted from aerial images.

Combining roofs and DTM will result in building vector models. These models could be refined by using terrestrial images taken with camcorders or still video cameras. Aerial images, terrestrial images, and digitized maps can all contribute to the texture part of the hybrid model. It is also well known that to a certain extent texture information can compensate for missing vector data.

Since fully automated extraction methods cannot cope with most of the aforementioned requirements, semiautomated photogrammetric methods are currently the only practical solutions of choice. For a review of semi-automated methods for site recording, see Gruen [7].

There are two semiautomated approaches that have made it into the commercial domain so far:

1. InJECT, a product of INPHO GmbH, Stuttgart. This approach was based on the fitting of elementary, volumetric building models or, in the case of complex buildings, building component models to image data. This concept, originally introduced at the Stanford Research Institute, Menlo Park, United States, was refined and extended at

the Institute of Photogrammetry, University of Bonn and has been available as a commercial software package for some time.

2. CyberCity Modeler (CC-Modeler) is a method and software package that fits planar surfaces to measured and weakly structured point clouds, thus generating CAD-compatible objects like buildings, trees, waterways, roads, and so on. Usually these point clouds are taken from aerial images, but it is also possible to digitize them from existing building plans. This product is marketed by CyberCity AG, Urdorf, Switzerland, a spin-off company of ETH Zurich.

In the following section, we focus on the description of CC-Modeler because of its dominating position in the professional domain.

14.4   CyberCity Modeler

For the generation of 3D descriptions of objects from aerial images, two major components are involved: photogrammetric measurements and structuring. In CC-Modeler, object identification and measurement are done in manual mode by an operator within a stereoscopic model on a stereo instrument, like a Digital Station. According to our experiences, stereoscopy is very crucial for object identification. In many complex situations in urban areas, monoscopic image interpretation will inevitably fail, because the roof structure cannot be interpreted. Thus, the human operator defines the level of detail of representation. He measures the key points of the roof landscape that describe the objects sequentially building by building. In the case of complex roof structures, a building may have to be broken up into several CC-Modeler units. The point cloud is then passed over to the actual "Modeler."

The structuring of the point clouds is done automatically with the CC-Modeler software. Structuring involves essentially the intelligent assignment of planar faces to the given cloud of points, or in other words, the decision making about which points belong to which planar faces. This problem is formulated as a consistent labeling problem and solved via a modified technique of probabilistic relaxation. Then, a least-squares adjustment is performed for all faces simultaneously, fitting the individual faces in an optimal way to the measured points and considering the fact that individual points are usually members of more than one face. This adjustment is amended by observation equations that model orthogonality constraints of pairs of straight lines. For the purpose of visualization, the system can also triangulate the faces into a TIN structure. Figure 14.3 shows the data flow and the procedures involved in the CC-Modeler.

A detailed description can be found at our homepage www.photogrammetry.ethz.ch (Research—Projects—CC-Modeler) and in Gruen and Wang [11]. With this technique, hundreds of objects can be measured in a day.

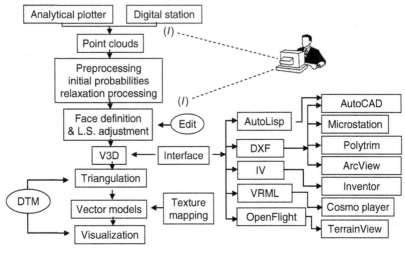

(*I*) ----Interactive functions

FIGURE 14.3
Data flow of CC-Modeler.

Although CC-Modeler generates a polyhedral world, objects with nonplanar surfaces can be modeled in sufficient resolution (compare Figure 14.4). A DTM, if not given a priori, can also be measured and integrated.

FIGURE 14.4
City model Zürich Oerlikon. Note the modeling of nonplanar surface objects.

Texture from aerial images is mapped automatically onto the terrain and the roofs, since the geometrical relationship between object faces and image patches has been established during the georeferencing and point cloud generation phases. The facade texture is produced semiautomatically via projective transformation from terrestrial images usually taken by camcorders or still video cameras.

Although many users are currently mainly interested in the visualization of the city models, there is also a clear desire to integrate the data into a GIS platform to use the GIS data administration and analysis functions. The commercial GIS technology is still primarily 2D-oriented and thus not really prepared to handle 3D objects efficiently. Therefore, we have developed in a pilot project a laboratory version of a hybrid 3D spatial information system (Figure 14.5), which is described in Wang and Gruen [8]. Figure 14.6 demonstrates a query function of our CC-Spatial Information System.

The CC-Modeler system and software are fully operational. Over 1,000,000 buildings at very high resolution have been generated already in cities and towns worldwide (see www.cybercity.tv). Figure 14.7 shows the integration of vector and image raster data into a joint model (images are mapped onto the DTM, roofs, and facades).

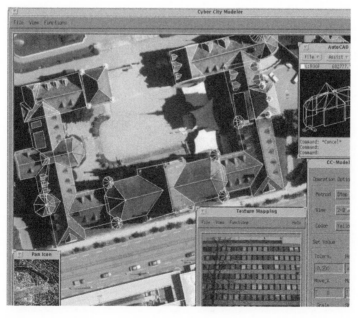

FIGURE 14.5
User interface of CC-Modeler.

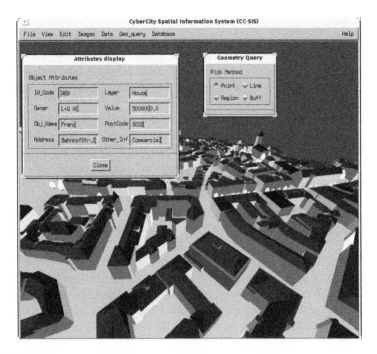

FIGURE 14.6
A geometrical query of the CC-Spatial Information System.

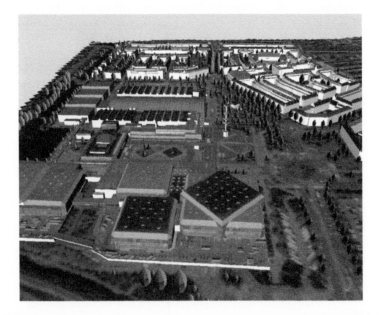

FIGURE 14.7
3D model of the Congress Center RAI, Amsterdam. Vector data, overlaid with natural texture.

14.5 Extensions to CC-Modeler

An important function in high-quality city modeling is the editing of the raw data. While the CC-Modeler was built to model the objects as close to their existing size and shape as possible, there arises sometimes the need to regularize the geometry. Under these constraints do fall the requests to make straight lines parallel and perpendicular where they are actually not, or to have all points of a group (e.g., eaves or ridge points) at a unique height. Another problem grew from the fact that CC-Modeler was designed to handle individual buildings sequentially and independent of each other. Building neighborhood conditions were not considered. The geometrical inconsistencies originating from that fact, like small gaps or overlaps between adjacent buildings (in the centimeter/decimeter range), are not dramatic and tolerable in many applications, especially those that are purely related to visualization. However, the topological errors constitute a serious problem in projects where the 3D model is subject to legal considerations or some other kind of analysis that requires topologically correct data.

Another significant extension refers to the precise modeling of building facades. Facades are usually not visible in aerial images, but available in cadastral maps. We combine this facade information with the roof landscape modeled with CC-Modeler to be able to represent the roof overhangs. We also show that we can model other vertical walls explicitly.

In the following sections, we will give a brief description of these new and practically important editing procedures. Figure 14.8 shows the flowchart of the processes mentioned earlier, which are executed after the face definition

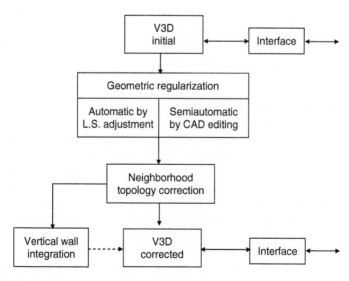

FIGURE 14.8
Flowchart of CC-Modeler extensions.

by probabilistic relaxation is done. For a more detailed description of these extensions, see Gruen and Wang [12].

14.5.1 Geometrical Regularization

Geometrical regularization refers to the task of modifying the geometry in such a way that regular structures are obtained. Measurements from images are always erroneous, although the errors may be very small. In addition, in particular with older buildings, the geometry deviates from regular patterns sometimes significantly. Edges are not parallel, intersections not perpendicular, roof faces not planar. We therefore have developed two strategies for regularization: a fully automatic adjustment based on least squares and a semiautomated approach of CAD editing. Both approaches are integrated in the software package CC-Edit.

The requirements for geometrical regularization are as follows:

1. Same height for groups of eave points, ridge points, and other structure points
2. Roof patches containing more than three points should form planar faces
3. Parallelism of straight edges
4. Right angles of intersecting roof edges
5. Collinearity of edge points

14.5.1.1 Automatic Regularization by Least-Squares Adjustment

We solve these requirements by formulating these geometrical constraints as stochastic constraints, that is, as weighted observation equations in a least-squares context. Details may be found in Gruen and Wang [12]. Figure 14.9 shows the result of such a regularization.

14.5.1.2 Regularization by CAD Editing

This is a semiautomated supervised procedure, which operates only in planimetry. Therefore, it requires that the equal height condition is observed during the point measurement phase. Then a grid of parallel construction lines is generated and overlaid to the measured lines. The measured lines are automatically adjusted to the direction of the grid. The grid's direction itself is derived from the average direction of the measured lines concerned. The selection of the concerned lines can be done automatically or manually. The overlay display is used for checking and manual editing if something goes wrong.

The right angle, collinearity, and the planar face constraints are automatically observed by that procedure. Since we use hard constraints here, the results are strict. An example is shown in Figure 14.10.

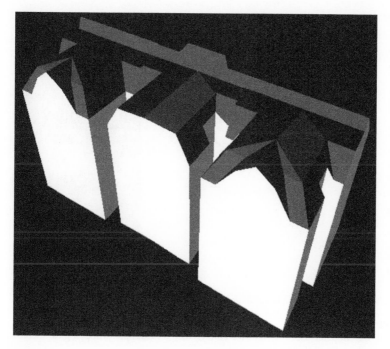

FIGURE 14.9
Correction of roof group by automated geometry regularization.

14.5.2 Topology Adjustment

Inconsistencies in topology between adjacent buildings may arise because of measurement errors and mutually overlapping roofs.

Figure 14.11 shows a typical topology problem, which may exist even after the previous geometry regularization. For its solution, we provide both an automated and a semiautomated procedure.

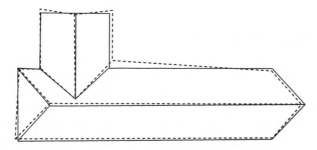

FIGURE 14.10
Line rectification (dotted line: before, solid line: after).

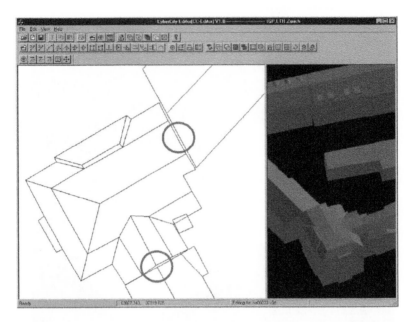

FIGURE 14.11
CC-Edit: CC-Modeler user interface for editing with an example of topological inconsistency between adjacent buildings.

In the automated mode, the system selects a reference border line, which is kept fixed and onto which the points of the other lines are projected perpendicularly. As reference line, the software selects the longest line (which is supposed to be the most stable). In the semiautomated mode, this reference line is selected manually.

The functioning of the automated and semiautomated procedures described here can be monitored by an operator within an editing window as shown in Figure 14.11. This has of course a certain similarity with a CAD interface. It actually contains many typical CAD functions, but also others that are unique to our system and application-related. An example of automatic topology correction is shown in Figure 14.12.

14.5.3   Building Facade Integration

The aim is a higher level of detail in building modeling. Since facades are in general not visible in aerial images we use digital cadastral maps, which show the outer walls of buildings as part of the legal definition of real-estate property. By integrating this information into the roof landscape, we are able to model the roof overhangs. What sounds like a simple problem at first sight turns out to be a formidable task to automate. In terms of structural detail, the roof landscape looks very different from the facade landscape. Sometimes the maps are outdated and the roofs do not match the map

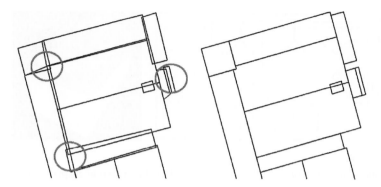

FIGURE 14.12
CC-Edit: Result of editing with an example of false (*left*) and automatically corrected topology (*right*).

content at all. Maps may also be inaccurate to an extent that the facade appears shifted and rotated with respect to the roof by a substantial amount. Facades can show a lot of additional, peripheral details, as for instance stairs and other add-ons (Figure 14.13).

Figure 14.14 shows a result of automated facade integration. The problem is not yet solved in general terms and still needs some manual interference in complex situations. We will report about technical details of our approach in another publication.

Beyond facade integration, also other vertical wall sections, as they may appear on parts of a building and not be available from maps, need to be explicitly modeled as faces. This holds for all vertical building sections that do not constitute the legal building boundary. We have also developed a solution for this problem based on the intersection of gutter point projections onto other building parts like roofs, balconies, and terraces.

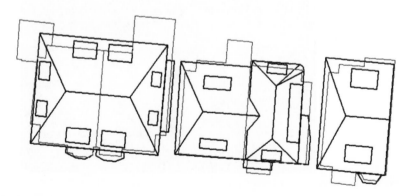

FIGURE 14.13
Plan view of roof landscape (dark) and the related facade representation from a map (light).

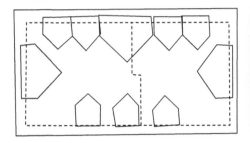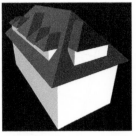

FIGURE 14.14
Automated facade integration. *Left*: Plan view of roofs and facades from map; *right*: integration
result.

14.6 Attribute Information and Connection to GIS

Wang and Gruen [8] proposed a solution for a 3D spatial information
system. It was developed as a prototype system for scientific investigations.
Within CC-Modeler (CC-Edit), attributes can be defined for geoencoding or
for adding attribute information for material, and so on. Figure 14.15 shows
this functionality.

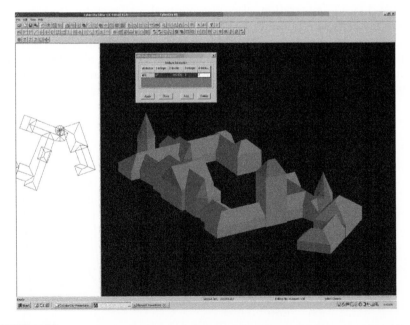

FIGURE 14.15
3D Spatial information system. Attributes can be entered within CC-Edit.

Additionally, CC-Modeler allows to compute attributes like volumes and areas for planning purposes. Even the volume of basements may be included optionally.

14.7 Texture Mapping and Visualization/Simulation

In most applications, images are used as realistic texture data. Photorealistic texturing applied to 3D objects gives the most natural representation of the real world. Texture presents details that are not modeled in the vector dataset and gives information about material properties. Therefore, for visualization purposes, images may even compensate for lack of modeling of fine details.

Generally, there are two types of data sources: aerial images and terrestrial images taken from street level. The former are usually used for mapping on terrain surfaces and roofs of buildings whereas the latter are for building facades and other vertical faces. From a data structure point of view, both kinds of images are expressed as 2D raster data, which can be stored or manipulated as a special layer in our 3D system. Visualization of 3D city models becomes a key issue when dealing with user requests. The best model is not worth too much if the user cannot look at it at reasonable speed. Here we must clearly distinguish between real-time and snail-time visualization requirements. Snail-time performance is acceptable, if images are produced for publications and the like. However, in most applications real-time capabilities are requested. This puts us into the somehow uncomfortable situation, that, although there are visualization programs available on the international market to the hundreds (compare www.tec.army. mil/TD/tvd/survey/survey_toc.html), only very few have real-time performance even with very large datasets. To make it clear, a large dataset in our applications starts with about 10,000 buildings of very fine detail and tenths of MB DTM vector data plus over GB of real-image texture. Synthetic image texture, although faster to handle, is of vanishing interest. We have seen datasets with more than 300,000 fine-detailed buildings.

For high-end performance level-of-detail (LoD) capabilities for both vector and image data are indispensable. LoD provides for on-the-fly switching between several resolution levels (three are mostly sufficient), depending on the viewing distance. With this functionality and sufficient host and graphics memory and an appropriate, but still standard graphics board, even laptops can handle very large datasets in real time.

In our group, we are using packages like Cosmo Player (for very small datasets), AutoCAD, Microstation, Inventor/Explorer (SGI), ERDAS VirtualGIS, Maya (Alias Wavefront), Terrainview (viewtec), Skyline (idc), and a variety of self-developed software.

Modern visualization software not only shows the "naked" model, but also allows for features like import of various standard data formats, preparation of interactive or batch-mode flyovers and walkthroughs, generation of videos, integration of text information, definition of various layer systems above terrain, search functions for objects, coupling of information in different windows, import of synthetic textures, integration and manipulation of active objects (e.g., clouds, fog, multiple light sources, cars, people, etc.), hyperlink functions for the integration of object properties, export via Internet/Intranet/CD/DVD, and so on.

New tools like Google Earth, Microsoft Virtual Earth, and NASA Wind have become available recently, making 3D landscape and city models of virtually the whole earth instantly accessible worldwide. Although high-quality 3D city models do not exist, yet on these platforms it is only a matter of time until they become available.

14.8 Generation of 3D City Models from Linear Array CCD Aerial Cameras

Linear array CCD-scanners are nowadays a prime technology for space and large format aerial image sensing. This new technology requires the development of novel sensor and trajectory models for precision processing. We have developed lately a suite of new algorithms and the related software for the processing of this kind of data (for details, see Refs. [13,14]). Here, we specifically address the issue of 3D city modeling. We combine the semi-automated object extraction method of CC-Modeler with the new sensor models of CCD-linear array cameras. In our examples, we use images from Three-Line-Scanner (TLS) and SI-200 (Starimager-200), both from STAR-LABO Corporation, Tokyo. This approach can also be applied to other sensors of similar type. We have interfaced TLS, SI-200 and IKONOS, QuickBird data with CC-Modeler functionality and have produced several datasets over Yokohama and Ginza, Tokyo, Japan, Izmir, Turkey, Phoenix, United States, and others. We report briefly about the current status of TLS and SI-200 functionality and we describe the related generation of city models. We show high-resolution phototextured models of Yokohama, including buildings and objects like street lamps, roads, waterways, parking lots, bridges and trees, and Ginza.

For our work with high-resolution satellite images, see Kocaman et al. [10]. The combination of two modern technologies from sensing and processing opens interesting perspectives for future applications in 3D virtual environment generation.

The TLS system is an aerial multispectral digital sensor system, developed by STARLABO Corporation, Tokyo. It uses the TLS principle to capture digital image triplets in along-strip mode. It can be used either in three-image panchromatic mode or in one-image multispectral (RGB) mode. The imaging system contains three times three parallel one-dimensional CCD focal plane arrays, with 10,200 pixels of 7 μm each. The TLS system produces seamless high-resolution images with usually 5–10 cm footprint

TABLE 14.1

TLS and SI200 Sensor Parameter

	TLS	SI-200
Focal length	60.0 mm	65.0 mm
Number of pixels per array	10,200	14,404
Pixel size	7 μm	5 μm
Stereo CCD arrays	3	3
Multispectral CCD lines	3 RGB	3 RGB + 1 infrared
Stereo view angle	21°	21°/30°[a]
Field of view	61.5°	68.0°
Scan line frequency	500 Hz	500 Hz

[a] Forward–nadir/nadir–backward stereo view angle.

on the ground with three viewing directions (forward, nadir, and backward). There are two configurations for image acquisition. The first configuration ensures the stereo imaging capability, in which the three CCD arrays working in the green channels are read out with stereo angles of about 21°. The second configuration uses the RGB CCD arrays in nadir direction to deliver color imagery. STARLABO has also developed another camera system, called SI-200 (STARIMAGER-200). This comes with an improved lens system and with 10 CCD arrays on the focal plane (3×3 work in RGB mode, one CCD array works in infrared mode). Each CCD array consists of 14,404 pixels at 5 μm size. All 11 channels can be read out simultaneously, producing threefold overlapping color imagery plus one IR channel. For the detailed sensor and imaging parameters, see Table 14.1.

In order to get highly precise attitude and positional data over long flight lines, a combination of a high local accuracy INS with the high global accuracy GPS is exploited. An advanced stabilizer is used to keep the camera pointing vertically to the ground to get high-quality raw-level images and outputs attitude data at 500 Hz. A Trimble MS750 serves as Rover GPS and collects L1/L2 kinematic data at 5 Hz and another Trimble MS750 serves as Base GPS on the ground. The rover GPS is installed on top of the aircraft and the INS and the TLS camera are firmly attached together.

14.8.1 Application Software Development

The application software has been developed by our group at the Institute of Geodesy and Photogrammetry, ETH Zurich. The processing modules include:

- *User interface and measurement system*: The user interface allows the display, manipulation, and measurement of images. It includes the mono and stereo measurement modules in manual and semiautomated mode. It employs large-size image roaming techniques to display the TLS forward, nadir, and backward (plus other channels if possible) view direction images simultaneously.

- *Triangulation*: For high-accuracy applications we recommend a previous triangulation. The related software is a modified bundle adjustment called TLS-LAB. We have developed a special TLS camera model and offer three different trajectory models (DGR, Direct Georeferencing Model; PPM, Piecewise Polynomial Model; and LIM, Lagrange Interpolation Model). For more details and results of several accuracy tests, see Gruen and Zhang [13–14]. The self-calibration technique for systematic error modeling has also been implemented.

- *Image rectification*: Rectification comes in two modes. The coarse version just uses the orientation elements as given (or derived from triangulation) and projects the raw images onto a predefined horizontal object plane. The refined version uses an existing DTM (of whatever quality) in replacement of the object plane. This latter method reduces the remaining y-parallaxes substantially.

- *DSM/DTM generation*: We have devised and implemented a new matching strategy for the automatic generation of Digital Surface Models. This strategy consists of a number of matching components (cross correlation, least-squares matching, multi-image matching, geometrical constraints, edge matching, relational matching, multipatch matching with continuity constraints, etc.), which are combined in particular ways to respond to divers image contents (e.g., feature points, edges, textureless areas, etc.). The matching module can extract large numbers of mass points by using multi-images. Even in nontexture image areas, reasonable matching results can be achieved by enforcing local smoothness constraints. For more details on matching, see Gruen and Zhang [13].

- *Orthoimage generation*: This is a special solution for fast derivation of orthoimages given the TLS/SI-200 geometry and images.

- *Feature and object extraction*: Building extraction is based on CC-Modeler. Another new technique, which operates at a higher level of automation, is under development. In addition, for road extraction we have different semiautomatic and fully automatic approaches at our disposal, which however still have to be adjusted to the particular sensor geometry of linear array images.

14.8.2   3D City Modeling with TLS/SI-200 Images and CyberCity Modeler

There are several advantages in using TLS/SI-200 imagery to derive a 3D city model. Firstly, very high-resolution seamless image data (3–10 cm ground resolution) can be obtained by installing the system on a helicopter. Several multispectral channels (RGB, infrared) are available simultaneously. Secondly, unlike with the traditional frame-based photography, the three-line geometry is characterized by nearly parallel projection in flight direction and perspective projection perpendicular to it (so-called line-perspective projection). This results in the following advantages compared with single-frame camera data:

1. More pixels per frame (larger area coverage)
2. Triple or more overlap (more redundancy)
3. Orthogonal in flight direction (minimal occlusions, true orthoimages are of better quality)
4. PAN and MS simultaneously acquired at same resolution
5. Building facades are better visible (impact on geometry and texture)

In the TLS/SI-200 system, a stabilizer is used to absorb the high-frequency positional and attitude variations of the camera during the flight to get high-quality raw-level images. Furthermore, the stabilizer always keeps the camera pointing nearly vertically to the ground. This results in minimal occlusions in the nadir view images. In addition, the image information of the building's facades are recorded in shortened form in the forward and backward view images (Figure 14.16).

With the TLS/SI-200 stereoscopic measurement software the buildings, roads, and other kinds of man-made objects can be measured manually or semiautomatically. The measurement procedure must follow the regulation of CC-Modeler, such that it can process TLS/SI-200 data directly.

Since the input data of CC-Modeler are just point clouds, it does not matter which sensor model is used to construct the 3D vector model.

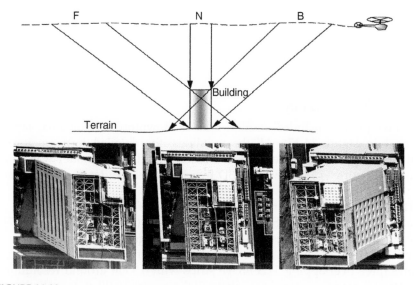

FIGURE 14.16
TLS principle: forward, nadir, and backward view images.

However, the sensor model must be identified if the full 3D model with texture mapping is required. In this case, the necessary modification of CC-Modeler is to extend the sensor model from the normal frame perspective projection to the line-perspective projection of the TLS/SI-200 system. The software "CC-TLSAutotext" does the texture mapping with TLS/SI-200 images as the original data source. In "CC-TLSAutotext," the procedure of texture mapping is to project the object faces from 3D space onto the TLS images (forward, nadir, and backward view) and take the image patch that has the best resolution. The best resolution is equivalent to the largest related image patch size. However, considering the possible occlusions between 3D objects, the best texture may not be contained in the patch with the best resolution. It could be the one with the highest amount of completeness or could be an image mosaic with texture patches that are from different TLS images. Therefore, an occlusion-checking procedure has to be involved. In case of occlusions, the user has three options: (1) paste partial patches from different images together, (2) use terrestrial images captured on the ground, and (3) randomly take artificial textures. In case of full occlusions, the procedure uses the artificial texture or manual texturing. Good natural texture mapping is one of the most demanding tasks in city modeling. There is much room for improvement and increase in efficiency.

14.8.3 Examples

We report here about two test projects, using TLS and SI-200 linear array sensor imagery, with the purpose of demonstrating the feasibility of our procedures:

1. Yokohama city, two selected subareas; TLS imagery
2. Ginza, Tokyo; SI-200 imagery

14.8.3.1 *Yokohama City*

The first project includes a small area in downtown Yokohama, Japan (Area 1 in Figure 14.17). All the buildings, the detailed infrastructures, main roads, and some trees were measured and a 3D model was constructed. Figure 14.18 shows the 3D hybrid model, rendered with Cosmo Player.

The second project represents a larger area of about 1.5 km², with a boulevard in front of the Shin-Yokohama Station (Area 2 in Figure 14.17). The whole model includes 2482 houses, 26 bridges, 20 road segments, 1 river, 129 trees, 8 electric power lines, and 170 street lights. Figure 14.17 shows three overlapping TLS image strips with 6.5 cm ground resolution. For texture-mapping purposes, about 100 terrestrial still video photos of some high buildings were used.

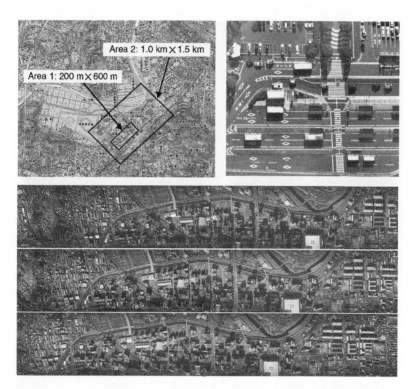

FIGURE 14.17
Experimental area of Yokohama (*Upper left*: area overview; *upper right*: zoom-in; *lower part*: three TLS strips).

After the triangulation procedure with several ground control points, a DTM was automatically generated with the TLS image matching module and some editing efforts, and a 0.25 m resolution orthoimage was generated with the TLS image rectification module.

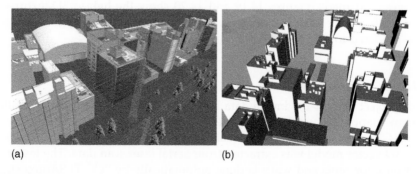

(a)   (b)

FIGURE 14.18
(a) Yokohama, area 2. View onto the reconstructed 3D model. Power lines are also modeled.
(b) Yokohama, area 2. Detailed roof structures of the 3D model.

FIGURE 14.19
Textured 3D model of a small area in downtown Yokohama.

We use the stereoscopic measurement module as the measurement platform to measure the point clouds, following the regulation of CC-Modeler. The integration of CC-Modeler and the stereo measurements is crucial for the 3D model generation. After measuring all the objects, CC-Modeler is employed to construct the 3D model. Figure 14.18a and b show views on the reconstruction results. The level of detail in reconstruction can be checked by the roof structures of Figure 14.18b.

Finally, the texture-mapping procedure for all measured objects is carried out with CC-Modeler's extended module "CC-TLSAutotext." In this procedure, the orthoimage mosaic is mapped onto the DTM, and the high-resolution image patches are mapped onto the 3D objects such as houses, bridges, and roads. Figures 14.19 and 14.20 show the hybrid 3D-textured models. In Figure 14.20, a background static image with sky and clouds is also rendered to achieve a more realistic effect.

14.8.3.2 Ginza, Tokyo

This project uses SI-200 image data and aims at creating a 3D model with texture mapping along the main street of downtown Ginza, Tokyo. The area of the Ginza project is 2.0 km^2 with about 600 houses and 15 road segments. The 3D vector model was extracted from aerial laser scan data. The texture mapping for roofs and walls is done automatically by "CC-TLSAutotext," using only aerial image data. This explains the somehow blurred impression of facade texture. Figure 14.21 shows a view onto the 3D model.

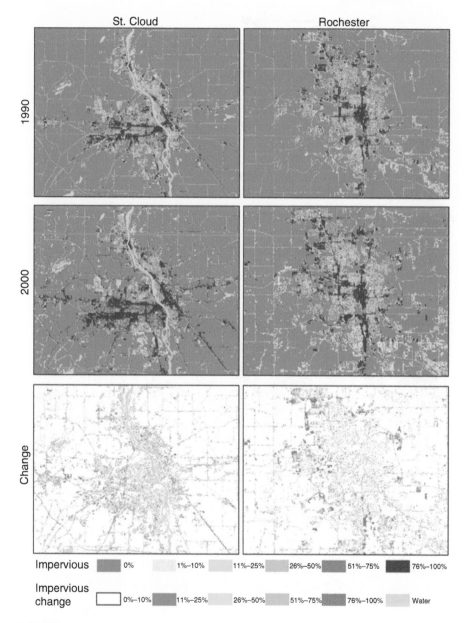

St. Cloud   Rochester

1990

2000

Change

Impervious ▮ 0% ▮ 1%–10% ▮ 11%–25% ▮ 26%–50% ▮ 51%–75% ▮ 76%–100%

Impervious
change ▢ 0%–10% ▮ 11%–25% ▮ 26%–50% ▮ 51%–75% ▮ 76%–100% ▮ Water

FIGURE 1.8
Impervious classifications of St. Cloud and Rochester for 1990 and 2000 and change maps.
(From Bauer, M., Loeffelholz, B., and Wilson, B., *Proceedings, Pecora 16 Conference, American Society of Photogrammetry and Remote Sensing*, October 23–27, 2005, Sioux Falls, South Dakota, 2005. With permission.)

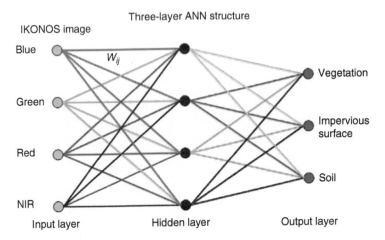

FIGURE 2.3
Artificial neural network structure.

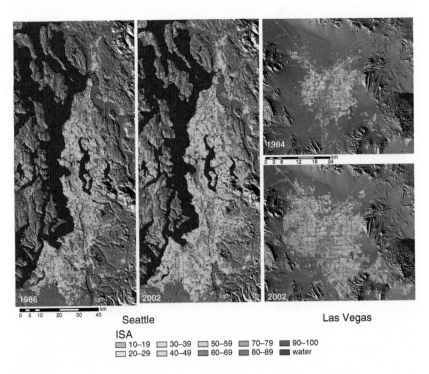

FIGURE 3.3
Impervious surfaces in 1986 and 2002 in Seattle, 1984 and 2002 in Las Vegas.

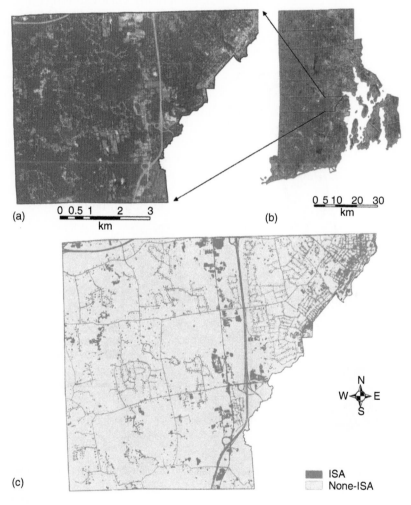

(a)

0 0.5 1   2   3
km

(b)

0 5 10 20 30
km

(c)

N
W ← → E
S

■ ISA
□ None-ISA

FIGURE 5.5
An example of true-color digital orthophoto data with 1 m spatial resolution for the East
Greenwich, Rhode Island (a, b) and the ISA extracted by MASC modeling (c).

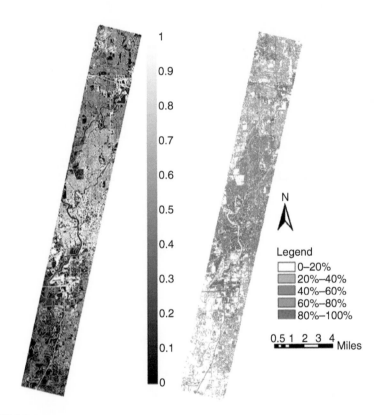

FIGURE 6.3
Impervious surface image derived from ALI image. The color figure to the *right* shows the distribution of impervious surface at four categories.

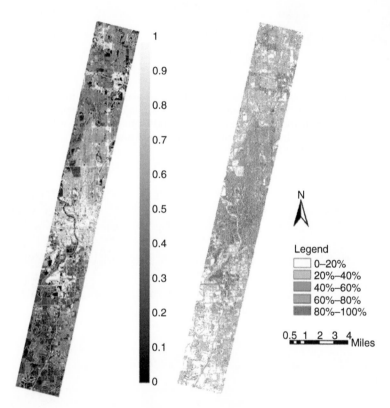

FIGURE 6.4
Impervious surface image derived from Hyperion image. The color figure to the *right* shows the distribution of impervious surface at four categories.

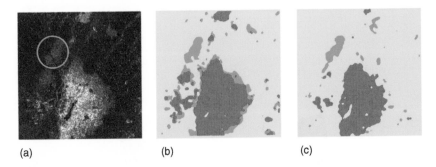

(a)   (b)   (c)

FIGURE 8.3
Classification maps for SAR data classification of human settlements in the area around the town of Al Fashir (Sudan). Three land-use classes are used: informal settlements (light green), formals settlements (red), rocks and bare soil (yellow). (a) Original SAR data with the tent camp highlighted by a green circle; (b) map from classification of SAR textures; (c) map from joint classification of SAR and SPOT textures.

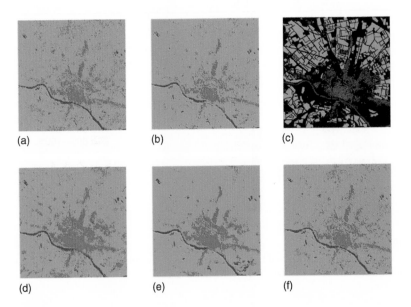

(a)   (b)   (c)

(d)   (e)   (f)

FIGURE 8.6
Classification maps of the three SAR available images obtained with ARTMAP (a) and with the MRF approach (b); the land-cover ground truth used to evaluate classification results (c); classification maps of the joint classification of the LANDSAT image of 21 June and the three SAR images with ARTMAP (d); and MRF (e); classification map of the three SAR available images obtained with spatial ARTMAP (f).

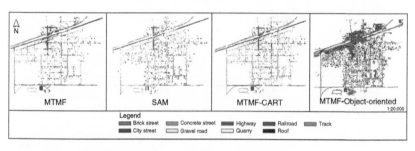

FIGURE 9.2

Comparison of classifications.

FIGURE 11.11

Final result after the extraction of intersections.

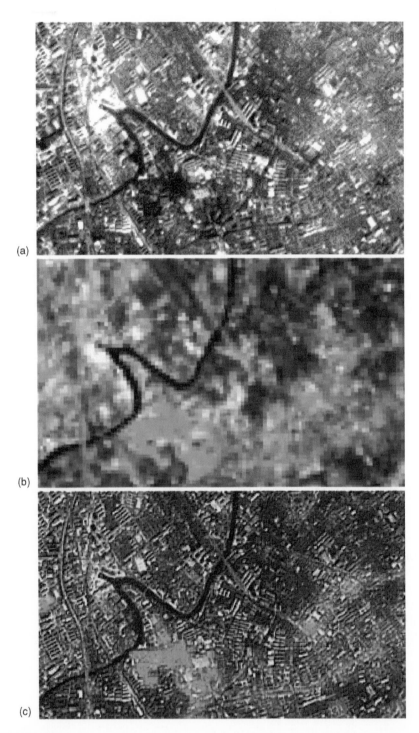

FIGURE 16.1
Medium-resolution satellite images for roof mapping (300 × 180 pixel section, 10 m pixel).
(a) SPOT Pan image; (b) TM bands 3, 4, and 7 in blue, green, and red; and (c) fused TM-SPOT
image.

FIGURE 16.6
QuickBird multispectral image of suburban scene in Oromocto (subset, 2.8 m resolution).

FIGURE 16.8
Pan-sharpened QuickBird image of the suburban scene produced from Figure 16.6 and the corresponding QuickBird Pan image (subset, 0.7 m resolution).

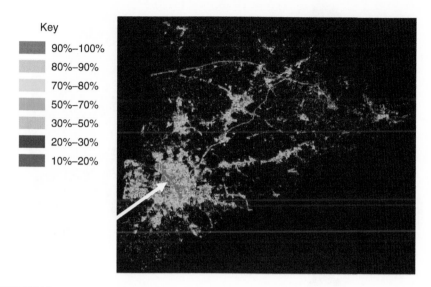

Key

- 90%–100%
- 80%–90%
- 70%–80%
- 50%–70%
- 30%–50%
- 20%–30%
- 10%–20%

FIGURE 17.5
Satellite-derived (Landsat) analysis of ISA for the Conestoga Watershed in eastern Pennsylvania for the year 2000. Arrow denotes the town of Lancaster referred to in Figure 17.6.

FIGURE 17.6
Photograph of the Conestoga River in the town of Lancaster, Pennsylvania (arrow in Figure 17.5) taken in the year 2004 near the location denoted by arrow in Figure 17.5.

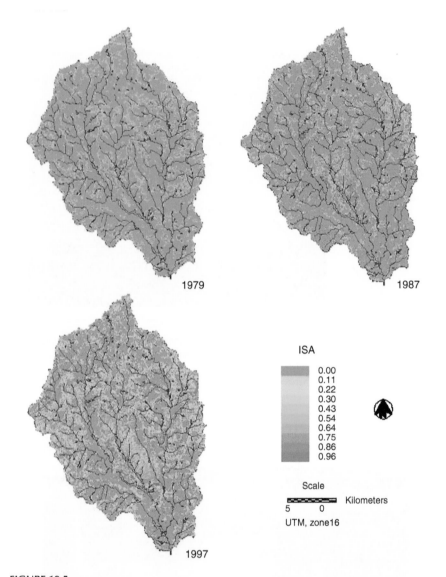

FIGURE 19.5

ISA (0–1 representing 0%–100%) maps for the entire Peachtree Watershed for the years 1979, 1987, and 1997. Color-coding represents the degree of ecological impact, as outlined by Schueler (1994). Overlaid in blue is the digital hydrography of the Peachtree watershed. (From Gillies, R.R. et al., *Remote Sensing Environ.*, 86, 441, 2003. Copyright 2003. With permission from Elsevier.)

FIGURE 14.20
An overview of the whole textured 3D model of the Yokohama project.

14.9 Conclusions

The efficient generation of 3D city models is an important task, both from a scientific and practical point of view. City models are already used in many

FIGURE 14.21
Textured 3D model of Ginza, Tokyo.

applications and we see a steady growth of users. City modeling is maturing from a niche application into a formidable market. The growth potential is immense. Synchronous with the broadening of the application base goes the need for higher resolution, both in terms of geometrical and textural detail.

Aerial photogrammetry plays an important role as a flexible and economic technique for data generation. Semiautomated photogrammetric techniques are available for efficient data production. Satellite images however, even in high-resolution mode, do not yet show enough detail for detailed and reliable modeling.

Terrestrial images, on the other hand, are already now relevant for facade texture generation, and in the future possibly also for facade geometry modeling and the recording of other objects that are not accessible from the air.

To what extent aerial laser scan data can be integrated into the city-modeling process has to be determined. Terrestrial laser scanners already play a significant role in the modeling of indoor scenes and also in high-resolution modeling of landmarks, for example, for 3D car navigation. Here, terrestrial laser scanning is combined with terrestrial images.

In the future, we will see more and more combined approaches, where images and laser scans are used, both from the air and in the terrestrial mode.

Much of the current discussion is centered around the generation of virgin databases. It would be appropriate however to discuss new methods for data maintenance. The fast pace with which our man-made environment is changing will also require innovative techniques for the updating of 3D city models.

References

1. Baltsavias, E. and Gruen, A., Resolution convergence—A comparison of aerial photos, LIDAR and IKONOS for monitoring cities, in *Remotely Sensed Cities*, Mesev, V., ed., Taylor & Francis, London, 2003, pp. 47–82.
2. *International Archives of Photogrammetry and Remote Sensing*, 31, B3, Proceedings of the XVIIIth Congress of the ISPRS, Vienna, 9–19 July 1996.
3. Foerstner, W. and Pluemer, L., *Semantic Modeling for the Acquisition of Topographic Information from Images and Maps*, Proceedings of the Workshop "SMATI'97," Birkhäuser Verlag, Bonn Bad Godesberg, 1997.
4. Gruen, A., Baltsavias, E., and Henricsson, O., *Automatic Extraction of Man-Made Objects from Aerial and Space Images (II)*, Proceedings Workshop, Monte Vérita, Birkhäuser Verlag, Basel, Switzerland, 1997.
5. Special Issue on Automatic building extraction from aerial images, in *Computer Vision and Image Understanding*, 72, 2, Gruen, A. and Nevatia, R., Guest eds., 1998.
6. Foerstner, W., 3D-City models: Automatic and semi-automatic acquisition methods, in *Proceedings Photogrammetric Week '99*, H. Wichmann Verlag, Heidelberg, 1999, p. 291.

7. Gruen, A., Semi-automated approaches to site recording and modeling, Invited Paper, in *International Archives of Photogrammetry, Remote Sensing and Spatial Information Sciences*, 33, 5/1, 2000, 309.

8. Wang, X. and Gruen, A., A hybrid GIS for 3-D City Models, *International Archives of Photogrammetry, Remote Sensing and Spatial Information Sciences*, 33, 4/3, 2000, 1165.

9. Baltsavias, E., Gruen, A., and van Gool, L., *Automatic Extraction of Man-Made Objects from Aerial and Space Images (III)*, Balkema, Lisse, 2001, p. 415.

10. Kocaman, S. et al., 3D City modeling from high-resolution satellite images, in *International Archives of Photogrammetry, Remote Sensing and Spatial Information Sciences*, 6, 1/W41, 2006, on CD-ROM.

11. Gruen, A. and Wang, X., CC-Modeler: A topology generator for 3-D City Models, *ISPRS Journal of Photogrammetry & Remote Sensing*, 53, 5, 1998, 286.

12. Gruen, A. and Wang, X., News from CyberCity Modeler, in *Proceedings Workshop "Automatic Extraction of Man-Made Objects from Aerial and Space Images (III),"* Monte Verita, Baltsavias, E., Gruen, A., and Van Gool, L., eds., Balkema, Lisse, 2001, p. 93.

13. Gruen, A. and Zhang, L., Automatic DTM generation from Three-Line-Scanner (TLS) images, in *International Archives of Photogrammetry, Remote Sensing and Spatial Information Sciences*, 34, 2A, Graz, Austria, 2002, 131.

14. Gruen, A. and Zhang, L., Sensor modeling for aerial triangulation with Three-Line Scanner (TLS) imagery, *Journal of Photogrammetrie, Fernerkundung, Geoinformation* (PFG), 2, 2003, 85.

15

SAR Images of Built-Up Areas: Models and Data Interpretation

Giorgio Franceschetti, Antonio Iodice, and Daniele Riccio

CONTENTS

15.1   Introduction and Motivations

Monitoring of the urban environment is a key issue within the framework of modern and efficient management of Earth resources. Urban areas represent a very critical part of our planet due to the extremely high density of population and the continuous change of its environment. Spaceborne remote sensing instruments provide frequently updated and relatively inexpensive data to monitor and possibly plan urban area development and optimal resources distribution [1,2]. In particular, synthetic aperture radar (SAR) images provide all-day, all-weather, synoptic views whose potentiality has been only partly explored. Accordingly, electromagnetic models, aimed to quantitatively explore the information content in SAR images of urban areas, are required [3–5] to provide a sound analytical background to devise interpretation tools able to recover value-added information from the available SAR images [6–9].

On qualitative basis, SAR images of urban areas are characterized by a very typical texture: roads and other not built-up areas are often clearly recognized because they appear as relatively dark features; conversely, buildings generate complicated features on the images with alternate extremely bright and dark areas. Models to understand the information content in building-free areas are available, as they are essentially coincident with those pertinent to natural areas. These areas are modeled as rough surfaces, where the classical SAR image formation mechanism takes place: then, each element of the SAR image exhibits a well-known dependence on a few surface geometric and electromagnetic parameters. Conversely, models to represent the built-up areas must deal with multiple-scattering phenomena as well as severe geometric distortion including shadowing and layover: in this case, each element of the SAR image exhibits a very complicated dependence on several surface geometric (building's dimensions and orientation, surrounding terrain classical or fractal roughness parameters) and electromagnetic (dielectric constants) parameters, and its geocoding turns out to be almost impossible.

For a quantitative, systematic, meaningful, and efficient approach, it is therefore convenient to consider a canonical urban scene composed of canonical elements. A list of these elements would include: flat (at the electromagnetic wavelength scale) surfaces, representing man-made pavements as well as buildings' flat walls and roofs; rough surfaces, representing terrains, as well as buildings' nonflat walls and roofs; dihedrals, each one representing an isolated building surrounded by a flat background; trihedrals, each one representing not isolated buildings whose walls are not aligned, as well as building balconies; canyons, representing not isolated buildings aligned along two sides of a street. Modeling all these elements would provide a complete handbook for quantitative analysis of SAR data relevant to urban areas. In this chapter, rationale and results for a sound electromagnetic SAR modeling of the dihedral canonical elements are detailed. The dihedral hereafter examined represents a single dielectric building surrounded by a dielectric background: the radar return from this canonical element is evaluated in closed form in terms of the scene geometric and radiometric parameters as well as the radar characteristics and the acquisition geometry.

A detailed derivation of results, even on this relatively simplified geometry item, is out of the scope of this chapter. We prefer to include all the relevant results, arranged in the form of a handbook, along with some useful comments to use it to clarify the applicability of the presented material. The quantitative discussion is referred to some meaningful cases presented in a section along with numerical results. Details to obtain these final results and their intermediate achievements can be found in the quoted literature.

To close this section, it is convenient to list the main characteristics of the electromagnetic models presented in this chapter.

The models discussed in this chapter are appropriate for airborne and the upcoming generation of high-resolution spaceborne SAR sensors: moreover, they make use of a description of the urban scene based on raster (topography and dielectric constants of the area) as well as vectorial (geometry and dielectric constants of the buildings) data; finally, they include relevant radar and mission parameters.

The presented *direct* scattering and radar models rely on a sound physical and mathematical basis, thus providing an innovative track and quantitative means to devise *inverse* algorithms for information retrieval from urban areas SAR data, as well as to identify the optimum SAR sensor configuration and operational mode (look angle, radar polarization, altitude, etc.), which maximizes the information content in the corresponding SAR images. With respect to any numerical approach, the presented radar and scattering models are easily amenable to efficient implementations in appropriate computer codes.

15.2 SAR Image Model

In this section, radar models are presented to highlight the dependence of the SAR image on the radar parameters.

The reference coordinate system makes use of x and r, the azimuth and slant range coordinates referring to the sensor trajectory (see Figure 15.1). This azimuth–slant-range coordinate system is the one that best matches the SAR sensor functional mode but does not easily cope with any geocoded representation of the SAR images. This representation would benefit by an azimuth–ground-range representation instead. As a matter of fact, SAR images are intrinsically affected by typical distortions: foreshortening, layover, and shadowing take place and, especially for images of urban areas, deeply affect the SAR data.

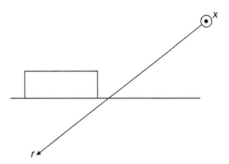

FIGURE 15.1
The azimuth–slant-range coordinate system.

By assuming a perfect focusing of the SAR raw data, the SAR complex image turns out to be expressed as [10]

$$i(x',r') = \iint \gamma(x,r)\exp\left(-j\frac{4\pi}{\lambda}r\right)\text{sinc}\left[\frac{\pi}{\Delta x}(x'-x)\right]\text{sinc}\left[\frac{\pi}{\Delta r}(r'-r)\right]\,dx\,dr,$$

(15.1)

where
- x' and r' are the independent coordinates employed for the SAR image
- $\gamma(x,r)$ is the scene reflectivity function, whose values are proportional to the coherent sum of all the fields backscattered by points located at azimuth x, and whose complete propagation path (sensor trajectory–scene–sensor trajectory, also including any multiple reflection on scene elements) is equal to $2r$
- λ and f are the carrier wavelength and frequency, respectively, of the transmitted signal
- $\text{sinc}\,(t) = [\sin\,(t)]/t$
- $\Delta x = \dfrac{L}{2}$ and $\Delta r = \dfrac{\lambda}{2}\dfrac{f}{\Delta f}$ are defined as the azimuth and range spatial resolutions, respectively, where L is the azimuth dimension of the sensor antenna and Δf is the bandwidth of the transmitted chirp
- Twofold integration is extended over the illuminated area

Determination of the image representation (Equation 15.1) assumes that the SAR raw data are continuously acquired as function of $x' = vt'$ and $r' = c(t' - t_n)/2$, where t' is the acquisition time variable, t_n is the time of transmission of the nth pulse, and c is the speed of light in the vacuum. This is not the case due to the pulsed operation of the SAR system. However, the representation (Equation 15.1) simplifies the notations and helps the interpretation of the results and is formally justified on the basis of the sampling theorem [10]. Accordingly, the quoted representation is the best choice for any theoretical analysis. However, it should be stressed that along all the processing chain, both the raw signal and the SAR image are appropriately sampled and quantized to get numerical representations amenable to computer processing and displaying via software codes and graphical tools. The quantized SAR image, normally coded on a gray-level scale, is usually obtained by considering the (square) modulus of the numerical version of the complex image (Equation 15.1). The SAR image is represented by superposition of sampled sinc($\cdot$) functions, each one centered over a resolution cell and holding an amplitude proportional to the radar cross section (RCS) of the cell as sensed by the SAR instrument.

15.3 Models for the Single Building Canonical Element in the SAR Image

According to Equation 15.1, the evaluation of the SAR image, including the previously mentioned canonical dihedral elements representing the buildings, can be performed as detailed hereafter.

First, the scene reflectivity map is evaluated as a function of the geometric and electromagnetic parameters of the elements constituting the scene; this step is accomplished by taking into account the SAR orbit and radar parameters. Then, the proper superposition of all the contribution from the elements of the scene is computed; this step is accomplished by taking into account the SAR resolutions.

The first step requires as input data the descriptions of sensor, orbit, terrain, and building data. Buildings are modeled as dielectric prisms lying on a rough surface, the latter being in principle of infinite extent. Accordingly, the scene geometric description is provided by the coordinates of vertices of the buildings' projection over the horizontal plane, the buildings' heights, and the coordinates of the terrain surrounding them; the electromagnetic description is provided by the permittivity and conductivity of the buildings' walls and the terrain surrounding them. The scene geometrical profile is represented by rectangular facets that are smaller than the SAR resolution, but much larger than wavelength; the reflectivity map of the scene, $\gamma(x,r)$, is obtained by appropriately summing up the radar returns from the illuminated facets within the scene. The facets can belong to the terrain surrounding the buildings, to the buildings' roof, or to the buildings' walls. A ray-tracing procedure is used to identify shadowed facets [11,12].

When a plane electromagnetic wave impinges on such a canonical structure, *single*-scattered contributions, shortly *single returns*, come back to the sensor, related to the backscattering properties of the ground, the buildings' walls and roof (see Figure 15.2a). It must be taken into account that facets belonging to the buildings' vertical walls usually appear almost smooth at

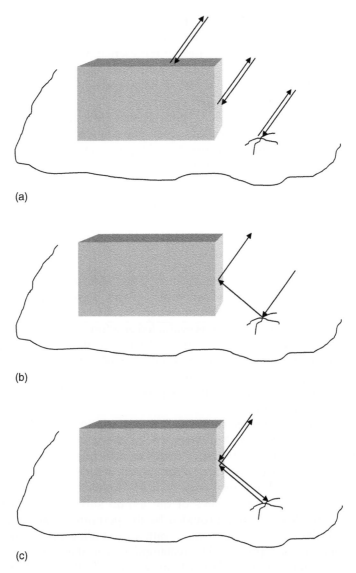

(a)

(b)

(c)

FIGURE 15.2
Ray path relevant to scattering contributions: (a) single; (b) double; (c) triple.

the radar frequencies ignoring any balcony that can be viewed as very small dihedral elements. Then, multiple reflections between each illuminated wall of any building and the terrain surrounding it can provide significant contributions to the electromagnetic field backscattered toward the SAR instruments. Part of the electromagnetic wave is *double*-scattered from the ground to the wall (and vice versa), and then sent back to the sensor, forming the *double returns* to the scattered field (see Figure 15.2b). Further contributions are by triple-scattering phenomena, the *triple returns* from walls and ground (see Figure 15.2c).

Closed-form solutions for the single-, double-, and triple-scattered components can be obtained in the phasor domain, that is, assuming time harmonic signals; for SAR employing pulsed radar [13], these closed-form solutions can be applied in an appropriate mixed slant range–phasor domain, which addresses the temporal issues related to the ray path of each electromagnetic wave contribution by distributing any radar return in the appropriate slant range cell [4].

A general consideration of the electromagnetic model's limits of validity is in order. Closed-form scattering solutions can only be obtained by employing appropriate approximations that are valid only under appropriate roughness regimes for the nonflat facets. In almost any practical application, SAR models relevant to urban areas can be safely dealt with using physical optics (PO) and geometrical optics (GO) scattering models.

The single returns are obtained by considering the electromagnetic field backscattered by each illuminated facet: they are evaluated as in Ref. [11] as a function of the mean facet slope, roughness, and electromagnetic parameters and by taking into account transmitting and receiving polarizations and the incidence angle. Depending on the facet surface roughness, geometrical optics or physical optics are employed [14]. Single returns are related to ray paths of different lengths and must be summed up in the appropriate slant-range cell.

The double returns are considered by referring to a ray that runs the path S-A-B-S, S indicating the radar position (see Figure 15.3); its optical path length is equal to S-O-S relevant to the ray that is backscattered at point O. This result is independent of the location of points A and B, the incidence angle and the building height. Hence, the returns corresponding to all the double-scattered rays hold the same time delay, equal to that of the single-scattering return from point O. Then, double-scattering contribution is added to the reflectivity map $\gamma(x,r)$, after being concentrated at a slant-range distance r_O.

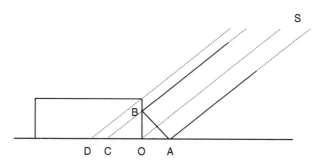

FIGURE 15.3
Optical ray path relevant to single, double, and triple returns.

The triple returns are considered by refering to a ray that runs the path S-B-A-B-S (see Figure 15.3); its optical path length is equal to S-C-S of one of the rays that would be backscattered by the point which is the specular counterpart of A with respect to the building's wall. Hence, the delays of the triple-scattered components are distributed over the interval spanning the delay corresponding to single scattering from O to the delay corresponding to single scattering from $D \equiv (x, r_D)$. Then, triple-scattering contribution is added to the reflectivity map $\gamma(x, r)$, after being evenly distributed over the range interval from r_O to r_D.

Superposition of first-, second-, and third-order returns fully represent the scattered field; higher-order mutual interactions do not provide any contribution to the backscattered field to the radar antenna because the wall surface is supposed to be flat.

In the case of random surfaces, the evaluation of the single-, double-, and triple-scattered contributions is obtained by generating a circular Gaussian complex random variable with mean and variance computed as presented in Sections 15.5 through 15.7.

A ground-range to slant-range projection is applied to model the fore-shortening and layover effects; moreover, a power-sharing approach [11] is employed to get a discretized version of the radar signals. Iterating this procedure for all the facets composing the scene allows the evaluation of the scene reflectivity map $\gamma(x, r)$.

15.4 Electromagnetic Methods to Evaluate the Backscattered Field

Electromagnetic models are required to evaluate in closed-form single and multiple returns for each element within the scene. Numerical techniques could be adopted: however, numerical methods are difficult to be used whenever random surfaces are in order; moreover, they do not show the functional dependence of the scattered electromagnetic field on the scene parameters, are not efficient, and are time-consuming. Conversely, analytical methods easily deal with random surfaces; explicitly account for any significant scene parameter in the analytically evaluated scattered field; and allow fast evaluation of the scattered field whenever closed-form solutions are available.

15.4.1 Scattering Methods

The proposed model is relevant to a building that is isolated (from the electromagnetic point of view) from any other man-made structure. The building model consists of a prism with smooth walls, lying over the ground (in principle of unlimited extent), modeled as a random rough surface;

building and ground surfaces are supposed to be made of dielectric materials. Each building wall forms a dihedral with the surrounding ground. The wall, which constitutes one face of the dihedral, forms a generic angle, φ, with respect to the sensor line of flight; the extension of the ground area interacting (according to the GO or PO rationale) with the building wall constituting the other element of the dihedral is not prescribed: the proposed method evaluates this area in terms of the radar geometry and ray-tracing procedures. The radar return from such a structure can be decomposed into single and multiple returns.

Single returns from the rough ground, the building roof, and vertical walls can be evaluated in closed form by employing physical optics or geometrical optics scattering methods, depending on surface roughness. PO approximation is used for surface slope variance much smaller than unity, whereas GO approximation is used for surface standard deviations much greater than the electromagnetic wavelength.

Multiple returns from the dihedral structures formed by vertical walls and ground must be included. As the wall roughness is often marginal at the wavelength typical of any SAR instruments, the multiple scattering between buildings and terrain can make use of the GO to evaluate the field reflected by the smooth wall toward the ground (first bounce) or the sensor (second or third bounce), and GO or PO (according to ground surface roughness) to evaluate the field scattered by the ground toward the wall (first or second bounce) or to the sensor (second bounce).

Note that GO and PO solutions are valid under the Kirchhoff approximation (KA), which assumes surfaces of infinite extent to evaluate the surface fields. To this extent, scattering contribution due to the currents actually flowing on surfaces of finite extent can be accounted for by including appropriate diffraction contributions from the building's horizontal and vertical edges (reference is made to the uniform theory of diffraction [UTD]) as well as the building's vertexes. The building's dimensions are very large in terms of wavelength at microwave frequency; then, building edge diffractions are expected to be small [15,16] with respect to reflection contributions, and errors caused by neglecting the diffractions are definitively lower than those caused by the assumed simple shape of the building. In addition, inclusion of these contributions would not allow obtaining analytical solutions to the scattering phenomena, and definitively less manageable and noninformative numerical solutions should be used.

The geometry of the wall–ground structure is presented in Figure 15.2b, in which wall height and length are h and l, respectively; the center of wall base onto the ground plane is (x_0, y_0); φ is the angle between x axis and wall base.

The structure can be modeled as an appropriate dihedral holding three relevant properties, which do not allow employing the classical theory of scattering from corner reflectors. First of all, the ground face area is in principle not limited because the portion of ground involved in electromagnetic double reflection depends on the radar off-nadir angle: it can vary from zero (radar illumination close to nadir) to infinite (radar illumination at near-grazing angles). Second, the material of the observed structure is not a perfect conductor. Third, the ground is a random rough surface, whose roughness may generate a nonnegligible double-scattering return for non-null values of φ. In the following the terrain profile is modeled by means of a Gaussian random process, $z(x,y)$, with the Gaussian correlation function whose variance and correlation length are σ^2 and l_c, respectively.

In the following text, we report only those formulae that are needed to present, in the remaining part of the chapter, the result for the scattered field relevant to single, double, and triple return contributions. Motivations to use GO and PO are added along with the rationale to use them. A discussion to distinguish between GO solutions as they apply to deterministic and random surfaces is also added to correctly apply GO to the smooth wall as well as to the stochastic ground.

15.4.2 Kirchhoff Approximation

A plane wave incident field, $\mathbf{E}_i$, is considered with amplitude E_i, polarization fixed by the unit vector $\hat{\mathbf{e}}_i$, and direction of propagation (in the plane $x = 0$), individuated by the wavevector $\mathbf{k}_i$:

$$\mathbf{E}_i = E_0 \hat{\mathbf{e}}_i \exp\left(-j\mathbf{k}_i \cdot \mathbf{r}'\right). \tag{15.2}$$

The electromagnetic field $\mathbf{E}_s$ scattered by a generic (possibly rough) surface S', which separates two media with different complex dielectric constants, is [14]

$$\mathbf{E}_s(\mathbf{r}) = \iint_{S'} \left\{-j\omega\mu G(\mathbf{r},\mathbf{r}') \cdot [\hat{\mathbf{n}} \times \mathbf{H}(\mathbf{r}')] + \nabla \times G(\mathbf{r},\mathbf{r}') \cdot [\hat{\mathbf{n}} \times \mathbf{E}(\mathbf{r}')]\right\} dS', \tag{15.3}$$

where
 $\hat{\mathbf{n}}$ is the unit vector normal to the surface pointing toward the source region;
 $G(\mathbf{r},\mathbf{r}')$ is the dyadic Green's function for the homogeneous space.

If the observation point is in the far-field region, then the dyadic Green's function simplifies and in the observation direction individuated by the unit vector $\hat{\mathbf{k}}_s$ the scattered field is

$$\mathbf{E}_s(\mathbf{r}) = -\frac{jk \exp\left(-jkr\right)}{4\pi r} E_0 \left(I - \hat{\mathbf{k}}_s\hat{\mathbf{k}}_s\right) \cdot \iint_{S'} \left\{\hat{\mathbf{k}}_s \times [\hat{\mathbf{n}} \times \mathbf{E}(\mathbf{r}')] + \eta[\hat{\mathbf{n}} \times \mathbf{H}(\mathbf{r}')]\right\}$$

$$\times \exp\left(-j\mathbf{k}_s \cdot \mathbf{r}'\right) dS', \tag{15.4}$$

where
 η is the intrinsic impedance in the source region
 $\mathbf{k}_s = k\hat{\mathbf{k}}_s$ is the wavevector of the scattered field.

In order to evaluate the surface fields, KA refers to the tangent plane. Accordingly, the scattered field is expressed as [14]

$$\mathbf{E}_s(\mathbf{r}) = -\frac{jk\exp(-jkr)}{4\pi r}E_0(\mathbf{I} - \hat{\mathbf{k}}_s\hat{\mathbf{k}}_s)\cdot \iint_{S'} \mathbf{F}(\hat{\mathbf{k}}_i, \hat{\mathbf{e}}_i, \hat{\mathbf{n}})\exp[j(\mathbf{k}_i - \mathbf{k}_s)\cdot \mathbf{r}']dS', \quad (15.5)$$

where

$$\mathbf{F}(\hat{\mathbf{k}}_i, \hat{\mathbf{e}}_i, \hat{\mathbf{n}}) = -(\hat{\mathbf{e}}_i \cdot \hat{\mathbf{q}}_i)(\hat{\mathbf{n}} \cdot \hat{\mathbf{k}}_i)\hat{\mathbf{q}}_i(1 - R_\perp) + (\hat{\mathbf{e}}_i \cdot \hat{\mathbf{p}}_i)(\hat{\mathbf{n}} \times \hat{\mathbf{q}}_i)(1 + R_{||})$$
$$+ (\hat{\mathbf{e}}_i \cdot \hat{\mathbf{q}}_i)[\hat{\mathbf{k}}_s \times (\hat{\mathbf{n}} \times \hat{\mathbf{q}}_i)](1 + R_\perp) + (\hat{\mathbf{e}}_i \cdot \hat{\mathbf{p}}_i)(\hat{\mathbf{n}} \cdot \hat{\mathbf{k}}_i)(\hat{\mathbf{k}}_s \times \hat{\mathbf{q}}_i)(1 - R_{||})$$
$$(15.6)$$

and

$$\hat{\mathbf{q}}_i = \frac{\hat{\mathbf{k}}_i \times \hat{\mathbf{n}}}{|\hat{\mathbf{k}}_i \times \hat{\mathbf{n}}|} \quad (15.7)$$

$$\hat{\mathbf{p}}_i = \hat{\mathbf{q}}_i \times \hat{\mathbf{k}}_i. \quad (15.8)$$

In Equation 15.6, $R_{||}$ and $R_\perp$ are (local) Fresnel reflection coefficients for locally parallel and perpendicular polarizations, respectively.

15.4.3 PO Solution

KA does not allow, in general, evaluation of the scattered field in closed form for nonplane surfaces due to the involved dependence of **F** on the surface profile, described by the local normal $\hat{\mathbf{n}}$. In the PO approximation, the surface is modeled as a microscopic roughness superimposed on a mean plane with normal unit vector $\hat{\mathbf{n}}_0$; in the small slope roughness hypothesis, **F** is expanded around $\hat{\mathbf{n}}_0$. The first term of the expansion provides the zero-order PO solution for the scattered field:

$$\mathbf{E}_s(\mathbf{r}) = -\frac{jk\exp(-jkr)}{4\pi r}E_0(\mathbf{I} - \hat{\mathbf{k}}_s\hat{\mathbf{k}}_s)\cdot \mathbf{F}(\hat{\mathbf{k}}_i, \hat{\mathbf{e}}_i, \hat{\mathbf{n}}_0)I_s, \quad (15.9)$$

where

$$I_s = \iint_{S'} \exp[j(\mathbf{k}_i - \mathbf{k}_s)\cdot \mathbf{r}']dS'. \quad (15.10)$$

In Equation 15.10, $\mathbf{r}'$ accounts for the surface microscopic random roughness so that I_s is a random variable.

By referring to horizontal and vertical polarizations, Equation 15.9 can be rewritten as

$$\begin{bmatrix} E_{sh} \\ E_{sv} \end{bmatrix} = -jk\frac{\exp(-jkr)}{4\pi r}\begin{pmatrix} S_{hh} & S_{vh} \\ S_{hv} & S_{vv} \end{pmatrix}\begin{bmatrix} E_{ih} \\ E_{iv} \end{bmatrix}I_s, \quad (15.11)$$

where the entries of the scattering matrix S are given by

$$S_{pq} = [(I - \hat{\mathbf{k}}_s\hat{\mathbf{k}}_s) \cdot \mathbf{F}(\hat{\mathbf{k}}_i, \hat{\mathbf{e}}_{ip}, \hat{\mathbf{n}}_0)] \cdot \hat{\mathbf{e}}_{sq}. \tag{15.12}$$

15.4.4 GO Solution

GO applies to scattering from both deterministic and random surfaces: however, in these two cases the GO formulation is obtained by following different rationales.

In the deterministic case, GO solution is the first term of the asymptotic solution to the Maxwell equations in the high-frequency regime $k \to \infty$. GO leads to electromagnetic propagation along ray paths. In a homogeneous medium, rays are straight lines and, when they hit a deterministic boundary surface, they are reflected in the specular direction [17]. Electromagnetic fields are locally plane waves: the evaluation of field polarization and amplitude is predicted by means of the transport equation [17].

GO evaluation of the scattered fields for rough surfaces can be obtained by asymptotically evaluating the integral in Equation 15.5. In this case, the main contributions to the scattered field arise from stationary phase points whose normal unit vector is $\hat{\mathbf{n}}_s$ implicitly defined as

$$\hat{\mathbf{k}}_s = \hat{\mathbf{k}}_i - 2(\hat{\mathbf{k}}_i \cdot \hat{\mathbf{n}}_s)\hat{\mathbf{n}}_s, \tag{15.13}$$

where the incident and the scattered direction of propagation satisfy the specular reflection condition. The GO solution is then written as [14]

$$\mathbf{E}_s(\mathbf{r}) = -\frac{jk \exp(-jkr)}{4\pi r} E_0(I - \hat{\mathbf{k}}_s\hat{\mathbf{k}}_s) \cdot \mathbf{F}(\hat{\mathbf{k}}_i, \hat{\mathbf{e}}_i, \hat{\mathbf{n}}_s)I_s. \tag{15.14}$$

The scattering matrix entries are in this case equal to

$$S_{pq} = [(I - \hat{\mathbf{k}}_s\hat{\mathbf{k}}_s) \cdot \mathbf{F}(\hat{\mathbf{k}}_i, \hat{\mathbf{e}}_{ip}, \hat{\mathbf{n}}_s)] \cdot \hat{\mathbf{e}}_{sq}. \tag{15.15}$$

Equations 15.14 and 15.15 are formally coincident with Equations 15.9 and 15.12, but for use of $\hat{\mathbf{n}}_s$ instead of $\hat{\mathbf{n}}_0$. The surface integral I_s is formally provided by Equation 15.10 and for random surfaces is a random variable: analytic evaluations in closed form of its mean and variance are of interest.

15.5 Single Return Contributions

The entries of the scattering matrix S, as well as mean and variance of the integral I_S, are hereafter evaluated in closed form for the different single return contributions.

Single returns to the radar are relevant to the terrain, the building's wall, or the roof. In the next section, we evaluate the single returns in closed form (see Figure 15.2a) by using both GO and PO.

15.5.1 Single Return from the Building's Walls

(a) GO solution:

For a plane deterministic surface, GO expresses the scattered field as a superposition of rays reflected along the specular direction. Then, the contribution backscattered by the wall, evaluated by the GO approach, is zero but for the case of $(\vartheta = \pi/2, \varphi = 0)$, which is never the case for the SAR sensor.

(b) PO solution:

For a plane deterministic surface, no expansion of $\mathbf{F}$ is necessary because the wall is plane and smooth. The scattering matrix entries can be evaluated by using Equation 15.12, in which $\hat{n}_0$ is replaced by $\hat{n}_w$. The incidence angle on the wall is

$$\psi = \cos^{-1}(\sin \vartheta \cos \varphi). \tag{15.16}$$

The scattering matrix entries are evaluated in terms of the Fresnel coefficients of the wall R_W:

$$S_{hh} = \frac{2 \sin \vartheta \cos \varphi}{\cos^2 \vartheta + \sin^2 \vartheta \sin^2 \varphi} \left[-R_{\perp W}(\psi) \cos^2 \vartheta \cos^2 \varphi + R_{\parallel W}(\psi) \sin^2 \varphi \right],$$

$$\tag{15.17a}$$

$$S_{hv} = S_{vh} = \frac{2 \cos \vartheta \sin \vartheta \cos^2 \varphi \sin \varphi}{\cos^2 \vartheta + \sin^2 \vartheta \sin^2 \varphi} \left[R_{\perp W}(\psi) + R_{\parallel W}(\psi) \right], \tag{15.17b}$$

$$S_{vv} = \frac{2 \sin \vartheta \cos \varphi}{\cos^2 \vartheta + \sin^2 \vartheta \sin^2 \varphi} \left[-R_{\perp W}(\psi) \sin^2 \varphi + R_{\parallel W}(\psi) \cos^2 \vartheta \cos^2 \varphi \right],$$

$$\tag{15.17c}$$

whereas the integral can be evaluated for a wall of height h and length l as

$$I_s = \exp(-j2ky_0 \sin \vartheta) \exp(-jkh \cos \vartheta) hl \cos \varphi \, \text{sinc}(kh \cos \vartheta) \text{sinc}(kl \sin \varphi \sin \vartheta). \tag{15.18}$$

15.5.2 Single Return from the Ground or from the Building's Roof

The electromagnetic field backscattered by a rectangular area of the rough ground with dimensions a and b is evaluated. This area can be set equal to that illuminated by the field reflected in the specular direction by one building's wall. This field component is a random variable, and its mean and variance are evaluated in terms of the statistical parameters used to model the surface roughness.

(a) GO solution:

In the backscattering case, the stationary phase points over the ground have a unit normal $\hat{n}_s$, which satisfy the following relation:

$$\hat{n}_s = \hat{k}_i,$$ (15.19)

which is satisfied in the case of normal incidence. Then, the scattering matrix entries can be evaluated by using Equations 15.15, thus getting

$$S_{hh} = -2R_{\perp}(0),$$ (15.20a)

$$S_{hv} = S_{vh} = 0,$$ (15.20b)

$$S_{vv} = 2R_{\|}(0).$$ (15.20c)

The surface integral I_s is a random variable. By using Equation 15.10, its mean value (i.e., its coherent component) is given by

$$<I_s> = \exp\left(-2k^2\sigma^2 \cos^2 \vartheta\right) ab \, \text{sinc}(ka \sin \vartheta)$$ (15.21)

and is negligible because the GO solution applies for $k\sigma \gg 1$; moreover, the surface integral mean square value is given by

$$<I_s I_s^*> = \frac{ab}{4k^2 \cos^4 \vartheta} \frac{1}{2\pi\sigma^2|C''(0)|} \exp\left[-\frac{\tan^2 \vartheta}{2\sigma^2|C''(0)|}\right],$$ (15.22)

where $C''(0)$ is the second derivative of the normalized correlation function $C(\rho)$ evaluated at $\rho = 0$.

(b) PO solution:

In this case, the scattering matrix entries can be evaluated by using Equation 15.12 in which $\hat{n}_0$ is replaced with $\hat{z}$ for the horizontal mean plane, thus getting

$$S_{hh} = -2R_{\perp}(\vartheta) \cos \vartheta,$$ (15.23a)

$$S_{hv} = S_{vh} = 0,$$ (15.23b)

$$S_{vv} = 2R_{\|}(\vartheta) \cos \vartheta.$$ (15.23c)

The surface integral I_s is a random variable. By using Equation 15.10, its mean value (i.e., its coherent component) is given by Equation 15.21, but is not negligible because $k\sigma$ is not necessarily high; its variance (i.e., the noncoherent component) is given by

$$\sigma_s^2 = <I_s I_s^*> - |<I_s>|^2$$

$$= \exp\left(-4k^2\sigma^2 \cos^2 \vartheta\right) ab\pi l_c^2 \sum_{m=1}^{+\infty} \frac{(2k\sigma \cos \vartheta)^{2m}}{m!m} \exp\left[-\frac{(2kl_c \sin \vartheta)^2}{4m}\right].$$

(15.24)

15.6 Double Return Contributions

The entries of the scattering matrix S, as well as mean and variance of the integral I_S, are hereafter evaluated in closed form for the different double return contributions. These take place whenever the incident field is first scattered by the building's vertical illuminated walls toward the ground and then scattered by the ground itself back to the SAR and vice versa (see Figure 15.2b). These backscattering mechanisms are referred to as *wall–ground reflections* and *ground–wall reflections*, respectively.

The field reflected by the wall is always evaluated via the GO solution according to its deterministic formulation: this situation is present in the first bounce of the wall–ground reflection, or in the second bounce of the ground–wall reflection. According to the ground surface roughness, the field scattered by the ground is evaluated by employing GO or PO: this happens in the first bounce of the ground–wall reflection, or in the second bounce of the wall–ground reflection. The cumbersome algebraic and vector intermediate calculations are skipped and only the final results are listed.

15.6.1 Wall–Ground Return

(a) GO–GO solution

The GO solution for a field reflected by a wall is a plane wave directed along the specular direction. Its amplitude and polarization are obtained by decomposing the horizontally or vertically polarized incident field into its orthogonal and parallel components with respect to the wall incidence plane and applying the proper Fresnel coefficient to each component. Scattering of this wave by the rough terrain toward the sensor must be evaluated.

The portion of ground invested by the field reflected by the wall is a parallelogram of area A_0 given by

$$A_0 = hl \tan \vartheta \cos \varphi; \tag{15.25}$$

the incidence angle on the wall is given by

$$\psi = \cos^{-1}(\sin \vartheta \cos \varphi), \tag{15.26}$$

whereas the incidence angle for the ground surface stationary points is given by

$$\zeta = \cos^{-1}\left(\frac{\cos \vartheta(\tan^2 \vartheta \sin^2 \varphi + 1)}{\sqrt{1 + \tan^2 \vartheta \sin^2 \varphi}}\right). \tag{15.27}$$

The scattering matrix entries are

$$S_{hh} = 2[-R_\perp(\zeta)R_{\perp W}(\psi)\cos^2\vartheta\cos^2\varphi$$

$$+ R_\|(\zeta)R_{\|W}(\psi)\sin^2\varphi]\cos\vartheta\sqrt{1+\tan^2\vartheta\sin^2\varphi}, \qquad (15.28a)$$

$$S_{hv} = S_{vh} = \sin 2\varphi[-R_\perp(\zeta)R_{\perp W}(\psi)$$

$$+ R_\|(\zeta)R_{\|W}(\psi)]\cos^2\vartheta\sqrt{1+\tan^2\vartheta\sin^2\varphi}, \qquad (15.28b)$$

$$S_{vv} = 2[-R_\perp(\zeta)R_{\perp W}(\psi)\sin^2\varphi$$

$$+ R_\|(\zeta)R_{\|W}(\psi)\cos^2\vartheta\cos^2\varphi]\cos\vartheta\sqrt{1+\tan^2\vartheta\sin^2\varphi}. \qquad (15.28c)$$

The surface integral I_{A_0} is a random variable. By using Equation 15.10, its mean value (i.e., its coherent component) is given by

$$<I_{A_0}> = \exp[-2(k\sigma\cos\vartheta)^2]\exp(-jk2y_0\sin\vartheta)\exp\left(jkh\sin^2\vartheta\frac{\sin^2\varphi}{\cos\vartheta}\right)$$

$$\sqrt{1+\tan^2\vartheta\sin^2\varphi}hl\tan\vartheta\cos\varphi\,\text{sinc}(kl\sin\vartheta\sin\varphi)\text{sinc}\left(kh\frac{\sin^2\vartheta\sin^2\varphi}{\cos\vartheta}\right). $$

$$(15.29)$$

Moreover, its mean square value is given by

$$<I_{A_0}I_{A_0}^*> = hl\tan\vartheta\cos\varphi\frac{1+\tan^2\vartheta\sin^2\varphi}{4k^2\cos^2\vartheta}\frac{1}{2\pi\sigma^2|C''(0)|}\exp\left[-\frac{\tan^2\vartheta\sin^2\varphi}{2\sigma^2|C''(0)|}\right].$$

$$(15.30)$$

(b) GO–PO solution

GO solution for the field reflected by the wall toward the ground is obviously equal to the one obtained in the previous section. Results for the PO solution to the field scattered by the area A_0 toward the sensor are presented hereafter. The scattering matrix entries can be evaluated by using Equation 15.12, in which $\hat{n}_0$ is replaced by $\hat{z}$, the incidence direction is that provided by the wall specular direction, and the incident polarization depends on the decomposition of the scattered field in the local (for the ground) horizontal and vertical polarization states.

The scattering matrix entries are

$$S_{hh} = A_h[-2R_\perp(\vartheta)\cos\vartheta\cos 2\varphi] + B_h[\sin^2\vartheta\sin 2\varphi + R_\|(\vartheta)\sin 2\varphi(1+\cos^2\vartheta)],$$

$$(15.31a)$$

$$S_{hv} = A_h[\sin^2\vartheta\sin 2\varphi + R_\perp(\vartheta)\sin 2\varphi(1+\cos^2\vartheta)] + 2B_h R_\|(\vartheta)\cos\vartheta\cos 2\varphi,$$

$$(15.31b)$$

$$S_{vh} = A_v[-2R_\perp(\vartheta)\cos\vartheta\cos 2\varphi] + B_v[\sin^2\vartheta\sin 2\varphi + R_{||}(\vartheta)\sin 2\varphi(1+\cos^2\vartheta)],$$
$$(15.31c)$$

$$S_{vv} = A_v[\sin^2\vartheta\sin 2\varphi + R_\perp(\vartheta)\sin 2\varphi(1+\cos^2\vartheta)] + 2B_vR_{||}(\vartheta)\cos\vartheta\cos 2\varphi,$$
$$(15.31d)$$

where

$$A_h = R_{\perp W}(\psi)\cos^2\vartheta\cos^2\varphi - R_{||W}(\psi)\sin^2\varphi, \tag{15.32a}$$

$$B_h = (R_{\perp W}(\psi) + R_{||W}(\psi))\cos\vartheta\cos\varphi\sin\varphi, \tag{15.32b}$$

$$A_v = -(R_{\perp W}(\psi) + R_{||W}(\psi))\cos\vartheta\cos\varphi\sin\varphi, \tag{15.32c}$$

$$B_v = -R_{\perp W}(\psi)\sin^2\varphi + R_{||W}(\psi)\cos^2\vartheta\cos^2\varphi. \tag{15.32d}$$

The surface integral I_{A_0} is a random variable. By using Equation 15.10, its mean value (i.e., its coherent component) is given by

$$<I_{A_0}> = \exp\left[-2(k\sigma\cos\vartheta)^2\right]\exp\left(-jk2y_0\sin\vartheta\right)\exp\left(jkh\sin^2\vartheta\frac{\sin^2\varphi}{\cos\vartheta}\right)$$

$$hl\tan\vartheta\cos\varphi\,\mathrm{sinc}(kl\sin\vartheta\sin\varphi)\mathrm{sinc}\left(kh\frac{\sin^2\vartheta\sin^2\varphi}{\cos\vartheta}\right).$$

$$(15.33)$$

Moreover, its variance is given by

$$\sigma_{A_0}^2 = \exp\left(-4k^2\sigma^2\cos^2\vartheta\right)hl\tan\vartheta\cos\varphi$$

$$\sum_{m=1}^{+\infty}\frac{(2k\sigma\cos\vartheta)^{2m}}{m!}\pi\frac{l_c^2}{m}\exp\left[-\frac{(2kl_c\sin\varphi\sin\vartheta)^2}{4m}\right]. \tag{15.34}$$

15.6.2 Ground–Wall Return

This contribution is equal to that of the wall–ground as far as the reflection over the wall is modeled by GO. As a matter of fact, among all waves scattered by the rough ground surface (evaluated by GO or PO), only the one propagating toward the wall that is successively specularly reflected by the wall toward the sensor is of interest. The length of the path followed by this wave, which first hits the ground and then the wall, is the same as the one followed by the corresponding wave, which first hits the wall and then the ground. Then, the two double-scattered contributions sum up coherently and the overall double-scattered field is twice the fields evaluated in Section 15.6.1, whereas the overall scattered power density is four times the powers evaluated in Section 15.6.1.

15.7 Triple Return Contributions

The entries of the scattering matrix S, as well as mean and variance of the integral I_s, are hereafter evaluated in closed form for the different triple return contributions. Two different triple-scattering contributions take place (see Figure 15.2c): the first one is obtained whenever the incident field is first scattered by the building's illuminated wall toward the ground; then, it is scattered by the ground itself back to the wall and finally it is scattered by the wall toward the SAR sensor. The second one is obtained whenever the field is first scattered by the ground toward the building's wall; then, it is scattered by the wall toward the ground and finally it is scattered by the ground toward the SAR sensor. These mechanisms are referred to as wall–ground–wall reflections and ground–wall–ground reflections, respectively.

The cumbersome algebraic and vector intermediate calculations are skipped in the following sections and only the final results are listed.

15.7.1 Wall–Ground–Wall Return

The wall–ground–wall contribution only arises in the case of rough terrain [13]. It can be evaluated using a full GO approach or considering the PO solution for the scattering over the ground.

(a) GO–GO–GO solution

The two GO reflections from the wall (first and third bounces) can be treated as they are in Section 15.6 for the double-scattering contributions, whereas the field backscattered in the second bounce by the ground can be evaluated as it is in Section 15.5.2a for the GO solution to the single backscattered contribution from the ground.

The scattering matrix entries are

$$S_{hh} = \frac{1}{(\cos^2 \vartheta + \sin^2 \vartheta \sin^2 \varphi)^2} 2R(0) \left\{ [R_{\perp W}(\psi) + R_{\parallel W}(\psi)]^2 \cos^2 \vartheta \cos^2 \varphi \sin^2 \varphi \right.$$
$$\left. + [R_{\perp W}(\psi) \cos^2 \vartheta \cos^2 \varphi + R_{\parallel W}(\psi) \sin^2 \varphi]^2 \right\}, \tag{15.35a}$$

$$S_{hv} = S_{vh} = \frac{1}{(\cos^2 \vartheta + \sin^2 \vartheta \, \sin^2 \varphi)^2} 2R(0)[R_{\perp W}(\psi) + R_{\parallel W}(\psi)]^2$$
$$\cos \vartheta \sin \varphi \cos \varphi (\sin^2 \varphi - \cos^2 \vartheta \cos^2 \varphi), \tag{15.35b}$$

$$S_{vv} = \frac{1}{(\cos^2 \vartheta + \sin^2 \vartheta \, \sin^2 \varphi)^2} 2R(0) \left\{ [R_{\perp W}(\psi) + R_{\parallel W}(\psi)]^2 \cos^2 \vartheta \cos^2 \varphi \sin^2 \varphi \right.$$
$$\left. + [R_{\perp W}(\psi) \cos^2 \vartheta \cos^2 \varphi - R_{\parallel W}(\psi) \sin^2 \varphi]^2 \right\}, \tag{15.35c}$$

where $R(0) = R_{\perp}(0) = -R_{\parallel}(0)$.

The surface integral I_{R_0} is a random variable. By using Equation 15.10, its mean value (i.e., its coherent component) is given by

$$<I_{R_0}> = \frac{1}{\cos \vartheta} \exp\left(-2k^2\sigma^2 \cos^2 \vartheta\right) \exp\left(-j2ky_0' \cos \varphi \sin \vartheta\right)$$

$$\exp\left(j2kx_0' \cos \varphi \sin \vartheta\right) \exp\left(-jkb \cos \varphi \sin \vartheta\right) ab \, \text{sinc}(kl \cos \varphi \sin \vartheta)$$

$$\text{sinc}(kl \sin \varphi \sin \vartheta), \tag{15.36}$$

and is negligible because the GO solution applies for $k\sigma \gg 1$; moreover, the surface integral mean square value is given by

$$<I_{R_0}I_{R_0}^*> = hl \tan \vartheta \cos \varphi \frac{1}{4k^2 \cos^4 \vartheta} \frac{1}{2\pi\sigma^2|C''(0)|} \exp\left[-\frac{\tan^2 \vartheta}{2\sigma^2|C''(0)|}\right]. \tag{15.37}$$

(b) GO–PO–GO method

The two GO reflections from the wall (first and third bounce) can be treated as done in Section 15.6 for the double-scattering contributions, whereas the field backscattered in the second bounce by the ground can be evaluated as done in Section 15.5.2b for the PO solution to the single backscattered contribution from the ground.

The scattering matrix entries are

$$S_{hh} = \frac{1}{(\cos^2 \vartheta + \sin^2 \vartheta \sin^2 \varphi)^2} \left[2R_\perp(\vartheta)R^2_{\perp W}(\psi)\cos^5 \vartheta \cos^4 \varphi\right.$$

$$+ 2\cos^3 \vartheta \cos^2 \varphi \sin^2 \varphi\left[-2R_\perp(\vartheta)R_{\perp W}(\psi)R_{\|W}(\psi) + R_\|(\vartheta)R^2_{\perp W}(\psi)\right.$$

$$\left.+ 2R_\|(\vartheta)R_{\perp W}(\psi)R_{\|W}(\psi) + R_\|(\vartheta)R^2_{\|W}(\psi)\right] + 2R_\perp(\vartheta)R^2_{\|W}(\psi)\cos \vartheta \sin^4 \varphi\right], \tag{15.38a}$$

$$S_{hv} = S_{vh} = \frac{1}{(\cos^2 \vartheta + \sin^2 \vartheta \sin^2 \varphi)^2} \left[2\cos^4 \vartheta \sin \varphi \cos^3 \varphi\right.$$

$$\left[-R_\perp(\vartheta)R^2_{\perp W}(\psi) - R_\perp(\vartheta)R_{\perp W}(\psi)R_{\|W}(\psi) + R_\|(\vartheta)R_{\perp W}(\psi)R_{\|W}(\psi)\right.$$

$$\left.+ R_\|(\vartheta)R^2_{\|W}(\psi)\right] + 2\cos^2 \vartheta \sin^3 \varphi \cos \varphi\left[R_\perp(\vartheta)R_{\perp W}(\psi)R_{\|W}(\psi)\right.$$

$$\left.\left. - R_\|(\vartheta)R^2_{\perp W}(\psi) - R_\|(\vartheta)R_{\perp W}(\psi)R_{\|W}(\psi) + R_\perp(\vartheta)R^2_{\|W}(\psi)\right]\right], \tag{15.38b}$$

$$S_{vv} = \frac{1}{(\cos^2 \vartheta + \sin^2 \vartheta \sin^2 \varphi)^2} \left[2R_\|(\vartheta)R^2_{\|W}(\psi)\cos^5 \vartheta \cos^4 \varphi\right.$$

$$+ 2\cos^3 \vartheta \cos^2 \varphi \sin^2 \varphi\left[2R_\perp(\vartheta)R_{\perp W}(\psi)R_{\|W}(\psi) + R_\perp(\vartheta)R^2_{\|W}(\psi)\right.$$

$$\left.- 2R_\|(\vartheta)R_{\perp W}(\psi)R_{\|W}(\psi) + R_\perp(\vartheta)R^2_{\perp W}(\psi)\right] + 2R_\|(\vartheta)R^2_{\perp W}(\psi)\cos \vartheta \sin^4 \varphi\right]. \tag{15.38c}$$

The surface integral I_{R_0} is a random variable. By using Equation 15.10, its mean value (i.e., its coherent component) is given by

$$<I_{R_0}> = \exp(-2k^2\sigma^2\cos^2\vartheta)\exp(j2k[x_0' - y_0\tan\varphi])$$
$$\exp(-j2ky_0'[\sin\vartheta\cos\varphi - \tan\varphi])$$
$$\exp(-jkh\tan\vartheta[\cos^2\varphi\sin\vartheta - \sin\varphi])\,hl\tan\vartheta\cos\varphi$$
$$\text{sinc}(kl\sin\vartheta\sin\varphi)\,\text{sinc}(kh\tan\vartheta[\cos^2\varphi\sin\vartheta - \sin\varphi]). \qquad (15.39)$$

Moreover, its variance is given by

$$\sigma_{R_0}^2 = \exp(-4k^2\sigma^2\cos^2\vartheta)\,hl\tan\vartheta\cos\varphi\pi l_c^2\sum_{m=1}^{+\infty}\frac{(2k\sigma\cos\vartheta)^{2m}}{m!m}$$

$$\exp\left[-\frac{(2kl_c\sin\vartheta)^2}{4m}\right]. \qquad (15.40)$$

15.7.2 Ground–Wall–Ground Return

Use of the GO solution for wall scattering implies that the field backscattered by the wall is zero so that the ground–wall–ground term does not contribute to the overall field backscattered to the SAR.

15.8 Numerical Examples

In this section, we numerically evaluate the RCS of the element of the urban structure described in previous sections as a function of the look angle ϑ, polarization, building orientation angle φ, building size hl, and surface roughness. The RCS is defined as

$$\text{RCS} = \frac{4\pi r^2 <|E_s|^2>}{|E_0|^2}. \qquad (15.41)$$

We separately compute the RCSs corresponding to single-, double-, and triple-scattering contributions. We assume the frequency $f = 9.6$ GHz ($\lambda = 0.031$ m), typical of X-band SAR systems and that both the building's wall and ground have a relative dielectric constant equal to $15 - j0.1$. The building is located on rough soil and both its height h and length l are 20 m.

We first consider very rough soil whose height standard deviation σ is 0.08 m and whose correlation length l_c is 0.3 m. In this case, the GO solution applies to the ground.

Single-scattering contribution from the wall is negligible except that for $\vartheta \cong \pi/2$ and $\varphi = 0$ and is not considered here. For the assumed ground roughness, the single-scattering contribution from the ground can be evaluated using GO, and the corresponding RCS is plotted in Figure 15.4 as a function of the look angle. Note that in this case hh and vv RCS are equal, and cross-polarized (i.e., hv and vh) RCS is zero. The latter value is related

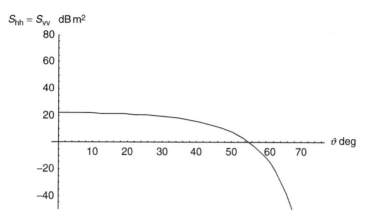

FIGURE 15.4

GO RCS vs. incidence angle ϑ for single scattering from a very rough terrain whose height standard deviation σ is 0.08 m and whose correlation length l_c is 0.3 m. Illuminated effective area is 400 m², frequency is 9.6 GHz ($\lambda = 3.125$ cm), and both wall and ground have a relative dielectric constant equal to $15 - j0.1$.

to the approximations employed by the KA, and in practice the cross-polarized RCS is not zero; however, it is certainly negligible with respect to the copolarized (i.e., hh and vv) one.

With regard to the double-scattering contribution, it can be properly evaluated by using the GO–GO solution. Corresponding RCS for different polarizations are plotted in Figure 15.5 as a function of the look angle ϑ for

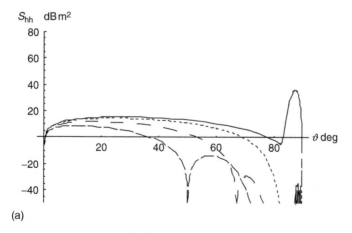

(a)

FIGURE 15.5

(a), (c), (e): GO–GO RCS vs. incidence angle ϑ for wall–ground double scattering. Wall orientation φ is 0° (solid line), 15° (short dashes), 30° (long-spaced dashes), and 45° (long dashes). (b), (d), (f): GO–GO RCS vs. wall orientation angle φ for wall–ground double scattering. Look angle ϑ is 23° (solid line), 45° (short dashes), and 60° (long-spaced dashes). For all plots, wall height h and length l are 20 m. Terrain roughness, frequency, and wall and ground relative dielectric constant are equal to those of Figure 15.4.

(*continued*)

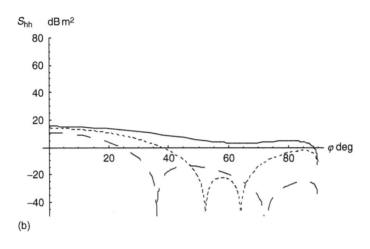

(b)

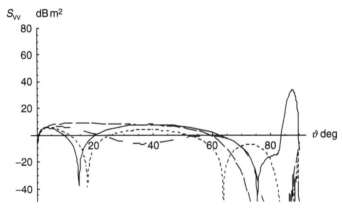

(c)

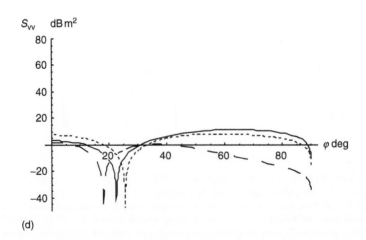

(d)

FIGURE 15.5 (continued)

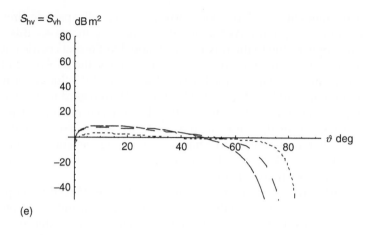

(e)

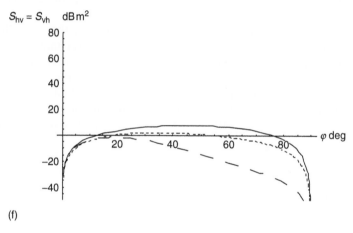

(f)

FIGURE 15.5 (continued)

different values of the wall orientation angle φ and also as a function of φ for different values of ϑ. In all cases, RCS is zero at $0°$ and $90°$, as expected. Note also that for ϑ near to $90°$ a peak appears: this is due to the field coherent component. As a matter of fact, the latter is always negligible due to the first exponential of Equation 15.29, except for the near-grazing angle incidence, in which case $\cos \vartheta$ is small. However, this is a spurious result due to the employed approximations: in fact, we recall that KA and hence GO and PO do not hold at near-grazing angle incidence. Let us now analyze the different polarizations. At hh polarization, RCS is almost constant with respect to the look angle for small values of the wall orientation angle. As the latter increases, the range of ϑ values over which the RCS remains constant is reduced. If the orientation angle is small, the diagram of RCS shows two valleys at vv polarization, corresponding to the pseudo-Brewster angles of wall and ground. As the orientation angle φ increases, this effect tends to disappear, due to the mixing of TE and TM Fresnel coefficient (see Equations 15.28). Finally, for cross-polarized channels, the RCS dependence on ϑ is weak, and, similarly to the hh case, the range of ϑ values over which RCS remains constant is reduced as φ increases. With regard to the dependence on φ, for copolarized channels, the dependence on the orientation angle is weak for small values of the incidence angle, except that for φ near to 90°. For cross-polarized channels, a strong dependence on φ is present also for small values of φ, when RCS rapidly changes from zero $(-\infty \text{ dB m}^2)$ to more than 0 dB m^2.

Triple-scattering contribution is plotted in Figure 15.6. By comparing Figures 15.5 and 15.6, we note that triple-scattering contribution is comparable to double scattering for hh polarization and small incidence angles, and in all other cases it is negligible with respect to double-scattering contribution.

Let us now move to consider the same building and the same illumination, but a much smoother ground, whose roughness is characterized by a height standard deviation σ equal to 0.001 m and by a correlation length l_c equal to 0.02 m. In this case, the PO solution applies for the ground.

In this case, single-scattering contribution must be evaluated by using PO and double-scattering contribution must be evaluated by using GO–PO. In Figure 15.7, single-scattering RCS is plotted vs. the look angle. At variance

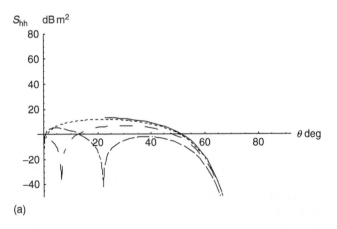

(a)

FIGURE 15.6
GO–GO–GO RCS vs. incidence angle θ for wall–ground–wall triple scattering. Wall height h and length l are both 20 m. Terrain roughness, frequency, and wall and ground relative dielectric constant are equal to those of Figure 15.4. Wall orientation φ is 0° (solid line), 15° (short dashes), 30° (long-spaced dashes), and 45° (long dashes).

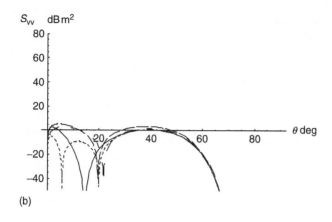

(b)

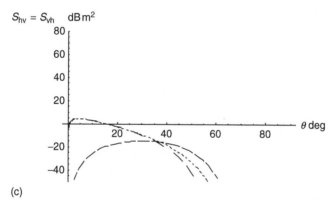

(c)

FIGURE 15.6 (continued)

with the previous example, in this case hh and vv RCS are not equal. In addition, in this case cross-polarized (i.e., hv and vh) RCS is zero.

Double-scattering RCS vs. the look angle is plotted in Figure 15.8. The main difference with respect to the previous example is the strong return for $\varphi = 0$ in the copolarized channels, due to the presence of the strong coherent component. In addition, for φ not near to zero, a stronger dependence on ϑ with respect to the previous example is shown, and hv and vh polarizations are not equal. In addition, in this case the effect of the pseudo-Brewster angle is visible on the vv channel, and the cross-polarized return is null for $\varphi = 0$. We also note that there is always a strong dependence on the orientation angle when the latter is small. These RCS fast variations are due to the strong coherent component. For orientation angles larger than about 10°, the dependence on φ is weaker, especially for small look angles. By comparing Figures 15.7 and 15.8, we note that double scattering is the main scattering

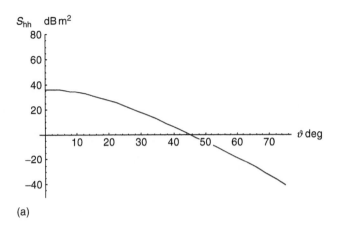

(a)

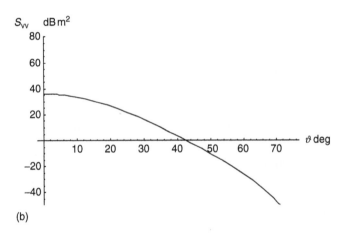

(b)

FIGURE 15.7
RCS vs. ϑ for single scattering from a moderately rough terrain whose height standard deviation σ is 0.001 m (i.e., $k\sigma = 0.314$) and whose correlation length l_c is 0.02 m. Illuminated effective area is 400 m^2, frequency is 9.6 GHz ($\lambda = 3.125$ cm), and both wall and ground have a relative dielectric constant equal to $15 - j0.1$.

mechanism for a wide range of scattering and orientation angles; a significant exception is the case of illumination at near vertical incidence for copolarized channels.

Triple-scattering contribution is plotted in Figure 15.9. By comparing Figures 15.8 and 15.9, we note that triple-scattering contribution is negligible with respect to that of double scattering, except that for small incidence angles and nonnull building orientation angle. As expected, triple-scattering contribution does not show a large coherent component for $\varphi = 0$, at variance with the double-scattering case.

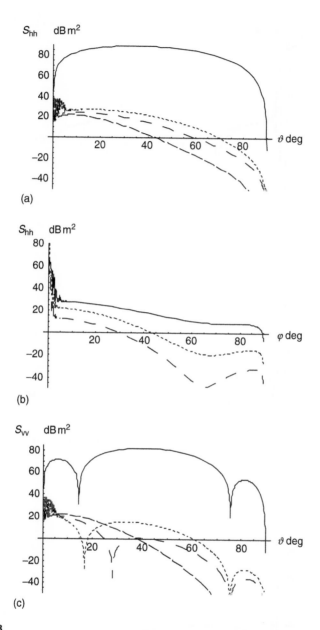

FIGURE 15.8
(a), (c), (e), (g): GO–PO RCS vs. ϑ for wall–ground double scattering. φ is 0° (solid line), 15° (short dashes), 30° (long-spaced dashes), and 45° (long dashes). (b), (d), (f), (h): GO–PO RCS vs. φ for wall–ground double scattering. ϑ is 23° (solid line), 45° (short dashes), and 60° (long-spaced dashes). For all plots, wall height h and length l are 20 m. Terrain roughness, frequency, and wall and ground relative dielectric constant are equal to those of Figure 15.7. Note that zones with very rapid oscillations, at very small incidence and orientation angles, are due to the coherent field component.

(*continued*)

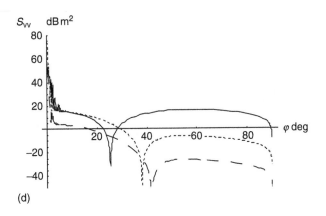

(d)

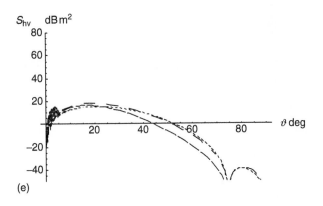

(e)

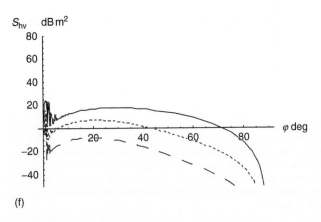

(f)

FIGURE 15.8 (continued)

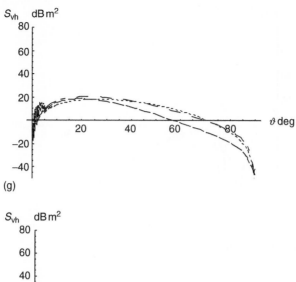

(g)

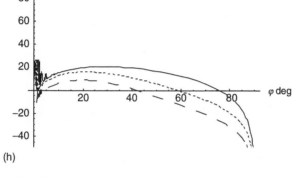

(h)

FIGURE 15.8 (continued)

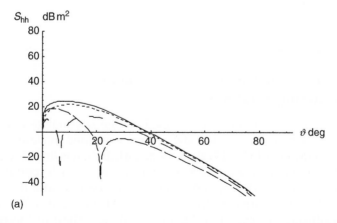

(a)

FIGURE 15.9
GO–PO–GO RCS vs. incidence angle ϑ for wall–ground–wall triple scattering. Wall height h and length l are both 20 m. Terrain roughness, frequency, and wall and ground relative dielectric constant are equal to those of Figure 15.7. Wall orientation φ is 0° (solid line), 15° (short dashes), 30° (long-spaced dashes), and 45° (long dashes).

(*continued*)

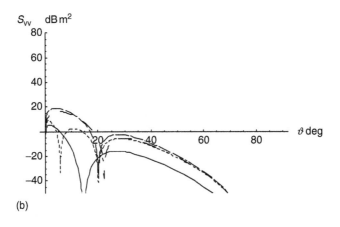

(b)

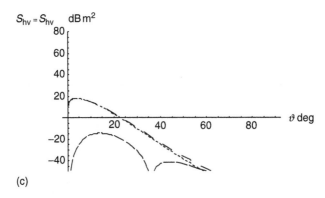

(c)

FIGURE 15.9 (continued)

15.9   Slant-Range Distribution of Canonical Scattering Solutions

The SAR signal formation is determined by the slant-range distribution discussed in Section 15.3 of the radar returns evaluated in closed forms in Sections 15.5 through 15.7. This slant-range distribution is linked to the time of arrivals of the pulses to the SAR. Time of arrival for each contribution, t_i, is proportional to the corresponding optical ray path length: then, single, double, and triple returns are located in the SAR image at different slant-range coordinates $r_i = ct_i/2$ (see Figure 15.10), according to the rationale hereafter presented.

In the time interval before t_1, the sensor receives only the field back-scattered from the ground and the reflectivity function value is determined just by the RCS of the rough terrain.

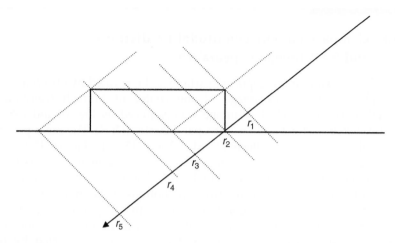

FIGURE 15.10
Relevant to the slant-range location of the single, double, and triple returns.

In the time interval (t_1, t_2), the field backscattered by the terrain is to be summed up to the return by part of the building roof and by the vertical wall; the layover phenomenon takes place and the reflectivity function increases with respect to the sole presence of the radar return from the ground.

At time t_2, all the double reflection contributions are received and summed up to the terrain, roof, and wall single-scattering contributions.

In the time interval (t_2, t_3), the SAR sensor receives the triple reflection contribution.

In the time interval (t_2, t_4), the SAR sensor receives the building roof return.

In the interval (t_4, t_5), no radar return is achieved and the shadowing phenomenon takes place.

After the time t_5, only the ground backscattering contributes to the radar return.

Note that for different building heights or widths or for different look angles, the presented scenario may change and the reflectivity map (and, as consequence, the SAR images) may even exhibit different patterns. For instance, if $t_3 < t_4$ triple reflection contribution is always received by the sensor together with the field backscattered from the building roof. Conversely, if $t_3 > t_4$ the triple reflection contribution is partly located in the shadow area. The condition $t_3 > t_4$ is verified if

$$h > \frac{L}{2} \sin 2\vartheta; \tag{15.42}$$

then, if $h > L/2$ the triple contribution term reduces the dimension of the shadowed area on the SAR image. In addition, it may occur that $t_4 < t_2$, hence the whole roof contribution is detected before the double reflection contribution.

15.10   Comparison between Model Predictions and SAR Image Appearance

Theoretical considerations reported earlier can be verified by observing the real high-resolution SAR image of the center of Munich (Germany), depicted in Figure 15.11. This image was acquired by the X-band E-SAR airborne system of the German Aerospace Center (DLR). The main parameters of the system are collected under Table 15.1. In Figure 15.11, the illumination is approximately from the bottom of the image. Let us focus attention on the building "Alte Pinakothek," indicated by an arrow in Figure 15.11. By moving upward from the bottom of the image, we notice first a gray area corresponding to the return from the ground; then, a very bright stripe corresponding to the southwest wall double reflection and layover; further above, a nonuniform dark gray area generated by the scattering from the roof (which is not flat and has some irregular objects on it, so that it is partly shadowed); and finally a black stripe corresponding to the building shadow. Moving up further, we find again a gray area representing the return from the ground. This sequence of stripes with different intensities is foreseen by the model presented in the previous sections and is a *signature* of the building geometry, so it can be used to derive building height and size [18]. But there is also an additional interesting feature that is not related to the building geometry: the bright southwest wall layover stripe is brighter on its right part, where some aligned,

FIGURE 15.11
SAR image of the center of Munich, Germany. (Courtesy of DLR.) This image was acquired by the X-band E-SAR airborne system. The arrow at the *bottom-right* part of the image points to the building of the "Alte Pinachotek." It also indicates the north direction.

TABLE 15.1

Main X-Band E-SAR System Parameters

Frequency [GHz]	9.60
Polarization	HH
PRF [Hz]	2000
Number of looks	8
Platform altitude [m]	3231
Look angle [degree]	About 50
Azimuth resolution [m]	3.0
Range resolution [m]	2.0

even brighter, points can be noticed. Such a difference is not visible on optical images and is not related to the wall geometry, which is the same in both parts of the wall. The reason is that the building was partly destroyed during the Second World War; the eastern part of the wall was reconstructed afterward with a different material (this explains the different brightness); and steel pipes were added (this explains the brighter points). Accordingly, this feature illustrates, at least qualitatively, the dependence of the building radar return on the wall complex dielectric constant, as also foreseen by the model presented in this chapter.

References

1. F.M. Henderson, Z. Xia, SAR applications in human settlement detection, population estimation and urban land use pattern analysis: a status report, *IEEE Trans. Geosci. Remote Sensing*, GE-35, 79–85, 1997.
2. Z. Xia, F.M. Henderson, Understanding the relationships between radar response patterns and the bio- and geophysical parameters of urban areas, *IEEE Trans. Geosci. Remote Sensing*, GE-35, 93–101, 1997.
3. Y. Dong, B. Forster, C. Ticehurst, Radar backscatter analysis for urban environments, *Int. J. Remote Sensing*, 18, 1351–1364, 1997.
4. G. Franceschetti, A. Iodice, D. Riccio, A canonical problem in electromagnetic backscattering from buildings, *IEEE Trans. Geosci. Remote Sensing*, GE-40, 1787–1801, 2002.
5. M. Quartulli, M. Datcu, Stochastic geometrical modeling for built-up area understanding from a single SAR intensity image with meter resolution, *IEEE Trans. Geosci. Remote Sensing*, GE-42, 1996–2003, 2004.
6. J.A. Benediktsson, M. Pesaresi, K. Arnason, Classification and feature extraction for remote sensing images from urban areas based on morphological transformations, *IEEE Trans. Geosci. Remote Sensing*, GE-41, 1940–1949, 2003.
7. F. Dell'Acqua, P. Gamba, G. Lisini, Improvemets to urban area characterization using multitemporal and multiangle SAR images, *IEEE Trans. Geosci. Remote Sensing*, GE-41, 1996–2004, 2003.

8. P. Gamba, B. Houshmand, M. Saccani, Detection and extraction of buildings from interferometric SAR data, *IEEE Trans. Geosci. Remote Sensing*, GE-38, 611–618, 2000.

9. M. Quartulli, M. Datcu, Information fusion for scene understanding from interferometric SAR data in urban environments, *IEEE Trans. Geosci. Remote Sensing*, GE-41, 1976–1985, 2003.

10. G. Franceschetti, R. Lanari, *Synthetic Aperture Radar Processing*, CRC Press, New York, 1999.

11. G. Franceschetti, M. Migliaccio, D. Riccio, G. Schirinzi, SARAS: a SAR raw signal simulator, *IEEE Trans. Geosci. Remote Sensing*, GE-30, 110–123, January 1992.

12. G. Franceschetti, M. Migliaccio, D. Riccio, SAR raw signal simulation of actual scenes described in terms of sparse input data, *IEEE Trans. Geosci. Remote Sensing*, GE-32, 1160–1169, November 1994.

13. G. Franceschetti, A. Iodice, D. Riccio, G. Ruello, SAR raw signal simulation for urban structures, *IEEE Trans. Geosci. Remote Sensing*, GE-41, 1986–1995, 2003.

14. L. Tsang, J.A. Kong, R.T. Shin, *Theory of Microwave Remote Sensing*, Wiley-Interscience, John Wiley & Sons, New York, 1985.

15. T. Griesser, C.A. Balanis, Backscatter analysis of dihedral corner reflectors using physical optics and the physical theory of diffraction, *IEEE Trans. Antennas Propagat.*, AP-35, 1137–1147, 1987.

16. R.G. Kouyomjian, P. Pathak, A uniform geometrical theory of diffraction for an edge in a perfectly conducting surface, *Proc. IEEE*, 62, 1448–1461, 1974.

17. G. Franceschetti, *Electromagnetics: Theory, Techniques and Engineering Paradigms*, Plenum Press, USA, 1997.

18. A.J. Bennett, D. Blacknell, The extraction of building dimensions from high resolution SAR imagery, in *Proceedings of the International Radar Conference 2003*, pp. 182–187, 2003.

16

Roof Mapping Using Fused Multiresolution Optical Satellite Images

Yun Zhang and Travis Maxwell

CONTENTS

16.1 Introduction

Automatic roof mapping using remote sensing imagery has always been a challenging research topic in remote sensing [1–4]. In general, high spatial resolution and multispectral information are two important information components for automatic classification or extraction of building roofs. The high spatial resolution provides geometric details for delineating individual objects, whereas the multispectral information supplies color information for differentiating objects with different spectral reflectance.

However, most remote sensors, such as Landsat ETM+, SPOT, IKONOS, QuickBird, and Z/I-DMC, collect high-resolution panchromatic (Pan) images and low-resolution multispectral (MS) images simultaneously. An optimal roof mapping result is usually difficult to achieve with the direct use of either the original Pan image or MS image.

For example, Landsat TM images contain a high level of spectral information, but their spatial resolution (30 m) is too coarse to interpret buildings. SPOT Pan images have higher spatial resolution (10 m), in which buildings can be interpreted visually, but their spectral information is not sufficient for digital classification. Similar to QuickBird MS and Pan images, for example, roofs of individual family houses can be identified in Quick-Bird MS images (2.8 m) according to color differences between roofs and their surroundings, but the roof edges cannot be clearly delineated. In QuickBird Pan images (0.7 m), however, roof edges can be clearly identified, but the gray value in the Pan images is not sufficient for classification.

To achieve accurate roof mapping results, it is, therefore, necessary to combine the spectral information of the MS image and spatial information of the Pan image of a given sensor in the building roof extraction. One efficient way to combine the spectral and spatial information is image fusion, that is, pan-sharpening the MS image using the Pan image.

This chapter introduces two different types of research studies on roof mapping using pan-sharpened remote sensing imagery: (1) pixel-based postclassification for small-scale roof mapping using fused Landsat TM and SPOT Pan images and (2) object-oriented classification for medium-scale roof mapping using pan-sharpened QuickBird images. Other types of techniques for roof mapping have also been reported [1,5,6].

16.2 Small-Scale Roof Mapping Using Fused SPOT Pan and Landsat TM Images

For detailed urban roof mapping, image fusion, multispectral classification, and spatial feature postclassification were performed in this study. The satellite images and the principles of each processing step are described in the following sections.

16.2.1 Satellite Images

The study area was the entire urban area of Shanghai, China, in the late 1980s. It covered more than 30×30 km^2. Most buildings in the center areas were small and close to each other. Bigger buildings (about 10–20 m in width) increased from the city center to the outskirts and they aligned relatively regularly.

Landsat TM, acquired on 18 May 1987, and SPOT Pan, acquired on 25 October 1989, were used for the study (Figure 16.1a and b). The TM bands 3, 4, and 7 were selected for building detection because these bands display building areas, built-up areas, green areas, and water areas well.

16.2.2   Fusion of Landsat TM and SPOT Pan

For detailed roof mapping using Landsat TM and SPOT Pan images, the synthetic variable ratio (SVR) image fusion method [7] was applied to the TM and SPOT Pan images (Figure 16.1c). The SVR method was developed based on the ratio techniques of Price [8] and Munechika et al. [9]. The SVR method preserves the spectral information of original TM images better than the widely used IHS method. In addition, its spatial resolution is as good as that of IHS fusion [10]. Hence, the SVR method leads to higher classification accuracy than the IHS method when it is applied to the fusion of SPOT Pan and Landsat TM images [11].

The principle of the SVR method can briefly be described with the following formula [7]:

$$XSP_i = Pan_H \times \frac{XS_{Hi}}{Pan_{HSyn}} \qquad (16.1)$$

$$Pan_{HSyn} = \sum \varphi_i XS_{Hi}, \qquad (16.2)$$

where
XSP_i is the gray value of the ith band of the fused TM-SPOT image
XS_{Hi} is the gray value of the ith band of the magnified TM image, which has the same pixel size as the SPOT pan image
Pan_H is the gray value of the SPOT Pan image
Pan_{HSyn} is the gray value of the synthetic panchromatic image simulated using XS_{Hi}
φ_i is the regression coefficient of the variable Pan_H and XS_{Hi}

16.2.3   Multispectral Classification

Since the unsupervised clustering method is better suited for classifying heterogeneous classes than supervised classification [12–15], the unsupervised ISODATA clustering method (ERDAS) was used for the spectral classification. In the unsupervised classification, the fused image was first subdivided into 50 clusters. The resulting classes were then extracted through additional interpretation of the 50 clusters. The classified buildings from the fused and original TM images are shown in Figure 16.2.

From Figure 16.2, it can be seen that visible big buildings (about 10–20 m in width) were extracted from the fused TM-SPOT image—although with significant noise (Figure 16.2b), while they could not be extracted from the original TM image (Figure 16.2a). The advantage of using a fused image for

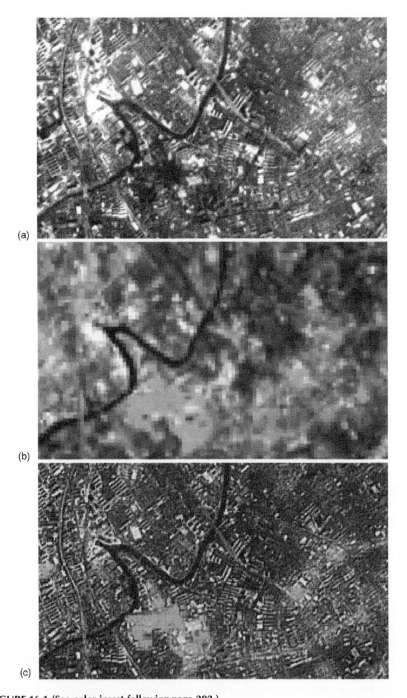

(a)

(b)

(c)

FIGURE 16.1 (See color insert following page 292.)
Medium-resolution satellite images for roof mapping (300 × 180 pixel section, 10 m pixel).
(a) SPOT Pan image; (b) TM bands 3, 4, and 7 in blue, green, and red; and (c) fused TM-SPOT
image.

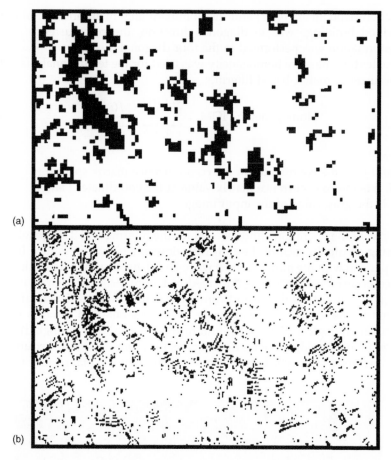

FIGURE 16.2
Multispectral classification of buildings. (a) From the original TM image and (b) from the TM-SPOT fused images.

roof mapping, instead of the original TM image, can clearly be seen by comparing the classification results in Figure 16.2.

16.2.4 Spatial Feature Postclassification

Despite the use of the fused image for the optimization of roof mapping, the accuracy of the classification result was still not satisfying for the detection of the housing development, because many nonbuilding objects were falsely classified as buildings (Figure 16.2b, also see Figures 16.4a and 16.5a). To remove the nonbuilding objects that have different spatial characteristics from buildings, a filtering algorithm based on the co-occurrence matrix technique was developed. Because the co-occurrence matrix is direction-dependent [16,17] and the dominant alignment orientation of the classified buildings was in diagonal direction, the co-occurrence matrix transformation was performed in the four diagonal directions.

The texture measure homogeneity was used for the texture analysis in the co-occurrence matrix-based filtering:

$$\text{Homogeneity: HOM} = \sum_{i=0}^{N-1} \sum_{j=0}^{N-1} \frac{f(i,j)}{1 + |i - j|}, \tag{16.3}$$

where

  i and j are the coordinates in the co-occurrence matrix space
  $f(i, j)$ is the co-occurrence matrix value at the coordinates i and j
  N is the gray value of the input image

The size of the operation window used for the filtering was 3×3 pixels. The filtering results in each of the four directions are shown in Figure 16.3. In comparison with the input image (Figure 16.4a), the direction dependence of the co-occurrence matrices is indicated in Figure 16.3a through d. In each direction, one side of the airport runway was fully filtered. The final result filtered in the four directions (Figure 16.4b) was obtained through logical combination (logical AND) of the four results (Figure 16.3a through d).

Common methods for noise filtering are, for example, simple texture analysis methods [18], which filter noise by directly calculating the texture

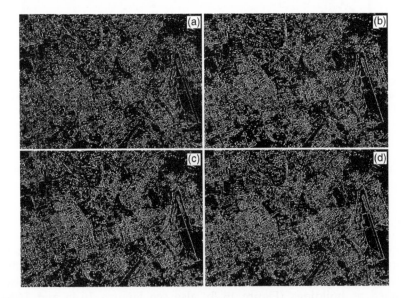

FIGURE 16.3
Results of the co-occurrence matrix-based filtering in four diagonal directions with a 3×3 operation window and homogeneity texture measure (540×380 pixel section, 10 m pixel). (a) North west, (b) north east, (c) south east, and (d) south west.

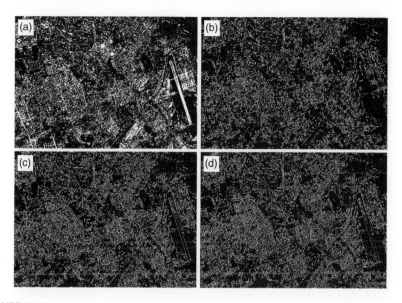

FIGURE 16.4
Comparison of buildings extracted using multispectral classification and buildings filtered using different methods (520 × 360 pixel section, 10 m pixel). (a) Multispectrally classified buildings, (b) result of the co-occurrence matrix-based filtering in the four diagonal directions with the texture measure homogeneity, (c) result of simple texture analysis filtering with the texture measure energy, and (d) result of simple texture analysis filtering with the texture measure homogeneity.

measures in original images. They are, however, insufficient for building filtering. Visual comparison of the co-occurrence filtered (Figure 16.4b) with the simple texture analysis filtered results (Figure 16.4c and d) shows clearly the advantages of co-occurrence matrix-based filtering: the airport (right), the railway station complex (bottom right), parts of streets (bottom left), and the gyms (top middle) were completely filtered in Figure 16.4b, whereas they were only partially filtered in Figure 16.4c and d.

Figure 16.5a through d show another section of the extracted buildings before and after the feature filtering. It can be seen that a significant amount of nonbuilding features (Figure 16.5c) exist in the multispectrally classified building roofs (Figure 16.5a). A clear impression about the improvement of the co-occurrence matrix-based filtering over the normal texture analysis filtering can be obtained by comparing Figure 16.5b and d. The same texture measure homogeneity was used in both methods. A detailed description of the filtering process can be found in the relevant publication [19].

16.2.5 Accuracy Assessment

A total of 400 random points were selected as reference pixels in each of two assessment areas (Area 1: Figure 16.4 and Area 2: Figure 16.5) for accuracy

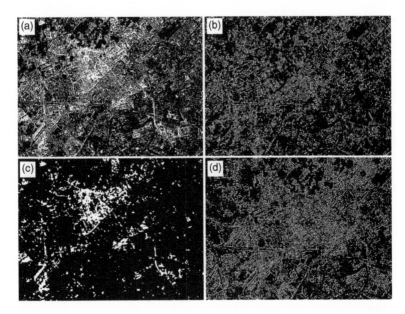

FIGURE 16.5
Comparison of the roof mapping results before and after performing a spatial feature filtering with a 3 × 3 operation window (500 × 360 pixel section, 10 m resolution). (a) Building roofs classified from the TM-SPOT fused image, (b) building roofs improved using the co-occurrence matrix-based filtering in four diagonal directions and the texture measure homogeneity, (c) nonbuilding feature removed by the co-occurrence matrix-based filtering in four directions, and (d) building roofs improved using a normal texture analysis filtering and the texture measure homogeneity.

assessment. The user accuracy and the kappa statistics of classified building roofs before and after the texture filtering are shown in Table 16.1.

The average user accuracy of the two areas has increased from 53.7% to 83.5% after the co-occurrence matrix-based improvement and the average kappa statistics has increased from 0.532% to 0.823%. Both the average user accuracy and the average kappa statistics increased by around 30% by using the co-occurrence matrix-based filtering.

TABLE 16.1

User Accuracy and Kappa Statistics of the r before and after the Texture Filtering

		User Accuracy (%)	Kappa Statistics
Multispectral classification	Area 1	48.57	0.4379
	Area 2	58.89	0.5261
Postprocessing with co-occurrence matrix-based filtering	Area 1	85.71	0.8480
	Area 2	81.25	0.7995
Postprocessing with normal texture analysis filtering	Area 1	68.75	0.6693

The accuracy improvement through normal texture analysis filtering with the texture measure homogeneity was also assessed in the first assessment area (Figure 16.5d) to compare the contributions of the co-occurrence matrix-based filtering and the normal texture analysis filtering. The assessment method was the same as earlier and the accuracy results are shown in Table 16.1. It can be seen that the accuracy of the co-occurrence matrix-based filtering is ~15% higher than that of the normal texture analysis filtering.

16.3 Medium-Scale Roof Mapping Using Fused QuickBird Pan and MS Images

For medium-scale roof mapping using high-resolution satellite images, new image fusion and classification techniques were used, because the traditional image fusion methods, such as IHS, PCA, Brovey, and wavelet fusion, introduce significant color distortion for the fusion of QuickBird (or IKONOS) Pan and MS images [20], and traditional pixel-based classification techniques cannot produce satisfactory results for high-resolution images [21,22].

The new image fusion method—Pansharp of PCI Geomatica—was used for the fusion of QuickBird MS and Pan images in this study, because the color distortion is significant when the SVR fusion method (used for the small-scale roof mapping earlier) is employed to fuse the MS and Pan images of the new sensors, such as Landsat ETM+, IKONOS, and QuickBird [20], and the PCI Pansharp produces superior fusion results [23].

The object-oriented classification of eCognition was applied to the roof mapping using pan-sharpened QuickBird MS images, because traditional per-pixel-based classification techniques are not capable of handling the gray-value variance within individual classes introduced by high spatial resolution, and object-oriented classification has demonstrated a promising direction for handling the high-resolution images [24].

16.3.1 Satellite Image Used

The images selected for this study consisted of two QuickBird scenes, both containing an MS (2.8 m) and a Pan (0.7 m) image. The first scene was a typical suburban area located in Oromocto, New Brunswick, Canada, covering 3.2 km by 2.6 km, which was collected on August 8, 2002. The second scene was a typical rural area, close to Fredericton, New Brunswick, Canada, covering 1.3 km by 1.2 km, collected on 26 July 2002. The reasons for choosing these two scenes were to evaluate the effectiveness of object-oriented classification for typical suburban areas and rural areas in Atlantic Canada. A subset of each of the two scenes is shown in Figures 16.6 and 16.7.

FIGURE 16.6 (See color insert following page 292.)
QuickBird multispectral image of suburban scene in Oromocto (subset, 2.8 m resolution).

16.3.2   Image Fusion

The image fusion method—PCI Pansharp—used in this study is a statistics-based fusion technique, which solved the two major problems in image fusion—color distortion and operator (or dataset) dependency. It is different from existing image fusion techniques in two principal ways [23]:

FIGURE 16.7
QuickBird multispectral image of rural scene close to Fredericton (subset, 2.8 m resolution).

1. It uses the least-squares technique to find the best fit between the gray values of the image bands to be fused and to adjust the contribution of individual bands to the fusion result to reduce the color distortion.

2. It employs a set of statistic approaches to estimate the gray value relationship between all the input bands to reduce the influence of dataset variation and to automate the fusion process.

The fused images of the two study scenes are shown in Figures 16.8 and 16.9, from which it can be seen that the spectral information of the fused images is well preserved, that is, the color of the fused images is almost identical to that of the original multispectral images (Figures 16.6 and 16.7).

16.3.3 Object-Oriented Classification

Object-oriented classification consists of two major processing components: (1) segmenting an image into meaningful segments and (2) classifying the segments into different classes.

Usually, segmentation is processed in two steps:

1. Performing initial segmentation to generate primitive subsegments

2. Selecting segmentation parameters to merge the subsegments (primitives) to form meaningful segments

FIGURE 16.8 (See color insert following page 292.)
Pan-sharpened QuickBird image of the suburban scene produced from Figure 16.6 and the corresponding QuickBird Pan image (subset, 0.7 m resolution).

FIGURE 16.9
Pan-sharpened QuickBird image of the rural scene produced from Figure 16.7 and the corresponding QuickBird Pan image (subset, 0.7 m resolution).

In the classification component, the following two processing steps are required:

1. Building a rule base for knowledge-based classification
2. Classifying the objects according to meaningful segments and rule base

16.3.3.1 Segmentation

First, a region-growing approach was applied to the segmentation of the image for producing homogeneous primitive subsegments (Figure 16.10).

The operator was then required to define three interrelated parameters—scale (f), shape weighting factor $(1-w)$, and smoothness weighting factor $(1-w_c)$—to merge the subsegments (primitives) to form meaningful segments [25]. The relationships between the three parameters $(f, 1-w,$ and $1-w_c)$ are described by the equations [25,26]:

$$f = w \times h_{\text{spectral}} + (1 - w) \times h_{\text{shape}}, \tag{16.4}$$

$$h_{\text{shape}} = w_c \times h_{\text{compact}} + (1 - w_c) \times h_{\text{smoothness}}, \tag{16.5}$$

where
  h_{spectral} stands for spectral heterogeneity—a measure of heterogeneity change after merging two adjacent primitives
  h_{shape} stands for shape heterogeneity—a measure of shape change after the merge of two adjacent primitives

FIGURE 16.10
Primitive subsegments obtained by performing initial segmentation.

$h_{compact}$ for compactness heterogeneity—a function of object perimeter and number of pixels within the object

$h_{smoothness}$ is smoothness heterogeneity—a function of object perimeter and the perimeter of the objects bounding box

Weights w, $1-w$, w_c, and $1-w_c$ are user-assigned weights associated with $h_{spectral}$, h_{shape}, $h_{compact}$, and $h_{smoothness}$, respectively.

The value f, so-called fusion value or scale parameter, is an indicator of the overall heterogeneity change for a potential merge between two primitives (subsegments). Once the fusion value (f) falls below a user-specified threshold, the merge between two objects will be accepted.

The definition of the three interrelated parameters—scale (f), shape weighting factor ($1-w$), and smoothness weighting factor ($1-w_c$)—is an iterative, trial-and-error process. Numerous combinations of f, $1-w$, and $1-w_c$ can be selected and tested according to the operator's experience. The best merging result is selected as the final segmentation of the image (Figure 16.11).

16.3.3.2 Classification

In object-oriented classification, a variety of features based on tone, texture, shape, and context can be taken into account for the classification of individual meaningful segments [27]. A classification scheme based on fuzzy logic is involved in the classification. In this study, rules for the classification

FIGURE 16.11
Meaningful segments obtained by merging primitive subsegments (Figure 16.10) through trial-and-error selection of the three parameters f, $1-w$, and $1-w_c$.

scheme were built using the previously mentioned features and user-defined membership functions. Using the meaningful segments as the basic processing units and according to the classification rules built by the operator, the segments were classified into different classes.

Because fuzzy logic is involved in the classification, it permits image segments (objects) to partially belong to any class. However, a defuzzification process is performed in the end to determine the highest membership value for a particular image segment and then to assign the segment to an appropriate class.

In this study, the pan-sharpened QuickBird images were classified into five classes: roofs of buildings or houses, pavements, bare soil, trees, and grass (Figures 16.12 and 16.13).

Further examples and detailed descriptions of roof mapping using fused QuickBird images can be found in the relevant publication [28].

16.3.3.3 Results Comparison

To assess the results of medium-scale roof mapping using pan-sharpened QuickBird MS images, the original QuickBird MS images of the same areas were also classified using the same classification method—object-oriented classification (eCognition)—under the same condition. Figure 16.14 illustrates the difference between the roof mapping results from pan-sharpened MS images and original MS images. It can be clearly seen that

FIGURE 16.12
Roof mapping result in a suburban area using object-oriented classification (eCognition) and pan-sharpened QuickBird MS image.

FIGURE 16.13
Roof mapping result in a rural area using object-oriented classification (eCognition) and pan-sharpened QuickBird MS image.

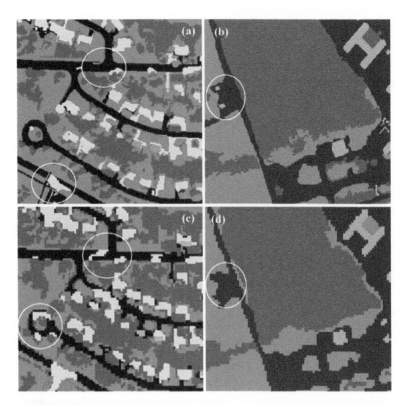

FIGURE 16.14
Comparison of roof mapping results from pan-sharpened MS images and original MS images.
(a) and (b) results from pan-sharpened QuickBird images; (c) and (d) results from original
QuickBird images.

1. The boundaries of individual objects are much more detailed in
   the results of pan-sharpened images (Figure 16.14a and b) than
   those of original MS images (Figure 16.14c and d) (compare the
   boundaries of roofs and roads).
2. The results from pan-sharpened images are more accurate than
   those from original MS images (compare the circled areas in
   Figure 16.14).

Quantitative assessment using randomly distributed points was also conducted. The average overall accuracy was 84% for the classification results
from pan-sharpened QuickBird MS images and 85% for those from the
original QuickBird MS images. Due to limited random points for the evaluation, the improvement at the edges of individual objects resulting from the
pan-sharpening could not be reflected in the statistic results.

However, when both the visual evaluation and statistic assessment are
considered, we can see that the contribution of pan-sharpening to the
classification is the improvement of object detail, including edge, shape,

and size, while the gray value variation introduced by the pan-sharpening into individual objects does not affect much the overall accuracy of the classification result.

16.4 Conclusions

From the two different types of research studies introduced in the chapter, we can see that an effective image fusion can significantly contribute to the improvement in accuracy of roof mapping.

When medium-resolution satellite images are used for small-scale roof mapping, it is impossible to extract building roofs by just using Landsat TM images. However, if the Landsat TM images are fused with corresponding SPOT panchromatic images, it becomes possible to extract big buildings. However, the results extracted using multispectral classification usually contain much noise due to other objects that have similar spectral reflectance, such as roads and other paved areas. A postclassification process is required to achieve better roof mapping accuracy.

When high-resolution satellite images are used for medium-scale roof mapping, traditional multispectral classification techniques are no longer adequate. Object-oriented classification produces better roof mapping results. However, if original high-resolution multispectral images, such as QuickBird MS, are directly used for roof mapping, building and house roofs can be extracted, but with coarse object boundaries. When pan-sharpened QuickBird images are used for roof mapping, the object boundaries can be significantly smoothened and the roof mapping accuracy can be increased. Some small houses that cannot be mapped using the original QuickBird MS images can be extracted from the pan-sharpened image.

From the two studies, it has also been recognized that even though pan-sharpened images can increase the roof mapping accuracy, the final roof mapping results still contain errors, irrespective of whether medium- or high-resolution satellite images are used. Such results may suit overall investigations of housing development. However, for detailed mapping purposes manual corrections are still required.

References

1. Peng, J. and Liu, Y.C., Model and context-driven building extraction in dense urban aerial images, *International Journal of Remote Sensing*, 26, 2005, 1289–1307.
2. Zebedin, L., et al., Towards 3D map generation from digital aerial images, *ISPRS Journal of Photogrammetry & Remote Sensing*, 60, 2006, 413–427.
3. Ruther, H., Martine, H.M., and Mtalo, E.G., Application of snakes and dynamic programming optimization technique in modeling of buildings in informal settlement areas, *ISPRS Journal of Photogrammetry & Remote Sensing*, 56, 2002, 269–282.

4. Mayunga, D., Coleman, D., and Zhang, Y., Mapping detailed informal settlement areas using high-resolution QuickBird images, *International Journal of Remote Sensing*, accepted March 2006, in press.

5. Tupin, F. and Roux, M., Detection of building outlines based on the fusion of SAR and optical features, *ISPRS Journal of Photogrammetry & Remote Sensing*, 58, 2003, 71–82.

6. Jung, F., Detecting building changes from multitemporal aerial stereopairs, *ISPRS Journal of Photogrammetry & Remote Sensing*, 58, 2004, 187–201.

7. Zhang, Y., A new merging method and its spectral and spatial effects, *International Journal of Remote Sensing*, 20, 1999, 2003–2014.

8. Price, J.C., Combining panchromatic and multispectral imagery from dual resolution satellite instruments, *Remote Sensing of Environment*, 21, 1987, 119–128.

9. Munechika, C.K., et al., Resolution enhancement of multispectral image data to improve classification accuracy, *Photogrammetric Engineering & Remote Sensing*, 59, 1993, 67–72.

10. Zhang, Y. and Albertz, J., Vergleich verschiedener Verfahren zur Kambination multisensoraler Satelliten-Bilddaten, *Photogrammetrie-Fernerkundung-Geoinformation (PFG)*, 5, 1998, 261–274.

11. Prinz, B., Wiemker, R., and Spitzer, H., Simulation of high resolution satellite imagery from multispectral airborne scanner imagery for accuracy assessment of fusion algorithms, in *Proceedings of the ISPRS Joint Workshop "Sensors and Mapping from Space" of Working Groups I/1, I/3 and IV/4*, Hannover, Germany, 1997, pp. 223–231.

12. Jensen, J.R., et al., Improved urban infrastructure mapping and forecasting for BellSouth using remote sensing and GIS technology, *Photogrammetric Engineering & Remote Sensing*, 60, 1994, 339–346.

13. Zhang, Y., *Aufbau eines auf Satellitenfernerkundung basierten Informationssystems zur städtischen Umweltüberwachung: Das Beispiel Shanghai*, PhD Dissertation, Berliner Geowissenschaftlichen Abhandlungen, Reihe C, Band 17, Berlin, 1998.

14. Csathó, B. and Schenk, T., Multisensor data fusion for automatic scene interpretation, *International Archives of Photogrammetry and Remote Sensing*, 32, 1998, 429–434.

15. Macleod, R.D. and Congalton, R.G., A quantitative comparison of change-detection algorithms for monitoring Eelgrass from remotely sensed data, *Photogrammetric Engineering & Remote Sensing*, 64, 1998, 207–216.

16. Haralick, R.M., Statistical image texture analysis, in Young, T.Y., Fu, K.-S. (Eds.), *Handbook of Pattern Recognition and Image Processing*, Academic Press, Orlando, 1986.

17. Bässmann, H. and Besslich, P.W., *Bildverarbeitung: ad oculus*. 2. Auflage, Springer-Verlag, Berlin, 1993.

18. Hsu, S.-Y., Texture-tone analysis for automated land-use mapping, *Photogrammetry Engineer & Remote Sensing*, 44, 1978, 1393–1404.

19. Zhang, Y., Optimization of building detection in satellite data by combining multispectral classification and texture filtering, *ISPRS Journal of Photogrammetry & Remote Sensing*, 54, 1999, 50–60.

20. Zhang, Y., Problems in the fusion of commercial high-resolution satellite images as well as Landsat 7 images and Initial solutions, in *International Archives of Photogrammetry and Remote Sensing (IAPRS)*, 34, Part 4, *ISPRS, CIG, SDH Joint International Symposium on "GeoSpatial Theory, Processing and Applications,"* Ottawa, Canada, July 8–12, 2002.

21. Carleer, A.P. and Wolff, E., Urban land cover multi-level region-based classification of VHR data by selecting relevant features, *International Journal of Remote Sensing*, 27, 2006, 1035–1051.

22. Puissant, A., Hirsch, J., and Weber, C., The utility of texture analysis to improve per-pixel classification for high to very high spatial resolution imagery, *International Journal of Remote Sensing*, 26, 2005, 733–745.

23. Zhang, Y., Highlight article: Understanding image fusion, *Photogrammetric Engineering & Remote Sensing*, 70, 2004, 657–661.

24. Hu, X., Tao, C.V., and Prenzel, B., Automatic segmentation of high-resolution satellite imagery by integrating texture, intensity and color features, *Photogrammetric Engineering & Remote Sensing*, 71, 2005, 1399–1406.

25. Definiens Imaging GmbH, *eCognition User Guide 4*, Munich, Germany, 2004.

26. Baatz, M. and Schape, A., Multiresolution segmentation—an optimization approach for high quality multi-scale image segmentation, in Strobl, J., et al (Ed.), *Angewandte Geographische Informationsverarbeitung XII*, AGIT Symposium, Salzburg, Germany, 2000, pp. 12–23.

27. Definiens Imaging GmbH, Whitepaper—eCognition Professional 4.0, Munich, Germany, 2004. [Internet: http://www.definiens-imaging.com/documents/brochures.htm]

28. Maxwell, T., *Object-Oriented Classification: Classification of Pan-Sharpened Quick-Bird Imagery and A Fuzzy Approach to Improving Image Segmentation Efficiency*, MScE Thesis, Department of Geodesy and Geomatics Engineering, University of New Brunswick, 2005, 169 p.

Part V

Impervious Surface Data Applications

17

Impervious Surface Area and Its Effect on Water Abundance and Water Quality

Toby N. Carlson

CONTENTS

17.1 Background

Impervious surface area (ISA) affects both water quality and water abundance through its influence on surface runoff. Runoff increases with increasing ISA, a fact that is well known to developers and watershed managers who must estimate the amount of surface runoff to streams and rivers resulting from precipitation events. Localized flooding occurs much more often now in urban areas than before because of development. Most serious urban flooding and surface runoff originate from commercially developed sites such as parking lots or shopping centers. Figure 17.1 shows a large shopping center with considerable impervious surface cover, punctuated by some scattered trees. Runoff from this site, which is a large parking lot, would be considerable.

Washout of city streets as well as runoff from agricultural and other developed surfaces brings with it elevated concentrations of nutrients (such as nitrogen and phosphorus) and sediment. Nitrogen and phosphorus in runoff are known to originate from farm practices (fertilizers) and pasturage, but also come from suburban and urban lawns and gardens. These and other constituents are swept into storm sewers and (ultimately) into streams and estuaries, where they foster the growth of algae. Increased

FIGURE 17.1
Photographs of Eagleview shopping center in Uwchlan Township, Pennsylvania (location C in Figure 17.3), illustrating an area of high ISA.

sediment loading in streams resulting from increased surface runoff erodes the land surface over which the runoff must travel to reach sewers and streams, where the additional volume of water causes increased bank erosion. Sediment itself adversely affects fish habitats, killing off trout and other game fish. Conversely, infiltration of storm water into the ground, rather than into streams, acts to purify the water, removing various toxic constituents such as metals, and ultimately to provide more clean water for plants, animals, and humans.

Fifty years ago, the chief worry in urban construction was to get rid of the rainwater as quickly as possible. Emphasis was on flood prevention, but not on water conservation. As a result, urban engineers developed very efficient methods using various types of channels to shunt the water away from where people congregate and deliver that water to the nearest stream. As urbanization gained momentum, an effort was made to reduce the runoff through various practices, the most common of which is a detention pond. Detention ponds result in the slowed release of water to streams but not in reducing the total surface runoff. Detention seemed to work in solving the urban flooding problem but not the greater problem of water abundance and water quality. Gradually, it dawned on people that such practices were contributing to the pollution of streams and rivers, regardless of how much city engineers were able to retard surface runoff. In effect the practice of getting rid of runoff was contributing to a loss of fresh water in the ground. Later, retention rather than detention became a more acceptable mode of storm water management. Retention strategies involve the creation of dry recesses or wetlands to trap and hold rainwater on the property.

The effect of ISA on both water quality and abundance is therefore of great economic importance, as well as a matter of public safety. Following the national Clean Water Act of 1972, states began to adopt their own plans for dealing with water pollution and abundance. For example, the Pennsylvania legislature passed Act 167 requiring municipalities to create their own Best Management Plans (BMP); other states have similar requirements. Currently, most of the BMPs are supposed to address the amount of increased surface runoff created by development as well as the peak flood rates engendered.

In principle, developers must account not only for the increased runoff from construction sites but they are also required to design those sites such as to retain the amount of increased runoff caused by the development. Even residential housing construction can produce significant increases in ISA and surface runoff. Figure 17.2 shows a typical residential neighborhood which, despite the abundance of trees and grass, has an ISA fraction of approximately 0.5, as measured by satellite (Hebble et al., 2001).

Knowledge of how much additional ISA is created by development and how much water will be discharged into storm sewers as a result of the development is crucial for building contractors and for the public, as the remedial efforts will result in greater costs to consumers such as home buyers. That knowledge is dependent on an accurate assessment of how building will change the landscape. At present, how to calculate the amount of additional storm water runoff created by development is as much an educated guess as it is science.

FIGURE 17.2
Photograph of typical residential area in Uwchlan Township, Pennsylvania (Acker Park in Figure 17.3).

17.2   Surface Runoff and Impervious Surface Area

One of the oldest methods involving the relationship between land use and surface runoff was developed by the Soil Conservation Service (SCS) more than 25 years ago (Bedient and Huber, 1992). The SCS method, sometimes referred to as TR-55, determines a quantity called the excess precipitation, Pe, which is the (cumulative) portion of the precipitation (e.g., in inches) that runs off at the surface directly during and just after a rainstorm. It is also the equivalent rainfall depth of the surface runoff, averaged over the area in question, for example, the area of a watershed basin. Surface runoff can arise from overland surface flow, flow within drainage pipes and sewers, or flow from the top, saturated layers of soil near the stream section. This direct runoff rate, qd, is customarily expressed as cubic feet per second or cubic meters per second. Rainwater that does not evaporate or is not abstracted into deep layers of the soil as ground water is left as surface runoff. The total flow rate, Qp, in a stream consists of this direct surface runoff plus that from ground water, the base flow, which continues at a much slower rate during and between rainstorms.

The ratio of Pe to the total precipitation P is sometimes referred to as a bulk runoff coefficient. Since qd can be estimated from hydrograph data obtained from stream gauges, it is customary to equate Pe to the total integrated volume of direct runoff Vd, that which reaches the stream as surface flow, using the relationship Vd/Ap = Pe where Ap is the effective drainage area. However, direct runoff seldom drains an entire basin, but rather a much smaller portion of the entire area; Ap is typically 10%–20% of the total basin area A (Sheeder et al., 2002). In general, the partial drainage area Ap is not known a priori.

The SCS equations governing the excess rainfall Pe can be written as

$$Pe = \frac{(P - KS)^2}{P + (1 - K)S},$$

(17.1)

where
   P is the total (cumulative) precipitation for a rain event
   S is a storage parameter related to land use and therefore to ISA and is directly a function of the so-called curve number (CN)
   K is an arbitrary coefficient that was originally determined from observations to be 0.2

The K value applies best in heavy rainstorms, although K is now thought to be smaller than 0.2 (possibly as small as 0.05), at least in light or moderate rainstorms (Sheeder et al., 2002). S represents a potential storage depth and is defined in terms of the land surface, vegetation, and soil characteristics, larger values representing more porous surfaces and smaller values representing more impervious surfaces including nominally pervious surfaces which, through compaction or lack of vegetation, tend to impede infiltration.

S is defined in terms of the curve number CN by the formula

$$S = 1000/CN - 10. \tag{17.2}$$

Note that the value of Pe may vanish for lighter rainstorms or for highly pervious surfaces (larger S) and approach the value of P for heavier rainstorms or for highly impervious surfaces (small S). It may even vanish for relatively heavy rainstorms over highly pervious surfaces such as a forest where S tends to be large. For typical urban and rural surfaces, S tends to range between values of 1 and 10 in.

Let us now look at the relationship, albeit highly empirical, between the land-use parameter S and the total volume of runoff, expressed as a volume in cubic feet, which would occur as the result of a 25 year rainstorm in central Pennsylvania, a precipitation amount approximately equal to 4.2 in. in 24 h. Table 17.1 is constructed for this event, assuming $K = 0.2$ and that only 20% of the watershed or basin area actually drains the surface precipitation, which is to say we let $Ap/A = 0.2$, as suggested earlier in this section.

Table 17.1 is schematic and meant to illustrate the importance of ISA on surface runoff. (The relationship between ISA and S is based on our own estimates derived from examining a wide range of published values.) Most such calculations in the literature assume that $A = Ap$ (the drainage area is the same as the entire watershed area), which, in this case, would yield values five times larger than those shown in the last column of Table 17.1. Such values appear excessive in our experience.

Let us consider an example for the case where a meadow is replaced by a dense housing development. In the case of a 25 year rainstorm a developer would, according to the table, be obliged to account for an increase of 879 ft^3 of water per acre of development resulting from the increase in surface runoff associated with the imposition of impervious surfaces. This amount of runoff would need to be contained by a pond with dimensions of approximately 20 ft × 20 ft × 2 ft for a 1 acre lot. Construction of

TABLE 17.1

S Value (inches), Percent Impervious for Typical Land Surfaces, Including Estimated Volume of Runoff (Cubic Feet Per Acre for the 25 year Rainstorm: 4.2 in. in 24 h)

Type of Surface	Percent Impervious	S Value (in.)	Volume of Runoff (25 year) Storm (ft³/acre)
Parking lot	100	0	3027
Urban center	95	1.1	2352
Dense housing ($\frac{1}{4}$ acre)	80	1.75	2069
Medium-density residential ($\frac{1}{3}$ acre lots)	50	3.3	1590
Sparse housing (1 acre lots)	20	4.3	1380
Meadow	0	5.4	1190
Woodland	0	8.2	860

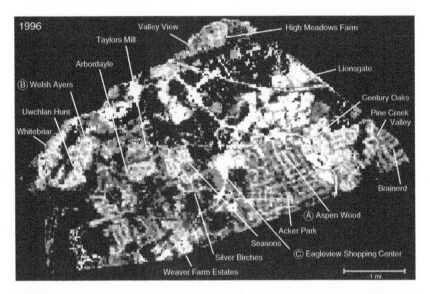

FIGURE 17.3
ISA map for Uwchlan Township for 1996, as determined from Landsat imagery. ISA values are represented from white to black, with completely impervious areas appearing white and areas with no ISA represented by black. Eagleview shopping center (Figure 17.1) is denoted by the letter C. Acker Park (Figure 17.2) is also labeled, along with a number of other housing developments.

medium-density housing would generate 400 ft^3 of additional surface run-off per acre. Clearly, such estimates translate to extra costs for the developer.

The pattern of such medium-density housing is easily recognizable on satellite imagery, expressed in the form of ISA maps. ISA analyses based on Landsat imagery for Uwchlan Township in Pennsylvania reveal the characteristic lattice structure for housing developments (Figure 17.3), whereas commercial centers and complexes appear as more solid white areas. Figure 17.3 shows various types of housing developments identified by name. Locations marked A or B and C refer, respectively, to the type of medium-density housing shown in Figure 17.2 and the Eagleview strip mall shown in Figure 17.1; the latter is a fair representation of the urban center category in Table 17.1.

Analyses of this township (Hebble et al., 2001) show a marked increase in Uwchlan township's overall ISA between 1986 and 1996; during the same time period, the area used for residential purposes increased about 20%. However, when examined according to lot size, it is apparent that most of the increase in residential ISA had come through growth of relatively low ISA neighborhoods. Nevertheless, the highest ISA values were found in commercial areas.

Residential ISA appears to be bounded by a threshold level of ISA. Hebble et al. (2001) found mostly commercial development with an ISA value on Landsat images greater than about 75%–80%, such as the Eagleview shopping center shown in Figure 17.1. Residential development generally corresponded in that study to ISA values between 30% and 70%, while the only residential areas to exceed this range were town house developments, which are similar in character to some Uwchlan Township business areas; these findings are in accord with those published by Arnold and Gibbons (1996).

17.3 Impervious Surface Area and Water Quality

Excess storm water—that which runs off as the result of urban development—carries with it increased loading of contaminants, including nutrients (phosphorus and nitrogen) and sediment. Arnold and Gibbons (1996) suggest a threshold value of 10% for the average ISA within a watershed, above which the principal stream can be classed as degraded, which is to say impaired for human use and aquatic life. A well-known diagram expressing the relationship between stream impairment and ISA was published by Schueler (1994) showing various threshold levels for stream impairment: 0%–5%, little impairment; 5%–10%, impairment begins; 10%–25%, runoff exceeds ability of stream to process; greater than 25%, severe impairment with little or no remedial methods available.

In fact, however, some stream degradation begins as soon as urban runoff begins. Even nonurban factors, such as agriculture and forest land, can affect the degree of impairment. Within the northeastern United States, and specifically the Chesapeake Bay Watershed, forest and agricultural land figure importantly in affecting the nutrient and sediment loading of storm runoff. Figure 17.4 shows relationship between phosphorus and percent woodland for approximately 40 watersheds in Pennsylvania. To

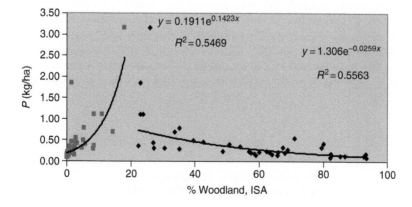

FIGURE 17.4

Phosphorus yield vs. percent woodland or percent ISA for approximately 40 watersheds in Pennsylvania. Square symbols pertain to the relationship between phosphorus yields and ISA and the small diamond symbols to the relationship between phosphorus yields and percent woodland. The two sets of data were used to create the regression equation (Equation 17.5).

compile these data, daily flow rates were obtained from the U.S. Geological Surface for a 10 year period from 1989 to 1999 (Evans et al., 2002), and nutrient concentration was obtained from the Pennsylvania Department of Environmental Protection (PaDEP). Historical water-quality data were compiled for either the 1987–1994 or 1990–1996 period, depending on the availability of the data collected. From these two datasets and the flow rate measurements, yearly nutrient and sediment yields (mass per unit area) were then computed.

Sheeder and Evans (2004) later evaluated these data in the context of stream impairment for 29 of the watersheds. In Pennsylvania, stream impairment is assessed using biological techniques, which are commonly applied to detect ecosystem impairment. Biological data are ideally suited to detect ecosystem impairment but do not provide information on the causes of that impairment. Their analysis yielded information on total, impaired, and unimpaired river mileage within each basin. These basins were then designated as impaired or unimpaired based on the percentage of total river mileage listed as impaired.

What Sheeder and Evans found was startling. Using the 95% confidence limit as a criterion, very sharply defined thresholds for impairment and nonimpairment existed for the nutrients and sediment yields. These thresholds are listed in Table 17.2.

These earlier results suggest that, while some deleterious effects of increased nutrient and sediment loads may gradually increase with increasing yields, a striking deterioration occurs within a very narrow range of nutrient or sediment yields as the latter increase.

All of the approximately 40 Pennsylvania watersheds were subjected to image analyses in which percent woodland and ISA were determined by methods described by Hebble et al. (2001). Regressing nutrient and sediment yields against ISA and percent woodland for each of the approximately 40 Pennsylvania watersheds produced relationships with a high degree of fit, as shown by Equations 17.3 through 17.5, respectively, for sediment (Sed), nitrogen (N), and phosphorus (P), expressed in units of kilogram/hectare/year as functions of percent woodland (%Wood), percent ISA, and the natural logs of the constituents (Ln). P values for the regressions were very small and the R^2 values for these three regression

TABLE 17.2

Estimated Nitrogen, Phosphorus, and
Sediment Unit Area Load Thresholds
(kg/ha/year)

Nitrogen	8.64
Phosphorus	0.30
Sediment	785.3

equations were all approximately 0.65. Values shown to the right of the equations represent one standard deviation in the measured data used to create the regressions.

$$\text{Ln (Sed)} = -7.184 + 0.1276 \times \text{Ln (\%ISA)}$$
$$-0.01459 \times \text{\%Wood } (\pm 110 \text{ kg/ha/year}), \tag{17.3}$$

$$\text{Ln (N)} = 3.6483 - 0.03115 \times \text{\%Wood } (\pm 1.57 \text{ kg/ha/year}), \tag{17.4}$$

$$\text{Ln (P)} = -0.375 + 0.20798 \times \text{Ln (\%ISA)}$$
$$-0.01556 \times \text{\%Wood } (\pm 0.2 \text{ kg/ha/year}). \tag{17.5}$$

It is somewhat curious that nitrogen exhibits no significant correlation with ISA. (Statistical treatment of the data is described by Haase et al. (2007)). This surprising result may be attributed to the fact that nitrogen depends mostly on agricultural sources and tends to be absorbed by trees and other vegetation.

One might question the absence of agricultural land as a regression predictor in these equations. Agricultural land area is not expressed here for a couple of reasons. First, when all the categories of classification are combined, the sum of agricultural land, woodland, and urban areas is almost equal to 100% of the area under consideration. This means that, given two of the categories, the third is essentially prescribed. Second, the choice of woodland as the nonurban predictor was made because agricultural land is inherently difficult to classify by satellite imagery; for example, bare soil may constitute plowed fields or a nonagricultural use while grassland may serve as pasture, be mistaken for a crop, or represent simply a meadow. Further justification for the preference for the choice of woodland over agricultural land, as pertains to the present dataset, is presented by Chang (2002). In any case, woodland is much easier to identify and classify on a satellite image than agricultural land. Yet, woodland in itself is more than a passive predictor—a category that is neither urban nor agricultural. Woodland and its attendant porous soils have the ability to filter and absorb nutrient and sediment before these constituents can reach the stream.

Table 17.2, in combination with Equations 17.3 through 17.5, enables one to assess stream impairment using percentages of woodland and ISA. The discrete nature of the impairment threshold, moreover, allows one to categorize impairment as either present or not present. In the former condition, one can tell how seriously degraded the stream is, whereas in the latter condition, one can assess the stream's relative absence of contaminants.

17.3.1 Evaluating the Regression Equations

Although a thorough discussion of the accuracy of these regression equations and how they were evaluated lies outside the scope of this chapter, a brief comparison of their output with independent measurements or estimates of nutrient yields is now presented. Two data sources were compared with the simple model (Equations 17.3 through 17.5): a set of observations and output from a complex model. Because Pennsylvania yield data are limited to that state alone, additional measurements were obtained for the region outside the state (USGS, 1998). Eighty-four nontidal stream sites were considered for the Chesapeake Bay Watershed; 36 of these within Pennsylvania and the rest in Virginia, Maryland, and New York. Output from a statistical model, SPARROW (Preston et al., 1998; SPARROW, 2006), was also obtained for additional comparison with the output of Equations 17.4 and 17.5 (sediment data not included). SPARROW requires a large range of different input parameters, some of which are quite specific and pertain to land cover, land use, and agricultural practices. Despite the practical difficulty in its application, this model is considered state-of-the-art and so can serve as a check on the validity of the regression equations, which treat ISA in a much more explicit manner than does SPARROW. In contrast, Equations 17.3 through 17.5 are more transparent in revealing the major factors in the loading of nutrients and sediment, at least in streams in the northeastern United States.

In order to compare the simple regression model with SPARROW—and therefore impart a semblance of credibility to the regression equations presented here—it was necessary to obtain additional ISA and land classification estimates for all of the Chesapeake Bay Watershed, as the number of data points for Pennsylvania was relatively limited. Land surface data were obtained from Landsat imagery, available on the Pennsylvania Spatial Data Site (PASDA) and as obtained from the University of Maryland (UMD) (Pittman, private communication). As a first step, estimates of ISA and percent woodland created at Penn State and those from the UMD were compared for both ISA and percent forest cover for overlapping areas in Pennsylvania. Values of both land-cover parameters were in close agreement, thereby allowing us to use either land surface classifications interchangeably and to expand the area over which SPARROW output could be compared with that from the simple regression model. No comparisons were made for sediment yields, however, because SPARROW output did not include sediment and because of large differences in magnitude between our sediment data (Evans et al., 2002) and those obtained published by USGS.

To facilitate comparison and also to allow the user to quickly assess the health of a particular stream, a stream health index (SHI) was created based on the relationship between the estimated or measured constituent yields as referenced to the impairment threshold values referred earlier. Realizing that the threshold range found by Sheeder and Evans (2004) was very narrow, an SHI value corresponding to the threshold was assigned an arbitrary value of zero, only when the estimated yield was within the narrow range of ±10% of the threshold. Nutrient or sediment yields >10% above the threshold but less than double the threshold value were assigned a value of −1. Yields equal to or greater than double the threshold value were assigned an SHI of −2. Similarly, yields ≤10% below the threshold but greater than half the threshold were assigned an SHI of +1. Yields equal to or less than half the threshold value were assigned an SHI of +2. Accordingly, the values of SHI for each constituent can range from −2 for seriously degraded streams to +2 for very clean streams. By summing the SHI of each of the three constituents, a stream can have a net SHI between −6 and +6.

The total (summed) SHI for two or for all three constituents is useful in that it accounts for the fact that each constituent may yield a different SHI value. For example, a stream may contain nutrient or sediment yields above the threshold for one constituent and below for another; typically, however, SHI for the three constituents tends to vary within ±1. The sum for all three constituents therefore yields a more meaningful gradation of values than for a single one. Moreover, in reducing the numerical values to discrete categories, interpretation of the data is facilitated because the data scatter is effectively minimized. For example, suppose an actual measured SHI was at a level five times the threshold and the estimated values from the regression equations indicated a yield only 20% above threshold. Despite the large difference in yields, the SHI would only differ by one category and yet convey the fact that the stream is suffering serious impairment. A fuller treatment of the comparisons between measurements or SPARROW and Equations 17.4 and 17.5 can be found on the web site, http://www.sharp.psu.edu.

SHI values were determined additionally for the measurements and for the SPARROW model output. Table 17.3 encapsulates the comparisons by

TABLE 17.3

Comparison of Output from Equations 17.3 and 17.4 with Measurements or Output from the SPARROW Model for Various Constituents, Showing the Percent of the Values That Differed by One SHI Category or Less

Constituent	Area	Comparison	Percent within One SHI Category
Nitrogen	Pennsylvania	PSU vs. SPARROW	86
Phosphorus	Pennsylvania	PSU vs. SPARROW	89
Nitrogen + Phosphorus	Pennsylvania	PSU vs. SPARROW	77
Nitrogen + Phosphorus	Pennsylvania	SPARROW vs. measurements	69
Nitrogen + Phosphorus	Pennsylvania	PSU vs. measurements	88
Nitrogen	Outside Pennsylvania	PSU vs. SPARROW	80
Phosphorus	Outside Pennsylvania	PSU vs. SPARROW	84
Nitrogen + Phosphorus	Outside Pennsylvania	PSU vs. SPARROW	71
Nitrogen + Phosphorus	Outside Pennsylvania	SPARROW vs. measurements	48
Nitrogen + Phosphorus	Outside Pennsylvania	PSU vs. measurements	34
Nitrogen + Phosphorus	Entire Chesapeake Bay Basin	SPARROW vs. measurements	32

Note: PSU in the comparison column refers to the output from Equations 17.1 and 17.2. Outside Pennsylvania refers to data within the Chesapeake Bay but not within Pennsylvania.

showing the percent of the cases for a particular pair of comparisons for output from Equations 17.4 and 17.5, measurements, and SPARROW. Listed are the percentages of cases for which the SHI differed by no more than one SHI category.

The table indicates that the PSU model (Equations 17.4 and 17.5) predicts SHI values close to those that would be obtained from SPARROW at least within Pennsylvania. In our opinion, poorer agreement between model and measurements outside Pennsylvania for both SPARROW and PSU is likely due to less reliable measurements rather than poorer model performance.

SHI presents a simple way of assessing stream health in the absence of direct measurements for any stream basin, however small, or part thereof, in Pennsylvania or within the Chesapeake Bay Basin. More to the point, Equations 17.3 and 17.5 underscore the importance of ISA within a stream basin in affecting stream health.

17.3.2 An Example

Let us consider an example of a seriously degraded watershed and the influence of ISA and woodland on the SHI. A satellite (Landsat)-derived image of an ISA analysis (Figure 17.5) for the year 2000 shows the Conestoga Watershed, which is located in Lancaster County, Pennsylvania. ISA is color-coded with red and orange generally pertaining to commercial structures (greater than 80%), and green and blue pertaining to residential areas

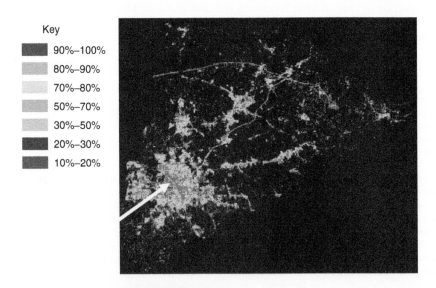

FIGURE 17.5 (See color insert following page 292.)
Satellite-derived (Landsat) analysis of ISA for the Conestoga Watershed in eastern Pennsylvania for the year 2000. Arrow denotes the town of Lancaster referred to in Figure 17.6.

FIGURE 17.6 (See color insert following page 292.)
Photograph of the Conestoga River in the town of Lancaster, Pennsylvania (arrow in Figure 17.5) taken in the year 2004 near the location denoted by arrow in Figure 17.5.

(20%–70%). The average ISA for the watershed is 8.4% and the SHI derived from Equations 17.3 through 17.5 is −6, the lowest possible category. A photograph of the Conestoga in Lancaster (Figure 17.6; refer to arrow in Figure 17.5), taken during 2004, shows a brown, muddy stream, clearly deserving of its low SHI. Although the ISA is large for an average over a sizeable watershed, it is not exceptionally so; the negative SHI is also associated with a relatively low woodland cover (20%) over this watershed.

17.4 Concluding Remarks

ISA is becoming an important parameter for water quality and surface runoff. Mitigation strategies now involve modification of the landscape and retaining water on developed property. Natural remedies consist of planting trees and conducting the runoff via pipe or grassy swale into vegetated recesses, which become ponds for a brief period after heavy rain. Indeed, a characteristic of more recent housing developments is that the developed property contains some form of storm water retention area (a grassy recess, a swale, or a wetland). In existing commercial areas, the planting of trees offers some reduction in storm water runoff, even if the tree simply shadows an underlying impervious surface.

In the future, it would be beneficial for assessing surface runoff and water quality (such as nutrient and sediment yields) to have a centralized database of land cover and ISA, and simple, widely accepted methodologies for estimating the impact of existing and future development on local streams anywhere in the United States. Such a facility must await not only the availability and easy access of ISA and other land-use data, but also further studies relating the land-cover parameters to the desired stream constituents for different types of terrain and climate regimes.

Large-scale databases of ISA are now being assembled at various institutions, albeit at differing resolutions and from various sources for part or all of the contiguous United States; for example, at approximately 30 m resolution from Landsat imagery for the Chesapeake Bay Basin at the UMD (Pittman, private communication), at Penn State for all of Pennsylvania for the years 1986 and 2000 (PASDA, 2004), and for the entire United States at 1 km resolution from NOAA satellite data (Elvidge et al., 2004). Some interesting statistics emerge from these analyses. The Pennsylvania analyses showed that an average ISA over the state in 2000 was 2.89%, covering approximately 1195 square miles and that total woodland cover was just under 60%. Elvidge et al. (2004) found that the total ISA over the contiguous USA is about the size of the state of Ohio.

On a smaller scale, developers and urban managers need to estimate the change in ISA that occurs with building and road construction and to estimate from changes in land use (including ISA) and the corresponding change (increase) in surface runoff that would occur in response to development. Enlightened communities now require that increased surface runoff be retained on developed properties, a requirement that adds extra cost to construction but, in the long run, is beneficial to the environment and to residents of new developments.

Assessing the effects of ISA and the resulting runoff on stream water quality is also important. At present, a simple regression model such as that described here, although best applied to the limited regions for which it was developed, can, at least, provide rough estimates of sediment and nutrient yields and stream health for any stream basin within Pennsylvania or the Chesapeake Bay Basin. This can be done with the aid of a simple Web tool called SHARP, which allows the user to delineate with the cursor the outline of a stream basin and instantly calculate yields of nitrogen, phosphate, sediment, ISA, and the SHI within the area circumscribed (http://www.sharp.psu.edu). Perhaps the most surprising result was the absence of a relationship between ISA and nitrogen, which alludes to the importance of rural (agricultural) effects on this constituent. These relationships as expressed in Equations 17.3 through 17.5 may change with time as measures are taken to retain water on developed properties and to treat surface runoff to streams in such a way as to minimize the washout of contaminants to streams. They may also vary with region. More studies are needed to determine similar relationships for other regions in this country.

References

Arnold, C.L. and C.J. Gibbons, 1996. Impervious surface coverage: The emergence of a key environmental indicator. *Journal of American Planning Association*, 62, 243–258.

Bedient, P.B. and W.C. Huber, 1992. *Hydrology and Floodplain Analysis* (Second Edition). Addison-Wesley Publishing Co., Reading, MA, 692 pp.

Chang, H., 2002. Spatial variations of nutrient concentrations in Pennsylvania Watersheds. *Journal of the Korean Geographical Society*, 37, 535–550.

Elvidge, C., C. Milisi, J.B. Dietz, B.T. Tuttle, P.C. Sutton, R. Nemani, and J.E. Vogelmann, 2004. U.S. constructed area approaches the size of Ohio. *EOS Bulletin*, 85, 15 June 2004, 2 pp.

Evans, B.M., D.W. Lehning, K.J. Corradini, G.W. Petersen, E. Nizeyimana, J.M. Hamlett, P.D. Robillard, and R. Day, 2002. A comprehensive GIS-based modeling approach for predicting nutrient loads in watersheds. *Journal of Spatial Hydrology*, 2, 1–18.

Hebble, E.E., T.N. Carlson, and K. Daniel, 2001. Impervious surface area and residential housing density: A satellite perspective. *Geocarto International*, 16, 13–18.

PASDA (Pennsylvania Spatial Data Access), Available at http://www.pasda.psu.edu/access/(Accessed in June 2004).

Preston, S.D., R.A. Smith, G.E. Schwartz, R.R. Alexander, and J.W. Brakebill, 1998. Spatially referenced regression modeling of nutrient loading in the Chesapeake Bay Watershed. In *Proceedings of the First Federal Interagency Hydrologic Modeling Conference*, Las Vegas, NV, 8 pp.

Schueler, T., 1994. The importance of imperviousness. *Watershed Prediction Technologies*, 3, 100–111.

Sheeder, S.A., T.N. Carlson, and J.D. Ross, 2002. Dual urban and rural hydrograph signals in three small watersheds. *Journal of the American Water Resources* (*JAWRA*), 38, 1027–1040.

Sheeder, S.A. and B.M. Evans, 2004, Development of nutrient and sediment threshold criteria for Pennsylvania TMDL assessment. *Journal of the American Water Resources Association*, 40, 881–888.

SPARROW, Available at http://www.neiwpcc.org/Index.htm?ne_sparrow.htm~mainFrame (Accessed in 2006).

USGS Report, September, 1998. Yields and trends of the nutrients and total suspended solids in nontidal areas of the Chesapeake Bay Basin, 1985–1995. In *Water-Resources Investigation Report 98-4192*. U.S. Department of the Interior, U.S. Geological Survey, 9 pp.

18

Impervious Surface Area Dynamics and Storm Runoff Response

Assefa M. Melesse and Xixi Wang

CONTENTS

18.1 Introduction

Impervious surface areas are characterized as surfaces that impede the natural infiltration of water into the soil and accelerate rainfall runoff processes and transport (USDA-SCS, 1986). Such surfaces are usually a result of urbanization and development, and they include roads, buildings, and parking lots. Urbanization, which converts pervious surfaces to impervious surfaces, is a global trend because of the population and transportation pressures. The impacts of this conversion on ecohydrological settings have been studied by a number of researchers (e.g., Hirsch et al., 1990; McCuen, 1998; Chin and Gregory, 2001; Rose and Peters, 2001; Booth et al., 2002).

The runoff from impervious surfaces raises serious environmental concerns because of its adverse impacts on the ecohydrology and water quality of receiving waters, as indicated by the degraded water quality and biodiversity (USEPA, 1983; Whalen and Cullum, 1989; Steuer et al., 1997). The impacts on the ecohydrology include the reduction of infiltration rate, increase in surface runoff volume and peak discharge, decrease in soil moisture compensation and groundwater recharge, decrease in flow duration and base flow drying up of wetlands, habitat degradation and fragmentation, and alteration of surface energy balance. As a result, the ecological functions and economic values of a watershed are likely to be greatly lowered. In addition, a combination of increased peak runoff volumes and decreased duration and hydraulic efficiency (Chow et al., 1988) results in more "erosive work" or hydraulic force acting on a stream channel, increasing risks of stream bank/bed erosion and loadings of sediment and its associated constituents (e.g., phosphorus, nutrient, and pesticide) (Driscoll, 1986; Chow et al., 1988).

Impervious surface areas also play an important role in determining the water quality of the receiving waters. During dry seasons, dry deposition of pollutants on such surfaces will be accelerated and washed out to the nearby streams and lakes in the wet seasons. The temperature of streams and rivers can rise in summer due to the heat conduction from the impervious surfaces and water washed into these waters. Nonpoint pollution sources associated with impervious surface areas are closely related to land management practices, and are a function of land-use type and intensity, such as the fraction of impervious surfaces, as well as climate conditions, such as the frequency and magnitude of storm events.

The conversion of pervious surfaces (e.g., grassland, agriculture, forest, and wetlands) to the corresponding impervious ones (e.g., roads, parking lots, and buildings) can substantially change the partition of the incoming solar radiation (USEPA, 1983). Solar radiation that reaches the Earth's surface is reflected, absorbed, and partitioned into sensible heat and latent heat (i.e., the energy used for water evapotranspiration). A very small percentage of the solar radiation is used in photosynthesis. Because of the absence of vegetation and soil moisture, impervious surfaces tend to have a lowered evapotranspiration. For pervious surfaces, the latent heat is used to evaporate the water from the soil and meet the transpiration demand of plants. However, the latent heat would be converted to the additional sensible heat in the absence of soil moisture and vegetation, which in turn, probably elevates the temperature and humidity of the ambient air above the impervious surfaces.

Further, the altered flow regime, water temperature, and elevated sediment and nutrient concentrations can have a substantial effect on aquatic species recruitment, age structure, taxa richness, and taxonomic composition (Poff and Ward, 1989; Kelsch and Dekrey, 1998). Both the aquatic and terrestrial habitats would be adversely impacted by the runoff and constituents from the impervious surfaces. Studies have indicated declines in biological integrity and habitat quality when the impervious fraction reaches 10%–20% (USEPA, 1983).

For a watershed, the characteristics of the overland runoff and its corresponding flow hydrograph are closely related to the land cover and are functions of imperviousness, defined as the ratio of the impervious areas to the inclusive watershed area (Schueler, 1987). The imperviousness determines the hydrologic processes of infiltration and surface runoff. Increased imperviousness results in less infiltration but larger surface runoff, and vice versa. In addition, because water is no longer infiltrating into the soil, the overland flow from impervious areas is mainly a result of infiltration excess, a rainfall runoff mechanism occurring when rainfall intensity exceeds the soil infiltration capacity. The infiltration excess runoff is also called Hortonian flow after Horton (1942), who developed a conceptual description of this runoff mechanism. This conceptual model has been widely used to estimate the overland flow for watersheds with a low soil infiltration capacity and urbanized areas where impervious surfaces hinder infiltration of water and accelerate the occurrence and transport of overland flow. In practice, the conceptual model is usually implemented using the Simple Method or the U.S. Department of Agriculture (USDA) Soil Conservation Service (SCS) Curve Number (CN) method (USDA-SCS, 1972).

The objectives of this chapter are to (1) provide an overview of these two methods; (2) introduce a method to quantify impervious areas based on the remotely sensed data; and (3) illustrate how to determine the runoff response to imperviousness in two watersheds with distinctly different climate conditions and development levels.

18.2 Estimation of Hortonian Overland Flow

18.2.1 Characteristics of Impervious Areas

Compared with rural areas, impervious surfaces in urban areas have an infiltration rate approaching zero. As a result, storm events tend to generate runoff processes with a larger volume and peak discharge from impervious areas than rural areas.

Impervious areas can be differentiated into two groups: (1) the area that is hydraulically connected to the drainage system (also called effective impervious area) and (2) the area that is not directly connected (USDA-SCS, 1986).

For description purposes, these two groups are designated as Group I and Group II impervious areas, respectively. The runoff generated from Group II impervious areas flows into their adjacent grass lands and is able to infiltrate into the soil, whereas, the runoff from Group I impervious areas is directly drained into the stream network by the drainage system. For the purpose of modeling the runoff processes, commercial, industrial, transportation, and institutional areas can be considered as Group I impervious areas. On the other hand, residential areas can be considered as Group II impervious areas.

18.2.2 Simple Method

The Simple Method was developed by Schueler (1987) to calculate pollutant loadings in urban runoff. The runoff volume, R, in cubic meters is calculated as

$$R = 0.01 \times P \times P_j \times R_v \times A, \tag{18.1}$$

where
  P is the annual precipitation (cm)
  P_j is the fraction of rainfall events that produce runoff
  R_v is the mean runoff coefficient
  A is the area of the watershed (ha)
  R_v is the proportion of rainfall converted to surface runoff and can be estimated as

$$R_v = 0.05 + 0.009 \times I, \tag{18.2}$$

where I is the percent imperviousness of the watershed.

P_j represents effects of interception, depression storage, and infiltration. Its value may be varied, depending on the topography, land use/cover, and soil property of the watershed. An empirical value of 0.9 can be used (Schueler, 1987) when the site-specific data are unavailable.

18.2.3 SCS-CN Method

The SCS-CN method was developed to provide a consistent basis for estimating the amounts of runoff under varying land use and soil types (USDA-SCS, 1972; Rallison and Miller, 1982). It is based on a hypothesized (empirical) relation between runoff and infiltration, which can be expressed as

$$\frac{F}{S} = \frac{Q}{P - I_a}, \tag{18.3}$$

where
  F is the actual retention of precipitation during a storm (cm)
  S is the maximum potential retention (cm)
  Q is the runoff (cm)

P is the precipitation (cm)

I_a (cm) is the initial rainfall abstraction, which represents the precipitation intercepted by vegetation, other surfaces (e.g., roof), and depression areas

F can be calculated as

$$F = P - Q - I_a. \tag{18.4}$$

An additional assumption is made that I_a is proportional to S, that is,

$$I_a = K \times S. \tag{18.5}$$

Substituting Equations 18.4 and 18.5 into Equation 18.3, we can get

$$Q = \frac{(P - K \times S)^2}{P - K \times S + S}. \tag{18.6}$$

S is related to a CN and is calculated as

$$S = \frac{2540}{CN} - 25.4. \tag{18.7}$$

CN is a function of land uses and soil types (hydrologic soil group). Its value can range from 1 to 100, but values of less than 35 and greater than 98 are seldom justified for practical applications. In addition, CN varies with antecedent soil moisture conditions. USDA-SCS (1972, 1985) developed the CN values associated with antecedent soil moisture conditions I, II, and III, which correspond to dry, average, and wet hydrologic conditions, respectively. The values for other moisture conditions are usually computed as a linear interpolation of that for those three conditions. Rallison and Miller (1982) describe in detail how to determine the values for CN.

The proportional coefficient, K, varies from storm to storm and watershed to watershed (Hawkins et al., 2002). For a watershed, K could have a value ranging from 0.0005 to 0.4910, with a median value of 0.0476. Nonetheless, a value of 0.2 was recommended by USDA-SCS (1985) and has been widely used. Lim et al. (2006) assessed the effects of different K values on the estimated runoff from a 70.5 km^2 Indiana watershed. This watershed has experienced significant urbanization, with about 50% urbanized land area in size in 1973 and about 68% in 1991. The study results indicated that the long-term surface runoff prediction had the best accuracy when a K value of 0.2 was used.

For Group I impervious areas, a CN value of 98 is usually used (USDA-SCS, 1986). However, two strategies are widely implemented to estimate the runoff generated from Group II impervious areas. Strategy I is to partition the impervious areas from their adjacent pervious areas and to apply different CN values to these areas. With this regard, the CN for the impervious areas has a value of 98, whereas, the CN for the pervious areas is determined as shown in Rallison and Miller (1982).

In contract, Strategy II is to calculate a composite curve number for Group II impervious areas using an equation developed by USDA-SCS (1986):

$$CN_c = \begin{cases} CN_p + imp_{tot} \times (CN_{imp} - CN_p) \times \left(1 - \frac{imp_{dcon}}{2 \times imp_{tot}}\right) & \text{when } imp_{tot} \leq 0.30 \\ CN_p + imp_{tot} \times (CN_{imp} - CN_p) & \text{when } imp_{tot} > 0.30, \end{cases}$$

(18.8)

where

CN_c is the composite curve number

CN_p is the pervious curve number

CN_{imp} is the impervious curve number and usually takes a value of 98

imp_{tot} is the fraction of the watershed area that is impervious (both Group I and Group II)

imp_{dcon} is the fraction of the watershed area that is impervious (Group II only)

The extra factor in Equation 18.8, $\left(1 - \frac{imp_{dcon}}{2 \times imp_{tot}}\right)$ when $imp_{tot} \leq 0.3$ is a value between 0 and 1 and it accounts for the reduction in the runoff. Imp_{dcon} is calculated as

$$imp_{dcon} = imp_{tot} - imp_{con},$$

(18.9)

where imp_{con} is the fraction of the watershed area that is impervious (Group I only).

Similarly, CN_c also varies with antecedent soil moisture conditions. The variation is because CN_p is a function of the antecedent soil moisture conditions. The values of CN_p for different soil moisture conditions are determined using the aforementioned procedure for CN.

18.2.4 Hydrologic Models

Among the commonly used hydrologic models, the Hydrologic Simulation Program—Fortran (HSPF) uses Strategy I to simulate the runoff from Group II impervious areas (Johanson et al., 1984), whereas, several USDA models, including the Soil and Water Assessment Tool (SWAT) (Arnold et al., 1998), the Agricultural Non-point Source Pollution Model (AGNPS) (Young et al., 1989), and the Water Erosion Prediction Project (WEPP) (Laflen et al., 1997), implement Strategy II. On the other hand, both Strategy I and Strategy II are provided by the Generalized Watershed Loading Functions (GWLF) (Haith et al., 1992), the Simulator for Water Resources in Rural Basins (SWRRB) (Arnold et al., 1990), and the Environmental Policy Integrated Climate (EPIC) (Mitchell et al., 1996).

Mandel et al. (1997) evaluated the performances of the Simple Method, HSPF, GWLF, and EPIC on predicting the average annual runoff in three gauged watersheds in which urban land uses were dominant, namely Rock Creek in Maryland and the District of Columbia, Beaverdam Run in Maryland, and Difficult Run in Virginia. The evaluation was conducted by: (1) estimating the average annual runoff from the stream gauge records using standard baseflow separation techniques; (2) predicting the runoff using the EPIC and GWLF models with both Strategy I and Strategy II for

Group II impervious areas and comparing the predicted runoff with the runoff estimates from baseflow separation; and (3) predicting the runoff using the Simple Method and the HSPF model and comparing the predicted runoff with the runoff estimates from baseflow separation. The results indicated a similar prediction performance, but the runoff predicted using Strategy II tended to be smaller than that using Strategy I.

In addition to these hydrologic models, the Storm Water Management Model (SWMM), developed by the U.S. Environmental Protection Agency (EPA), is also widely used for single event or long-term (continuous) simulation of runoff quantity and quality from primarily urban areas (Rossman, 2005). The runoff component of SWMM operates on a collection of subcatchment areas, which receive precipitation and generate runoff and pollutant loads. SWMM uses Strategy II to calculate the runoff from Group II impervious areas. The routing portion of SWMM transports this runoff through a system of pipes, channels, storage/treatment devices, pumps, and regulators. SWMM tracks the quantity and quality of runoff generated within each subcatchment and the flow rate, flow depth, and quality of water in each pipe and channel during a simulation period comprising multiple time steps.

While using different strategies to estimate the runoff from impervious areas, both the Simple Method and the CN-based hydrologic models require a common input of imperviousness and classifications of impervious areas. In practice, this input can be derived from the remotely sensed data.

18.3 Impervious Surface Area Estimation

Estimating the imperviousness in urbanized watersheds is an important task of water resources managers and planners. This information is needed for designing flood control systems, canals, and drainage network structures, as discussed earlier, and is an important input of hydrologic models. Various techniques have been developed and used to estimate the imperviousness and delineate the extents of impervious surface areas. Among these techniques, remote sensing-based techniques have proven to be superior to the others for studying watersheds with a large size and mixed land uses, because they are capable of quantifying impervious surfaces at a regional or watershed spatial scale.

One of the commonly used remote sensing techniques is based on the scaled normalized difference vegetation index ($NDVI_s$) (Price, 1987; Che and Price, 1992; Carlson and Arthur, 2000). $NDVI_s$ reflects the temporal change of the vegetation cover. It can be determined using remote sensing images with different scenes and dates as

$$NDVI_s = \frac{NDVI - NDVI_{low}}{NDVI_{high} - NDVI_{low}},\qquad(18.10)$$

where $NDVI_{low}$ and $NDVI_{high}$ are values for bare soil and dense vegetation, respectively.

Carlson and Ripley (1997) suggested a functional relationship between the fractional vegetation cover (FVC) and $NDVI_s$. FVC has a range between 0 and 1. This relationship can be expressed as

$$FVC \approx (NDVI_s)^2. \qquad (18.11)$$

Ridd (1995) and Owen et al. (1998) developed an equation for estimating the fractional impervious surface area (FIS) based on FVC. The equation can be expressed as

$$FIS = 1 - FVC. \qquad (18.12)$$

FIS is equivalent to the imperviousness required by the Simple Method, variable I in Equation 18.2. In addition to this composite parameter, the extents of various land uses can be accurately delineated using this remote sensing-based technique. In particular, this technique can distinguish Group I impervious areas from Group II ones, depending on the resolution of the images used. Further, supplemented by the ground-surveying data, this technique has the capability to partition the impervious areas and their adjacent pervious areas for Group II impervious areas. As discussed earlier, the resulting data on land-use classifications are common inputs of the hydrologic models.

18.4 Example Studies on Runoff Response to Impervious Surface

The studies were conducted in the Red River of the North Basin, located in North Dakota and Northwestern Minnesota and the Simms Creek watershed, located in Florida. These two study areas have distinctly different climate conditions.

18.4.1 Study Areas

The Red River of the North meanders about 880 km from its headwater in South Dakota along the state border between North Dakota and Minnesota to the international border between United States and Canada (Figure 18.1). It has a very large sinuosity as indicated by a shorter line distance of 456 km, and it drains 100,480 km^2 U.S. lands. In the U.S. portion, about one-third of the population lives in the two North Dakota–Minnesota metropolitan cities, namely Grand Forks–East Grand Forks and Fargo-Moorhead. These two cities are experiencing urbanization and noticeable changes on land use in the recent years. The climate of the Red River of the North basin is continental, with a dry/subhumid condition in the North Dakota side to a

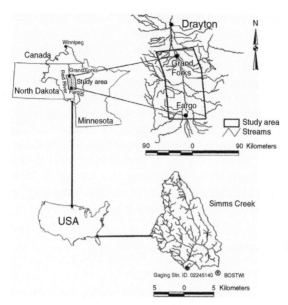

FIGURE 18.1
Map showing the locations and boundaries of the study watersheds.

subhumid/humid condition in the Minnesota side. The average annual precipitation varies from about 457 mm at the northwestern areas of the basin to about 686 mm at the southeastern portion. About 75% of the annual precipitation falls in the months between April and September. The average annual temperatures vary from 2.8°C to 6.1°C, and the monthly temperatures have a wider range, from −18.3°C in January to 21.7°C in July.

The study was conducted using the Landsat and meteorological data spanning the years 1974–2002 (Melesse, 2004) in the Fargo-Moorhead and Grand Forks–East Grand Forks metropolitan areas (Figure 18.1). The landuse/cover changes were analyzed using the Landsat imagery by making comparisons between the study years. In addition, the extents of impervious surface areas were delineated and the imperviousness was estimated.

In contrast, the Simms Creek watershed, located in the Etonia subbasin, Florida (Figure 18.1), has a drainage area of about 114 km². The Simms Creek flows eastward along with the Rice and Etonia Creeks to join the St. Johns River. The Simms Creek watershed is mainly covered with forest, wetlands, agriculture, and urban areas. It has a moderate topographic relief, with elevations ranging from 3.6 to 72 m above mean sea level. The average annual rainfall is about 142 cm, with ∼60% of rainfall occurring in the months between June and September. The rainfall is affected by the frontal and convective climate. The frontal precipitation usually occurs when a cold front from the north results in lifting of air masses and can have a long duration but a small intensity. It usually occurs during the dry season, that is, the months between November and March. In contrast, the convective rainfall, which usually occurs during the wet season, that is, in the months between May and September, is characterized by a short duration but fairly high rainfall intensity.

The average temperature of the Simms Creek watershed is 22°C, with the minimum temperature of 14.8°C in February and a maximum temperature of 29°C in June. The annual average potential evapotranspiration is estimated to be 116 cm. The dominant soils of the Simms Creek watershed are Myakka and Tavares, which constitute more than 60% of the watershed area in size. These soils have a very low permeability and thus a high runoff potential and are classified as Hydrologic Soil Group D. Melesse and Shih (2002) showed that infiltration excess rather than saturation excess was the dominant runoff generation mechanism for this watershed.

18.4.2 Red River of the North Basin

18.4.2.1 Land-Cover Change Analysis

Figure 18.2 shows the land-cover changes determined using the Landsat images that were taken from 1974 to 2001. The analysis used the land-use/land-cover classification system developed by Anderson (1976). The details on the techniques used to estimate the land-cover classes and extents can be found in Melesse (2004). The results indicated that the urban areas were increased by 19% from 1974 to1984 and 19% from 1974 to 1992. A larger percent increase of 54% from 1974 to 2001 indicates an intensive urbanization since 1992. Further, the results indicated that more than 50% of the study area was covered by cropland and forest and that the rangeland acreage was decreased by 29% between 1974 and 2001.

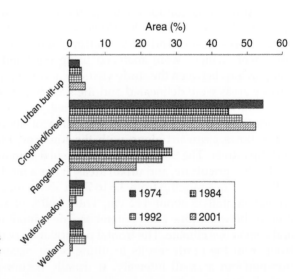

FIGURE 18.2
Land-cover analysis results from Landsat images for the Red River of the North Basin.

18.4.2.2 Imperviousness Dynamics

Using Equation 18.12, the FIS or imperviousness was estimated and mapped for the portions of the basin classified as urban areas. Figure 18.3 shows the spatiotemporal dynamics of the FIS for the Grand Forks–East Grand Forks and Fargo-Moorhead metropolitan areas. It can be seen that the impervious areas have been noticeably increasing, which was probably caused by the rapid urbanization and development activities along the Red River Valley in recent years. The increased imperviousness was likely to alter the storm runoff processes in the basin. For example, the change in FIS between 1992 and 2001 than that between 1974 and 1984 corresponds to

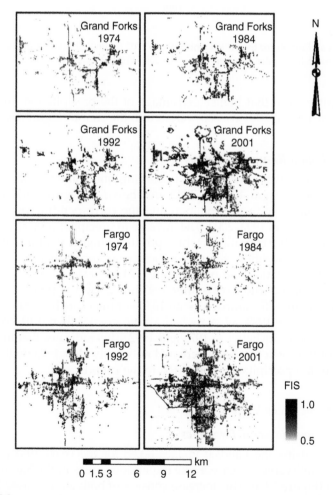

FIGURE 18.3
Impervious surface area analysis results for the urbanized areas in the Red River of the North Basin. (From Melesse, A.M., *Phys. Chem. Earth*, 29, 795, 2004. With permission.)

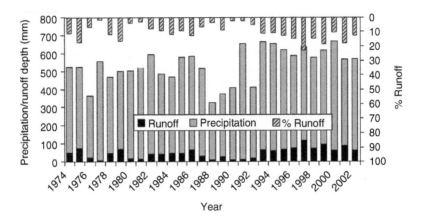

FIGURE 18.4
Precipitation and runoff for the period from 1974 to 2002 in the Red River of the North Basin.
(From Melesse, A.M., *Phys. Chem. Earth*, 29, 795, 2004. With permission.)

a consistent pattern that higher percentage of the precipitation during the period from 1992 to 2001 was converted into the runoff (Figure 18.4).

18.4.2.3 Runoff Response Analysis

The data on monthly precipitation (including both rainfall and snowfall) and runoff depth from 1974 to 2002 are shown in Figure 18.4. The plot indicates an increase in precipitation and the corresponding discharge of the river in the years 1993–2002. The increase in precipitation was driven by the regional weather conditions, which produced higher volumes of rainfall and snowfall. In addition, the plot indicates an approximately consistent increase in precipitation and runoff from 1974 to 2002.

About 40% of the study years after 1993 had a ratio of runoff to precipitation larger than 0.15, compared with 10.5% of the years between 1974 and 1992. Further, after 1993, all years had a ratio greater than 0.10, compared with 35% of the years between 1974 and 1992. This indicates that the runoff response of the basin has been increased noticeably after 1993, with greater than 10% of the precipitation converted to the runoff.

18.4.3 Simms Creek Watershed

18.4.3.1 Imperviousness Dynamics

The analysis using the Landsat images taken for the years 1984 and 2000 indicated a noticeable increase in the impervious areas in the Simms Creek watershed (Figure 18.5). This increase consists of both an increased FIS and expanded extents of the impervious areas. Compared with 1984, more areas in 2000 had an FIS value of 1, that is, 100% imperviousness, indicating an intense land-use alteration. Further, as shown in Figure 18.5, these areas are

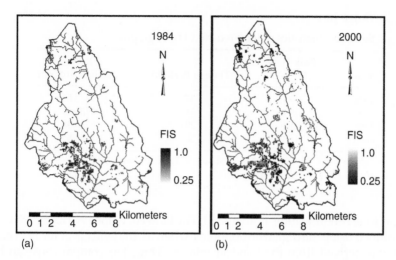

FIGURE 18.5
Impervious surface areas in the Simms Creek watershed in (a) 1984 and (b) 2000. (From
Melesse, A.M., Graham, W.D., and Jordan, J.D., *J. Spatial Hydrol.*, 3, 2003. With permission.)

scattered across the watershed, but mostly distributed in the areas adjacent
to the stream channels.

18.4.3.2 Hydrograph Comparison

In order to detect effects of land-use changes on the hydrograph characteristics, namely shape, peak discharge, and times to peak and recession end, an
arbitrary rainfall intensity of 12 mm/h with a 4 h duration was simulated.
Because the major purpose of this analysis was to assess the effects of the
land-use changes on the hydrograph characteristics, the excess rainfall was
computed using the SCS-CN method and taken as the sole variable. Using
the Spatially Distributed Direct Hydrograph (SDDH) method (Melesse
and Graham, 2004), hydrographs were generated using the SCS-CN number
method for the years 1984, 1990, 1995, and 2000. In addition, the
analysis varied the Manning's roughness coefficients in accordance with
the determined land-cover distribution. Manning's roughness coefficient is
a land-cover-dependent coefficient in Manning's formula for estimating open
channel flow velocity (Chow, 1959). It is a function of friction along the flow in
open channel flow. Flow over high-friction materials such as trees and shrubs
has higher Manning's coefficient than low-friction covers such as grass, sand,
and gravel. In this study, Manning's coefficient was used from published
materials (Brater and King, 1976; Montes, 1998) for each land-cover class.

The results indicated that time to peak was 37, 31, 29, and 31 h in 1984,
1990, 1995, and 2000, respectively (Table 18.1). The peak discharges for
1984, 1990, 1995, and 2000 were predicted to be 0.83, 0.94, 1.04, and
0.98 m^3/s, respectively. However, these 4 years were predicted to have a

TABLE 18.1

Summary Statistics of the Predicted Hydrographs

Year	Peak Flow (m³/s)	Time to Peak (h)	Runoff Volume (10³ m³)
1984	0.83	37	152.81
1990	0.94	31	153.69
1995	1.04	29	153.98
2000	0.98	31	153.59

Source: From Melesse, A.M., Graham, W.D., and Jordan, J.D., *J. Spatial Hydrol.*, 3, 2003. With permission.

similar total runoff volume (Table 18.1). The land cover for 1990 and 1995 was determined using the digital orthophotos, whereas, the land cover for 1984 and 2000 was derived from the Landsat images. Thus, fair comparisons can be made between the results for 1990 and 1995 and that for 1984 and 2000, as the procedures and data sources are identical. On the other hand, comparisons of the results for 1990 and 1995 with that for 1984 and 2000 would be biased because of the incomparability of the analysis procedures and data sources.

The comparison of the predicted hydrographs indicated an increase in peak discharge by 6.5% and reduction in time to peak by 10.6% in 1995 from that in 1990 (Figure 18.6). This was probably a result of the increased urban built-up areas in 1995, which increased the runoff and reduced the flow travel time (Melesse et al., 2003). This is one typical effect of urbanization on the shape of the hydrograph and time to peak (Chow et al., 1988). Similarly, increased urban built-up areas in 2000 from that in 1984 increased peak discharge by 16.2% and reduced time to peak by 18.1% (Figure 18.6).

18.5 Summary

This chapter elaborated the effects of increased imperviousness on the ecohydrology and water quality of receiving water bodies in urbanized watersheds. It provided an overview of the Simple Method and SCS-CN-based hydrologic models, which are widely used to predict the effects of urbanization on precipitation runoff processes. In addition, this chapter introduced a remote sensing-based technique for determining the imperviousness and extents of impervious areas in the watershed. Further, example studies were used to demonstrate the runoff responses to the increased impervious areas in the Red River of the North Basin and Simms Creek watershed, which have distinctly different climate conditions.

In contrast with pervious surfaces, impervious surfaces tend to absorb more incoming solar radiation, that is, the energy of the sensible heat

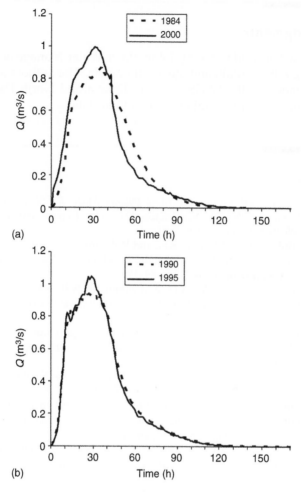

FIGURE 18.6

Predicted flow hydrographs using a synthetic rainfall event with an intensity of 12 mm/h for (a) 1990 and 1995 and (b) 1984 and 2000. (From Melesse, A.M., Graham, W.D., and Jordan, J.D., *J. Spatial Hydrol.*, 3, 2003. With permission.)

reduces the latent heat, which is likely to make the surface and ambient atmosphere warmer. This effect of impervious surfaces is due to the absence of vegetation and soil moisture. Accordingly, the warmer surface would provide a great opportunity to transfer more heat energy to the contacting runoff and thus to increase water temperatures of streams and other water bodies. In addition, the altered flow regime and increased sediment and its associated constituents are likely to have an adverse impact on aquatic ecosystems, leading to degraded water quality and biodiversity (Price and Waddington, 2000).

Acknowledgments

The authors acknowledge the St. Johns River Water Management District for providing GIS, rainfall, and stream flow data. The authors also extend their appreciation to the USGS of North Dakota and North Dakota State Water Commission for providing some of their data.

References

Anderson, E.A., *A Point Energy and Mass Balance Model for Snow Cover*. NOAA Technical Report NWS 19. Washington, D.C.: U.S. Department of Commerce, National Weather Service, 1976.

Arnold, J.G., Srinivasan, R., Muttiah, R.S., and Williams, J.R., Large-area hydrologic modeling and assessment: Part I. Model development. *Journal of the American Water Resources Association*, 34(1), 73–89, 1998.

Arnold, J.G., Williams, J.R., Nicks, A.D., and Sammons, N.B., *SWRRB: A Basin Scale Simulation Model for Soil and Water Resources Management*. College Station, TX: Texas A & M University Press, 1990.

Booth, D.B., Hartley, D., and Jackson, R., Forest cover, impervious-surface area, and the mitigation of stormwater impacts. *Journal of the American Water Resources Association*, 38, 835–845, 2002.

Brater, E.F. and King, H.W., *Handbook of Hydraulics for the Solution of Hydraulic Engineering Problems*. New York: McGraw-Hill, 1976.

Carlson, T.N. and Arthur, S.T., The impact of land use–land cover changes due to urbanization on surface microclimate and hydrology: a satellite perspective. *Global and Planetary Change*, 25, 49–65, 2000.

Carlson, T.N. and Ripley, A.J., On the relationship between fractional vegetation cover, leaf area index and NDVI. *Remote Sensing of Environment*, 62, 241–252, 1997.

Che, N. and Price, J.C., Survey of radiometric calibration results and methods for visible and near-infrared channels of NOAA-7, -9 and -11 AVHRRs. *Remote Sensing of Environment*, 41, 19–27, 1992.

Chin, A. and Gregory, K.J., Urbanization and adjustment of ephemeral stream channels. *Annals of the Association of American Geographers*, 91, 595–608, 2001.

Chow, V.T., *Open-Channel Hydraulics*. New York: McGraw-Hill, 1959.

Chow, V.T., Maidment, D.R., and Mays, L.W., *Applied Hydrology*. New York: McGraw-Hill Series in Water Resources and Environmental Engineering, McGraw-Hill, 1988.

Driscoll, E., Long normality of point and non-point source pollution concentrations, in *Engineering Foundation Conference*, New York, June 19–27, 1986.

Haith, D.A., Mandel, R., and Wu, R.S., *GWLF: Generalized Watershed Loading Functions*, Version 2.0, User's Manual. Ithaca, NY: Department of Agricultural and Biological Engineering, Cornell University, 1992.

Hawkins, R.H., Jiang, R., Woodward, D.E., Hjelmfelt, A.T., and Van Mullem, J.A., Runoff curve number method: examination of the initial abstraction ratio, in

Proceedings of the Second Federal Interagency Hydrologic Modeling Conference, Las Vegas, Nevada. Lakewood, CO: U.S. Geological Survey, CD-ROM, 2002.

Hirsch, R.M., Walker, J.F., Day, J.C., and Kallio, R., The influence of man on hydrologic systems, in *Surface Water Hydrology*, Wolman, M.G., and Riggs, H.C. (Eds.), Vol. 0–1. Boulder, CO: Geological Society of America, 1990, pp. 329–359.

Horton, R.E., Remarks on hydrologic terminology. *Transactions of American Geophysical Union*, 19(2), 479–482, 1942.

Johanson, R.C., Imhoff, J.C., Kittle, J.L., Jr., and Donigian, A.S., *Hydrologic Simulation Program—FORTRAN (HSPF)*: Users Manual for Release 8.0. Athens, GA: U.S. Environmental Protection Agency, 1984.

Kelsch, S.W. and Dekrey, D., *Effects of Environmental Factors on Stream-Fish Assemblages*. Bismarck, ND: North Dakota Game and Fish Department, Final F-2-R-43 Project Report, 1998.

Laflen, J.M., Elliot, W.J., Flanagan, D.C., Meyer, C.R., and Nearing, M.A., WEPP: Predicting water erosion using a process-based model. *Journal of Soil and Water Conservation*, 52(2), 96–102, 1997.

Lim, K.J., Engel, B.A., Muthukrishnan, S., and Harbor, J., Effects of initial abstraction and urbanization on estimated runoff using CN technology. *Journal of the American Water Resources Association*, 42(3), 629–643, 2006.

Mandel, R., Caraco, D., and Schwartz, S.S., *An Evaluation of the Use of Runoff Models to Predict Average Annual Runoff from Urban Areas*. Maryland: The Interstate Commission on the Potomac River Basin (ICPRB), ICPRB Report #97-7, 1997.

McCuen, R.H., *Hydrologic Analysis and Design*. Englewood Cliffs, NJ: Prentice Hall, 1998, p. 814.

Melesse, A.M., Spatiotemporal dynamics of land surface parameters in the Red River of the North Basin. *Physics and Chemistry of Earth*, 29, 795–810, 2004.

Melesse, A.M. and Graham, W.D., Storm runoff prediction using spatially distributed travel time concept utilizing remote sensing and GIS. *Journal of the American Water Resources Association*, 40(4), 863–879, 2004.

Melesse, A.M., Graham, W.D., and Jordan, J.D., Spatially distributed watershed mapping and modeling: GIS-based storm runoff response and hydrograph analysis part 2. *Journal of Spatial Hydrology*, 3(2), 2003.

Melesse, A.M. and Shih, S.F., Spatially distributed storm runoff depth estimation using Landsat images and GIS. *Computers and Electronics in Agriculture*, 37, 173–183, 2002.

Mitchell, G., Griggs, R.H., Benson, V., and Williams, J., *EPIC User's-Guide, Version 5300*. Temple, TX: Agricultural Research Service Grassland, Soil and Water Research Laboratory, 1996.

Montes, S., *Hydraulics of Open Channel Flows*. Reston, VA: ASCE Press, 1998.

Owen, T.W., Carlson, T.N., and Gillies, R.R., Remotely-sensed surface parameters governing urban climate change. *International Journal of Remote Sensing*, 19, 1663–1681, 1998.

Poff, N.L. and Ward, J.V., Implications of streamflow variability and predictability for lotic community structure: A regional analysis of streamflow patterns. *Canadian Journal of Fish and Aquatic Science*, 46, 1805–1817, 1989.

Price, J.C., Calibration of satellite radiometers and the comparison of vegetation indices. *Remote Sensing of Environment*, 21, 15–27, 1987.

Price, J.S. and Waddington, J.M., Advances in Canadian wetland hydrology and biogeochemistry. *Hydrological Processes*, 14, 1579–1589, 2000.

Rallison, R.E. and Miller, N., Past, present, and future SCS runoff procedure, in *Proceedings of the International Symposium on Rainfall Runoff Modeling: Rainfall-Runoff Relationship*, V.P. Singh (Ed.), May 18–21, 1981. Mississippi: Mississippi State University, 1982, pp. 353–364.

Ridd, M.K., Exploring a V-I-S (vegetation-impervious surface-soil) model for urban ecosystem analysis through remote sensing: Comparative anatomy for cities. *International Journal of Remote Sensing*, 16, 2165–2185, 1995.

Rose, S. and Peters, N., Effects of urbanization on streamflow in the Atlanta area (Georgia, USA): A comparative hydrological approach. *Hydrological Processes*, 15, 1441–1457, 2001.

Rossman, L.A., *Storm Water Management Model User's Manual*, Version 5.0. Cincinnati, OH: Water Supply and Water Resources Division, National Risk Management Research Laboratory, 2005.

Schueler, T.R., *Controlling Urban Runoff: A Practical Manual for Planning and Designing Urban BMPs*. Washington, D.C.: Metropolitan Washington Council of Governments, 1987.

Steuer, J., Selbig, W., Horewer, N., and Prey, J., *Sources of Contamination in an Urban Basin in Marquette, Michigan and an Analysis of Concentrations, Loads, and data Quality*. Washington, D.C.: U.S. Geological Survey, Water-Resources Investigations Report 97-4242, 1997.

USDA-SCS, *National Engineering Handbook*. Part 630, Hydrology, Section 4. Washington, D.C.: U.S. Government Printing Office, 1972.

USDA-SCS, *National Engineering Handbook*. Section 4, Hydrology. Littleton, CO: Water Resources Publications, 1985.

USDA-SCS, *Urban Hydrology for Small Watersheds*. Washington, D.C.: U.S. Department of Agriculture, Technical Release 55, 1986.

USEPA, *Results of the National Urban Runoff Project (Final Report)*. Washington, D.C.: U.S. Environmental Protection Agency, 1983.

Whalen, P. and Cullum, M., *An Assessment of Urban Land Use/Stormwater Runoff Quality Relationships and Treatment Efficiencies of Selected Stormwater Management Systems*. Florida: South Florida Management District Resource Planning Department, Water Quality Division, Technical Publication 88–9, 1989.

Young, R.A., Onstad, C.A., Bosch, D.D., and Anderson, W.P., Agricultural non-point source pollution model for evaluating agricultural watersheds. *Journal of Soil and Water Conservation*, 44(2), 168–173, 1989.

19

Growth of Impervious Surface Coverage and Aquatic Fauna

Robert R. Gillies

CONTENTS

19.1 Introduction

Human activity in the shape of land-cover and land-use change, in particular urbanization, is often cited in the literature as a major factor contributing to a staggering loss in ecological biodiversity. So much so, that The National Academy of Sciences has noted the ecological impact of land-cover and land-use change (NRC, 2001). Urbanization is persistent in the sense that it is a more lasting type of land-use change than are other types of habitat loss; moreover, it continues to expand geographically with resulting effects on the environment (Benfield et al., 1999). In the United States, the data on species loss reveal themselves in the increasing number of plants and animals added to the endangered species list. As an indicator, however, it is misleading and frequently understated, as knowledge of extinctions along with inventory lists are better for some species than others. In the context of aquatic environments, for example, the knowledge base for freshwater fish and bivalves* is quite extensive (USEPA, 2002; Nilsson et al., 2003) while those for other water settings (i.e., bed landscapes of water bodies), which harbor abundant biota, are sorely lacking.

There is a plethora of landscape indicators that are used in determining the biological integrity of ecosystems and in the assessment of the biological diversity of aquatic flora and fauna (Gergel et al., 2002). Of note, however, is one particular indicator approach—that of total impervious surface area (ISA)—that among other uses has been applied in the study of fish, insect, and invertebrate diversity (Klein, 1979; Schueler and Galli, 1992). What follows in this chapter is a background review of ISA and aquatic ecosystems, followed by a detailed look at a case study, and in particular the techniques, that was first published by Gillies et al. (2003) computing and examining ISA growth and expansion with associated subsequent fresh water mussel loss and extinctions in the Peachtree watershed in Atlanta. Lastly, some more recent studies on ISA and aquatic ecosystems are presented and discussed, as they exemplify further complexities that should be considered.

19.2   Background

The generalized thinking that one reads in the literature presently with respect to ISA and aquatic ecosystems is that by replacing pervious surfaces with artificial (i.e., impervious) ones, any precipitation that falls will be summarily transported to streams and rivers. While, in the process of transfer, the flowing water picks up any detritus that is part of the urban fabric, thus degrading the water body properties and aquatic biota through physical (increased stream flow with resultant sediment transport), chemical (pollutants), and biological (bacteria) changes. Such thinking is exemplified by the now ubiquitous observation of a negative correlation between stream condition and ISA (Beach, 2001; Gergel et al., 2002) and that originally described by Klein (1979) but in terms of fish species diversity as shown in Figure 19.1.

The imposition of ISA on aquatic ecosystems is harder to define than that of terrestrial ecosystems. This is due primarily to a lack of inventories for aquatic species, especially for bottom sediments where arguably more abundant biota exist—not surprising given the abundance of species that exist in disparate aquatic environments (e.g., lakes, streams, ponds, marshes, ground waters, and wetlands); for a good account, see Palmer

---

*Bivalves are mollusks belonging to the class Bivalvia. The class has 30,000 species including mussels.

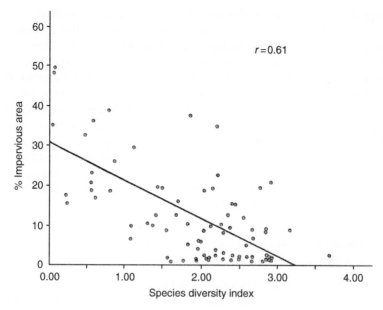

FIGURE 19.1

Montgomery/Prince George's County fish collections vs. watershed imperviousness. Note that the line of regression crosses the species diversity index of 2.00, which separates the good to fair range at the 12% imperviousness point. (From Klein, R.D., *Water Resour. Bull.*, 15 (4), 948, 1979. With permission.)

et al. (1997) for estimations as to the extent of biota (i.e., aquatic invertebrates, algae, protozoa, and bacteria) that exist globally. It is therefore understandable that information on aquatic species extinction is sparse. However, this is only one facet of the problem since lack of knowledge at an inventory level is similarly compounded by lack of understanding of the biological linkages that exist between the flora and fauna aquatic ecosystems. Palmer et al. (2000) describe such a state of affairs.

19.3 A Case Study of ISA and Aquatic Fauna—The Line Creek Watershed, Atlanta, Georgia, United States

19.3.1 Introduction

The case study described here serves two purposes. The first is a stepwise description of ISA derivation from first-order (integer digital numbers) satellite data. The second is to show how the ISA data field was manipulated in a fashion that permitted its application towards an assessment of the impacts on an aquatic ecosystem that had been severely disturbed due to rapid urbanization over a period of time.

Before describing steps in the process of ISA computation from derived fractional vegetation cover, it is worthwhile to set out the assumptions inherent in its derivation. Quoting from Gillies et al. (2003), "The theory, adopted behind the satellite determination of *ISA* is based upon the observation (Jennings and Jarnagin, 2000; Ridd, 1995; Klein, 1979; Leopold, 1973) that *ISA* is inversely related to vegetation cover in urban areas, i.e., non-vegetated surfaces in urban areas are almost entirely impervious in North American cities. In making this assumption it is also assumed that no impervious surfaces reside in areas that are not developed." Moreover, when it comes to the health of ecological systems and associated biodiversity of species that comprise them, the fractional vegetation cover approach to calculating ISA has, arguably, more direct meaning from a biophysical point of view than other satellite- or GIS-based methods.

19.3.2 Derivation Techniques and Algorithms to Determine ISA Delineated by Watershed

A technique for determining ISA from satellite data is described in detail with reference to the Landsat 7 platform (ETM+ data). The methods are ubiquitous in the sense that they are essentially the same for TM and MSS data—all that is required is the appropriate technical documentation. For Landsat ETM+ data, the relevant information is found in the Science Data Users Handbook (http://ltpwww.gsfc.nasa.gov/IAS/handbook/handbook.htmls/chapter11/chapter11.html/). Moreover, the rationale and steps involved are germane to other platforms (e.g., ASTER, MODIS) although, for certain platforms, the fractional vegetation cover may constitute one of the derived data products.

Step 1 Digital Numbers to Radiance
The so-called media output for Landsat 7 (i.e., the 1G product) supplies pixel data for an image scene as digital numbers (DN—expressed as integers). The parameters required are generically spectral reflectance, which necessitates a series of steps referred to as calibration. The first step in the calibration procedure is to render the data to units of radiance for which Equation 19.1 is the starting point:

$$L_\lambda = \text{Gain} \times \text{QCAL} + \text{Offset}, \tag{19.1}$$

where
L_λ is the spectral radiance in a spectral interval/band (λ) as $\text{Wm}^{-2}\,\text{sr}^{-1}\,\mu\text{m}^{-1}$ at the sensor's aperture.

The gain (rescaled gain) and offset (rescaled bias) are expressed in $\text{Wm}^{-2}\,\text{sr}^{-1}\,\mu\text{m}^{-1}$ as

$$\text{Gain} = \frac{\text{LMAX}_\lambda - \text{LMIN}_\lambda}{\text{QCALMAX} - \text{QCALMIN}}, \tag{19.2}$$

and

$$\text{Offset} = LMIN_\lambda, \tag{19.3}$$

and QCAL (unitless) is the quantized calibrated pixel value in DN, which is given as

$$\text{QCAL} = \text{QCAL} - \text{QCALMIN} \text{ (0 for NLAPS product, otherwise 1).} \tag{19.4}$$

The other variables represented are defined accordingly as follows:

$LMIN_\lambda$ and $LMAX_\lambda$ (both unitless) are the spectral radiances that are scaled to QCALMIN and QCALMAX, respectively. These values change slowly over time and are supplied by the responsible agencies; for example, for Landsat 7, as mentioned earlier, they are found in the Science Data Users Handbook.

Finally, QCALMIN and QCALMAX (both unitless) are, respectively, the minimum and maximum quantized calibrated pixel values (corresponding to $LMIN_\lambda$ and $LMAX_\lambda$) in DN (generally 0 and 255).

Step 2 Radiance to Apparent Reflectance
The following formula is applied to compute apparent reflectance:

$$\rho_a = \frac{\pi L_\lambda d^2}{E_{sun_\lambda} \cos \theta_s}, \tag{19.5}$$

where

ρ_a is the apparent reflectance (unitless)*
L_λ is the spectral interval radiance at the sensor's aperture
d is the Earth–Sun distance in astronomical units (obtained from a nautical handbook or interpolated from tables—usually supplied in the agency's technical notes)[†]
E_{sun_λ} is the mean solar exoatmospheric irradiance for the spectral interval (tabulated by the agency) $(Wm^{-2} \mu m^{-1})$
θ_s is the solar zenith angle in degrees[‡]

Step 3 Fractional Vegetation Cover
To obtain fractional vegetation cover, first compute the normalized difference vegetation index (NDVI). The NDVI is defined as

$$\text{NDVI} = \frac{\rho_{a_{ir}} - \rho_{a_r}}{\rho_{a_{ir}} + \rho_{a_r}}. \tag{19.6}$$

---

*You may find this referred to as planetary reflectance or planetary albedo in remote sensing science literature.
[†] Some simply default this to one.
[‡] Two things to note here: (1) The solar elevation angle is what is generally supplied in the header file that accompanies the image data so as to compute the zenith (the complementary angle, one computes $90 - \theta_s$). (2) Most trigonometric functions supplied in image-processing packages default units to radians.

The NDVI is calculated from the "reflectance" ρ_{a_r} in the red band (typically 0.6–0.7 μm) and the reflectance $\rho_{a_{ir}}$ in the near-infrared band (typically 0.8–1.0 μm). The notation used here is important as it indicates apparent reflectance; there is often confusion on this front. In theory, NDVI should be calculated using surface reflectance values, as it is this NDVI that is directly correlated with the amount of photosynthetically active radiation (PAR). Moreover, the apparent reflectances contain an atmospheric scattering component that can vary considerably depending on the atmospheric properties at the time of measurement. To account for any atmospheric addition or subtraction to ρ_{a_r} and $\rho_{a_{ir}}$ requires a so-called atmospheric correction, the details of which are not necessary here but generally involve the use of a radiative transfer model like MODTRAN. However, a straightforward atmospheric correction (Carlson and Ripley, 1997) is accomplished simply by normalizing the NDVI with respect to a bare soil (NDVI$_o$) and full vegetation (NDVI$_s$) NDVI reference points that correspond to a bare soil pixel and a completely vegetated pixel, as follows:

$$N^* = \frac{\text{NDVI} - \text{NDVI}_o}{\text{NDVI}_s - \text{NDVI}_o}. \qquad (19.7)$$

There are various ways (e.g., Gillies and Carlson, 1995) to determine NDVI$_s$ and NDVI$_o$, which vary in complexity; however, using the tails of the NDVI frequency distribution for the image scene is a good first approximation.

There are some subtleties to the computation of NDVI of which one should be cognizant. Theoretically, the index can range from −1 to 1; in particular, negative values of NDVI occur when ρ_{a_r} is greater than $\rho_{a_{ir}}$ and usually crop up in measurements of rather low reflectance in both bands. Such conditions will manifest when the image scene contains water "contamination" such as clouds, water bodies, and snow, or if the soil is particularly wet, each of which may fully or partially represent a pixel. To facilitate analysis, negative values of NDVI should be filtered from the image scene.

The fractional vegetation cover (Fr), the derivation of which is detailed in Gillies et al. (1997), is subsequently calculated as

$$\text{Fr} = N^{*2}. \qquad (19.8)$$

Step 4 Image Scene Classification
The corresponding piece that is coupled with the fractional vegetation cover (to infer ISA) is the knowledge as to which pixels are urban/built-up, that is, those that are developed as artificial impervious surfaces. The derivation of "developed" pixels through the process of image classification is arguably the most technically difficult and time-consuming part of the process. There are numerous classification techniques detailed in the literature. It is dependent upon the user's knowledge and skill-set to apply any of the various classification procedures (i.e., unsupervised, supervised, hybrid, fuzzy) that many of the current image-processing systems now make available.

Image-processing systems also evaluate the classification accuracy of each cover type, which is valuable information for assessing the success or failure of the classification scheme and whether further refinement might be considered necessary.

Step 5 ISA

A single new field of ISA (Carlson and Arthur, 2000) is generated by applying the following formulation:

$$ISA = (1 - FR)_{DEV}, \tag{19.9}$$

where the subscript DEV denotes that ISA is computed for all pixels classified as developed (i.e., urban/suburban).* Implicitly, all other pixels are excluded because they are other land-use/cover categories (e.g., water, vegetation types, etc.).

Equation 19.9 is evaluated heuristically in an image-processing system that contains a programmatic module (e.g., in Imagine, it is called the spatial modeler). Hence, the fractional vegetation cover digital layer, computed from basic DN via Equations 19.1 through 19.8, is coupled with the thematic classification layer to determine ISA for those pixels that are defined as "developed."

19.3.2.1　Watershed Delineation

In studies that pertain to the consequence of ISA on aquatic systems, it is often desirable to delineate a particular watershed or split a watershed into its subdivisions. Several software packages are available to demarcate a watershed by specifying certain criteria within a digital elevation model (DEM). At a superficial level the results from each are, or can be made, similar. Of those out there, examples include TauDEM (http://hydrology. neng.usu.edu/taudem/), ArcHydro (http://support.esri.com/index.cfm? fa = downloads.dataModels.filteredGateway&dmid = 15), and RiverTools (http://www.rivertools.com/). In each case, one has to make good choices in activities like picking the threshold for delineating channels. However, TauDEM does suffer from memory limitation problems limiting the DEM size that can be handled relative to ArcHydro and RiverTools, both of which have commercial support to implement efficient memory management. Depending on the application area, one might consider using preexisting results that have been quality controlled, for example, NHDPlus (http://www.horizon-systems.com/nhdplus/) or national elevation dataset derivatives (http://edna.usgs.gov/). One further digital data field that is particularly useful is the watershed digital hydrography. The digital hydrography can be overlaid to visually georeference the DEM of the watershed, identifing subwatersheds from which one can set down reference points to start the process of channel delineation.

---

*Depends on the specifics of your classification.

19.3.2.2 Data Masking

A data mask is a simulated construct that allows one to filter redundant/ unwanted data. In the context of image processing, an image consisting of pixels is stored as a raster file, a grid of x and y coordinates and is illuminated as such on a display space. In the case of a delineated watershed of ISA (often referred to as an area of interest (AOI) in image-processing parlance), the raster file contains data that are not actual ISA data but values that constitute the raster grid. One might think of it in this way. Consider the fact that ISA values ranging from 0 to 1 (representing fully vegetated and 100% ISA, respectively) would appear on the display as black (0) and bright white (1). However, if the system stores those values of the grid outside the AOI as zero, a bulk export of the data for statistical analysis would be totally misrepresentative. To remove the extraneous zeroes from the file, a mask is created that assigns a unique value for the AOI pixels and for the rest of the grid, which when run is used to export only those ISA values within the AOI. Hence, true values of zero (those fully vegetated) are retained for subsequent analysis. One has to construct an algorithm to perform this task. Figure 19.2 gives a visual impression of the data one wishes to export to an ASCII file—in this case, only the ISA data associated with the sub-watershed delineated within the AOI.

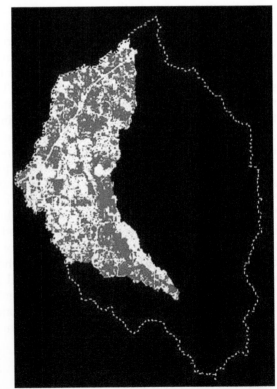

FIGURE 19.2
Delineated watershed (gray shading) for Line Creek (derived ISA). The Peachtree watershed (AOI) is denoted by the dotted line. To extract only the data associated with Line Creek subwatershed requires the use of a data mask. (From Gillies, R.R. et al., *Remote Sensing Environ.*, 86, 441, 2003. Copyright 2003. With permission from Elsevier.)

19.3.3 Peachtree Watershed ISA and Freshwater Mussel Diversity

As mentioned previously, the knowledge base for bivalves is relatively extensive. In the case of the Peachtree Watershed, the mussel data records were in part surveyed but also collated from historical records (Brim Box and Williams, 2000; Gillies, et al. 2003). Figure 19.3 indicates the survey sites that lie in the Peachtree watershed. There are three drainages to the Peachtree watershed—namely, Line Creek (where sites H, G, J, A, B, N, and E lie), Flat

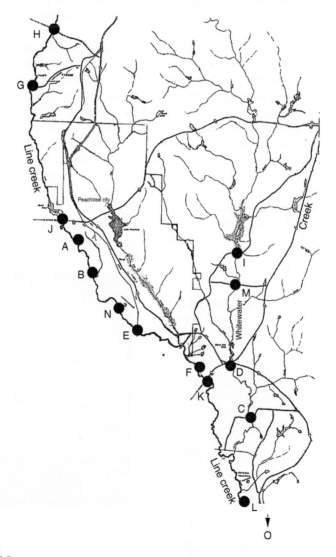

FIGURE 19.3

Site map showing study sites (A through L) on the Line and Whitewater Creeks, near Atlanta, Georgia. Sites O, further downstream, and P, to the east of Peachtree City, are located outside this map. (From Gillies, R.R. et al., *Remote Sensing Environ.*, 86, 441, 2003. Copyright 2003. With permission from Elsevier.)

Creek (water source: Lake Peachtree), and Whitewater (where sites I, M, D, and C lie). Sites F and K reside downstream from the confluence of Line and Flat Creeks and L, O, and P are sites where all drainages contribute. In terms of the ISA analysis, three drainages exist and were delineated as and named Line, Line plus Flat, and Whitewater Creeks. The Line Creek ISA breakout of the subwatershed is shown in Figure 19.2. All three subwatersheds along with the time series of ISA (using three Landsat images (1979 MSS, 1987 TM, and 1997 ETM+) for the entire Peachtree watershed were generated following the procedures outlined earlier. As a point of clarification, high- and low-density urban classes (step 4) were created through a hybrid classification procedure (Yang and Lo, 2002).

Again quoting from Gillies et al. (2003), "We were interested in examining the entire Peachtree watershed in terms of overall urban development and ecological impact. In addition, the sub-watersheds of the combined Line/Flat and Whitewater creeks might offer further insights from two perspectives: First, the urban development around the Line plus Flat sub-watersheds as compared to the Whitewater Creek watershed is markedly different in extent and magnitude, so we might realistically find differences in aquatic decline between the two sub-watersheds. Second, the confluence of Line and Flat creeks (where study sites F and K reside) is where we might reasonably expect to observe a greater ecological impact due to component flows from these two creeks (Morisawa and LaFlure, 1979)."

The results of the ISA derivation are shown in the following two figures. Figure 19.4 is the ISA calculated for the entire drainage area and expressed

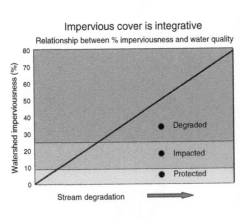

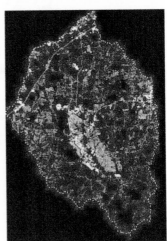

FIGURE 19.4

Entire Peachtree watershed expressed as a gradation of gray associated with degree of ISA. The shades correspond to stream quality conditions from unpolluted and natural (<10%) to polluted (10%–20%) and degraded (>25%). Zero ISA (100% vegetation) appears as black in this representation. (From Gillies, R.R. et al., *Remote Sensing Environ.*, 86, 441, 2003. Copyright 2003. With permission from Elsevier; Prisloe, S. et al., in *Proceedings of the American Society of Photogrammetry and Remote Sensing*, St. Louis, MO, April 23–27, 2001. With permission.)

in terms of a stream quality classification system as originally proposed by Schueler (1994) and later modified by Arnold and Gibbons (1996). It divides ISA into three groupings, as indicated, that serve as general guidelines for stream quality: less than 10%, 10% to 25%, and above 25%.

Figure 19.5 is the time series equivalent of Figure 19.4 and shows ISA expansion throughout the years. The color-coding corresponds to the

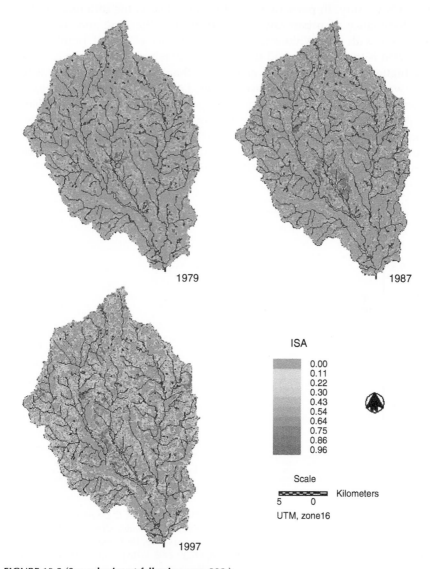

FIGURE 19.5 (See color insert following page 292.)
ISA (0–1 representing 0%–100%) maps for the entire Peachtree Watershed for the years 1979, 1987, and 1997. Color-coding represents the degree of ecological impact, as outlined by Schueler (1994). Overlaid in blue is the digital hydrography of the Peachtree watershed. (From Gillies, R.R. et al., *Remote Sensing Environ.*, 86, 441, 2003. Copyright 2003. With permission from Elsevier.)

protected (green zone), impacted (orange zone), and degraded (red zone) categories of Figure 19.4.

The information enclosed in Figure 19.5 indicates that a remarkable degree of urbanization has occurred over the 18 year period and, in particular, urban growth in the vicinity of Peachtree City. As noted by Gillies et al. (2003), industrial, commercial, and shopping center complexes (ICS) along with residential development mark the urbanization in the city's vicinity. In particular, what is very clear is the difference in type and dominance of urbanization in and around the Line and the Flat as compared with the Whitewater basin. In terms of actual numbers, the indication is that the Line and the Flat subwatersheds exhibited a considerably higher percentage of ICS complexes than that in Whitewater where there was no observable increase in high ISA values related to ICS complexes from 1987 to 1997. These findings were the result of a statistical analysis* that was performed on the exported ASCII data, the details of which can be found in Gillies et al. (2003) and from which, as an example, Figure 19.6 is taken. Figure 19.6, as stated in Gillies et al. (2003), "...shows percentiles (85th to 100th for 1979 and 1987, and 70th to 100th in 1997; in

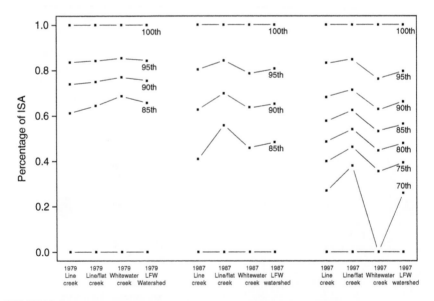

FIGURE 19.6
Percentiles of ISA percentages (0–1 representing 0%–100%), based on all pixels, for the different watersheds (Line, Line/Flat (indicating Line plus Flat), Whitewater Creeks, and the entire drainage area representing Line Creek (LFW)) for the years 1979, 1987, and 1997. (From Gillies, R.R. et al., *Remote Sensing Environ.*, 86, 441, 2003. Copyright 2003. With permission from Elsevier.)

---

*S-Plus (Mathsoft, 2000) Version 5.0 Release 3.0 for SunSPARC, SunOS 5.5.

steps of 5) of ISA percentage, based on all pixels," from which Gillies et al. (2003) note that "...the Whitewater sub-watershed has far less than 10% of its ISA values above 70% (in fact, the highest 10% of the ISA values are in the range 63% to 100%). For the Line Creek sub-watershed and the Line/Flat Creek (Line plus Flat) sub-watersheds, the highest 10% of the ISA values are above 68% and 71%, respectively, indicating a considerably higher percentage of ICS complexes for the latter sub-watersheds than for Whitewater." The indication of growth in ICS complexes is concomitant with growth in residential areas and is reported as such in Gillies et al. (2003).

The nature of the mussel data (Table 19.1) meant that only empirical evidence, in terms of aquatic species richness, was available to assess the relative impact of changes in ISA. Quoting some summary statistics from Gillies et al. (2003): "Two things are apparent when looking at the four sites on Line Creek. (1) Downstream sites historically contained more species than the upstream sites (e.g., K versus J). (2) There was a progression of species loss going from upstream to downstream sites. For example, site G on Line Creek (well upstream) did not lose any mussel species. Progressing downstream, sites J, K, and L each lost over 50% of their mussel species and the loss was progressively greater going downstream. For example, from the 10 species that were present at site J in 1995 or before (i.e., in 1985 and 1992 to be specific), only 5 were observed in 1995, which relates to a loss of 50%. At site K, 54% of species disappeared over time. Site L, which historically had the most species of any site, experienced the greatest loss. Eleven of 15 species (73%) that were initially present at that site had disappeared by 1999." On the other hand, site H along with site G (as already noted), both well upstream of the main ISA growth areas, had no major species losses.

Further, "Two of the six sites surveyed over multiple years occurred in the Whitewater Creek sub-watershed. Species losses at those sites were less than those on Line Creek. For example, site C lost no mussel species between 1995 and 1997 and site D lost 30%. It is possible that species losses at Site D were due, in part, to the reservoir directly above that site. Declines in mussel species below dams in other drainages have been well documented" (e.g., Bates, 1962; Williams et al., 1992).

19.4 Further Studies of ISA and Aquatic Ecology

A further refinement to ISA (the total ISA defined for a catchments' area) is the effective ISA, although first mentioned back in the late 90s (Booth and Jackson, 1997), which is now being scrutinized as a more insightful landscape indicator. The effective ISA is defined as those impervious surfaces with direct hydraulic connection to the downstream drainage system. In a sense, effective ISA is a more esoteric variable since it is not directly measured and is simply a crude surrogate of ISA (Wang et al., 2001; Stepenuck et al., 2002)—in other words, no direct information on actual drainage

TABLE 19.1

Sites vs. Species vs. Year (Body) for the Line Creek/Flat Creek (Line plus Flat)/Whitewater Sampling Sites near Atlanta, GA.

| | Site Location | | | | | | | | | | | | | | | | |
| | Line Creek | | | | | | | L/F | | Whitewater | | | | Other[a] | | | |
Species	H	G	J	A	B	N	E	F	K	I	M	D	C	L	O	P	Status
Elliptio complanata	+	+	+	+	+		+	+	+	+	+	+	+	+	+		cs
Villosa vibex	+	+	+	+	+				+			−	+	●			cs
Toxolasma paulus	+	+	●	+	+	+			−	+		+	+	●			cs
Villosa lienosa	+	+	●	+					−	+		+	+	●			sc
Quincuncina infucata		+	+	+	+	+			+	+	+	−	+	−	+		cs
Megalonaias nervosa							+		+			+	+	+	+	#	cs
Utterbackia imbecillus			+	+					●					●			cs
Pyganodon grandis			●	+					●			+					
Lampsilis subangulata			●						+				+	+		#	e
Elliptio icterina									+					+			cs
Elliptio arctata			●						●					●			sc
Lampsilis claibornensis									●								sc
Villosa villosa									●								sc
Uniomerus caroliniana					+									−			cs
Anodontoides radiatus																#	e

		#		e
Alasmidonta triangulata	•	+		e
Medionidus penicillatus	•			x
Lampsilis binominata	•			t
Elliptoideus sloatianus	•			cs
Elliptio crassidens			+	cs
Utterbackia peggyae			+	cs
Lampsilis teres			−	

Source: From Gillies, R.R. et al., *Remote Sensing Environ.*, 86, 441, 2003. Copyright 2003. With permission from Elsevier.

Note: The status indicates how critical the condition of a species is in general (cs = currently stable, sc = special concern, e = endangered, t = threatened, x = extinct). The top row shows sites, labeled according to Figure 19.3, and the first column shows mussel species. Sites have been arranged from upstream to downstream locations. Watersheds are separated by an additional space. Mussel species have been arranged in such a way that species that have been observed at similar sites are listed close to each other.

Key:

•   Observed in 1992 or before, but not afterward.

+   Observed in 1995 and later if that site was visited.

−   Observed in 1995, but not afterward (G, D, and C: 1997; K and L: 1999).

Historic data from 1966, no later visits to this site.

L/F   Sampling sites with feeds from Line and Flat Creeks.

[a] Site L is on Line Creek (with feeds from Flat and Whitewater Creeks) whereas sites O and P are situated on the Flint River.

connection is contained within the variable. It is simply inferred through other means, usually for a given land use. However, Walsh et al. (2004) have taken a more exacting approach to determining effective ISA and have applied it within the Melbourne metropolitan area, Victoria, Australia. The result of this study and an analogous study (Walsh, 2004; Walsh et al., 2004, respectively) indicated that drainage connection is significant and was, in point of fact, the strongest independent correlate (Walsh, 2004). Walsh's conclusion is perhaps more salient because the connection of ISA to streams by pipes is a more likely determinant of taxa loss than ISA in itself.

In the previous paragraph, the importance of representing the urban system more completely was emphasized. However, a recent study by King et al. (2005) elucidates another aspect of the inherent complexity of the system when it comes to assessing aquatic impacts by virtue of urban development, that is, the spatial arrangement may be an important modulator of watershed land-cover effects on stream ecosystems. Furthermore and perhaps more significantly, their work, from which Figure 19.7 is taken, indicates that a threshold effect is in place that comes into play when between 21% and 32% of the land is developed—at least for the watershed studied.

Once this point is reached, the macroinvertebrate assemblage composition plummets dramatically. The authors did not remark on whether the observed decline in diversity was irreversible. While the land-cover data

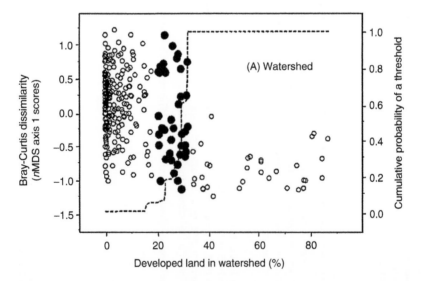

FIGURE 19.7
Scatterplot of the threshold effect of developed land in the watershed on the macroinvertebrate assemblage composition (Bray-Curtis dissimilarity expressed as nonmetric multidimensional scaling *n*MDS Axis 1 scores). Notice that there is essentially a 100% probability (right-hand scale) of a sharp decline in taxonomic composition beyond 32% development. (From King, R.S. et al., *Ecol. Appl.*, 15(1), 137, 2005. With permission).

(percentage developed land) used in the analysis are not strictly equivalent to percentage ISA, the corollary of using ISA is enticing as to where the biotic thresholds might lie, and one is inclined to remark here that the range 21%–32% is likely well within the range of ISA, which would be defined "demonstrable and probably irreversible, loss of aquatic system function" (Booth and Reinelt, 1993).

An additional study worthy of mention is that by Synder et al. (2005). In this study, landscape metrics were generated in part with the use of ISA and tree cover and used in a series of logistic regression models to predict stream health. The ISA turned out to be consistently the most important predictive variable in the models while tree cover in the watershed or in the riparian zones was the second strongest predictor. One of their concluding statements is somewhat revealing though—"The importance of impervious surface areas was likely influenced by the existence of storm drains that bypass buffer zones and effectively connect the stream to the built environment." The significance of "effectively" should not be lost on the reader given the substance of the discussion that has gone before. Nonetheless, in closing the authors' note that the ISA conveyed more information than those landscape metrics derived using it, which, if nothing else, entertains the notion that the generalized relationship that has been around for quite some time (Figure 19.4) has distinct merit to it with the proviso of a distinction being made with respect to effective ISA.

In point of fact, the results of the Synder et al. (2005) study indicated that watersheds in excellent health averaged <8% ISA, watersheds in good health averaged <10% ISA, those rated fair averaged <20% ISA, and those with a poor health ranking exceeded 29% ISA. These numbers agree with the bounds of those originally proposed and diagramed (Figure 19.4) by Schueler (1994) and customized later by Arnold and Gibbons (1996).

19.5 Discussion and Conclusions

A good part of this chapter was used to detail a methodology for computing ISA as well as giving some direction as to how one can use GIS and image-processing technologies to delineate and extract the derived ISA data within a geographical framework—in this case, at the watershed and subwatershed levels—to see how ISA (1) grew and evolved over time as well as (2) was linked comparatively with the disappearance or reduction of mussel species diversity. The conclusion with respect to (1) is that there was significant urbanization over time and that this occurred predominantly in the high-end development exhibiting high values of ISA. Moreover, there was distinct disparity at this level between the Line plus Flat drainages as compared with the Whitewater drainage—observed visually and corroborated statistically. As to (2), aquatic species richness data could only be used to assess the relative impact of changes in ISA. The empirical evidence is quite compelling. First, little impact (in terms of species loss) was observed in Whitewater (where there was marginal ISA growth); the exception is at site D (30% lost) and can be explained in that there was a dam immediately upstream. Second, mussel data on the Line Creek indicated that upstream sites (H and G), where ISA growth was minimal, had no major mussel loss whereas those sites further downstream (e.g., J and K with over 50% loss) were distinctly impacted. Along the reach (from J to E), observations in themselves reveal major growth in ISA in extent and close proximity to the creek, which is not the case further upstream at sites H and G. Third, the effect of combined ISA growth in both the Line and the Flat drainages is perhaps evidenced by the fact that the site where most species were lost (site L with a 73% loss) is downstream of the confluence (and hence combined runoff) of these creeks. Marked physical evidence of this is apparent at this site, which shows that there has been marked widening of the creek channel; this is consistent with streams heavily affected by urbanization. The discussion should, however, not preclude other sites as not all show such marked declines. What it shows is that confounding factors may lie in any or all of (1) the completeness of the mussel inventories, (2) the sensitivity of a species to tolerate changes in its physical environment, (3) a particular habitat's susceptibility to erosion as well as perhaps (4) the representation of total ISA rather than effective ISA. Gillies et al. (2003) recognized such a state of affairs and remarked: "We certainly do not want to claim that an increase in ISA 'causes' the loss of mussel species. However, as shown in this paper, ISA (an aggregate of many factors, some of which may 'cause' the loss of species) is a good indicator that is 'associated' with the loss of mussel species. Obtaining ISA percentages (based on remotely sensed data) for larger regions than in this study may quickly focus urban planners and environmentalists to regions where aquatic species are severely endangered—and to implement corrective actions that prevent future habitat degradation and aquatic species losses."

The issue of effective ISA is a significant qualification that should be considered wherever possible but may not be realistic in all instances. The use of either effective ISA or ISA will become site-specific depending on resources available or the problem that is addressed. Perhaps the most important point to come from Walsh's work on effective ISA is one of future design of drainage connections to a water body.

Several instances of departures from expectations of biodiversity as might be inferred from the ISA have pointed toward ecosystem complexity as an issue that is poorly understood or where the knowledge base is incomplete. While the discussion has not included the impact of ISA on terrestrial ecosystems, it must be realized that while they are distinct systems in themselves they are not mutually exclusive but inextricably linked with aquatic ecosystems. The biotic processes and patterns of adjacent terrestrial habitats are connected to overlying water bodies and vice versa. As an example of this, consider insects that spend part of their life cycle on land

and in the water. If a particular insect is lost, to what extent will this disrupt or intensify the interactions and linkages that exist across the aquatic, riparian, and terrestrial domains?

An alluring aspect of the work of King et al. (2005) is the observation of a threshold that in many ways is in the same category as the current debate on climate change. Each (an ecosystem or the Earth's climate) is inherently a complex system that has been observed and can shift from one state to another when a so-called "tipping-point" is reached and similarly may be irreversible (certainly without significant effort and energy) once initiated.

There is no doubt that a generalization of ISA as a measure of disturbance, as well as a heuristic tool, for aquatic ecosystems is a practical first-order indicator of expectations of aquatic ecosystem degradation of biotic assemblages. In fact, the preponderance of research, including those cited here, indicates that the ISA categories for stream quality conditions as originally set up by Schueler in 1994 still hold fast as representative, if not key, categories of environmental impact for aquatic ecosystems. This has directed efforts to the conservation of aquatic ecosystems, which are mostly centered on limiting a catchment's ISA to a very small level (Arnold and Gibbons, 1996; Beach, 2001).

References

Arnold, C.L. and Gibbons, C.J. 1996. Impervious surface coverage: the emergence of a key environmental indicator. *Journal of the American Planning Association* 62 (2), 243–258.

Bates, J.M. 1962. The impact of impounds on the mussel fauna of Kentucky Reservoir, Tennessee River. *American Midland Naturalist* 68, 232–236.

Beach, D. 2001. *Coastal Sprawl. The Effects of Urban Design on Aquatic Ecosystems in the United States.* Pews Ocean Commission, Arlington, Virginia.

Benfield, K., Raimi, M., and Chen, D. 1999. *Once There were Greenfields; How Urban Sprawl is Undermining America's Environment, Economy, and Social Fabric.* Natural Resources Defense Council.

Booth, D.B. and Jackson, C.R. 1997. Urbanization of aquatic systems—degradation thresholds, stormwater detention, and the limits of mitigation. *Journal of American Water Resources Association (JAWRA)* 33, 1077–1090.

Booth, D.B. and Reinelt, L.E. 1993. Consequences of urbanization on aquatic systems—measured effects, degradation thresholds, and corrective strategies. In *Proceeding Watershed'93. A National Conference on Watershed Management*, March 21–24, Alexandria, VA, pp. 545–550.

Brim Box, J. and Williams, J.D. 2000. Unionid mollusks of the Apalachicola Basin in Alabama, Florida and Georgia. *Bulletin of the Alabama Museum of Natural History* 21, 1–143.

Carlson, T.N. and Arthur, S.T. 2000. The impact of land use–land cover changes due to urbanization on surface microclimate and hydrology: a satellite perspective. *Global and Planetary Change* 25, 49–65.

Carlson, T.N. and Ripley, D.A.J. 1997. On the relationship between NDVI, fractional vegetation cover and leaf area index. *Remote Sensing of Environment* 61, 241–252.

Gergel, S.E., Turner, M.G., Miller, J.R., Melack, J.M., and Stanley, E.H. 2002. Landscape indicators of human impacts to Riverine Systems. *Aquatic Sciences* 64, 118–128.

Gillies, R.R., Brim-Box, J., Symanzik, J., and Rodemaker, E.J. 2003. Effects of urbanization on the aquatic fauna of the Line Creek watershed, Atlanta—a satellite perspective. *Remote Sensing of the Environment* 86, 441–422.

Gillies, R.R. and Carlson, T.N. 1995. Thermal remote sensing of surface soil water content with partial vegetation cover for incorporation into climate models. *Journal of Applied Meteorology* 34, 745–756.

Gillies, R.R., Carlson, T.N., Cui, J., Kustas, W.P., and Humes, K.S. 1997. A verification of the 'triangle' method for obtaining surface soil water content and energy fluxes of the Normalized Difference Vegetation Index (NDVI) and surface radiant temperature. *International Journal of Remote Sensing* 18, 3145–3166.

Jennings, D. and Jarnagin, S.T. 2000. Impervious surfaces and streamflow discharge: a historical remote sensing perspective in a northern Virginia subwatershed. In *ASPRS Annual Conference Proceedings*. Washington, D.C. May 22–26, CD-ROM.

King, R.S., Baker, M.E., Whigham, D.F., Weller, D.E., Jordan, T.E., Kazyak, P.F., and Hurd, M.K. 2005. Spatial considerations for linking watershed land cover to ecological indicator in streams. *Ecological Application* 15 (1), 137–153.

Klein, R.D. 1979. Urbanization and stream water quality impairment. *Water Resources Bulletin* 15 (4), 948–963.

Leopold, L.B. 1973. River channel change with time: an example. *Geological Society of America Bulletin* 84, 1845–1860.

Morisawa, M. and LaFlure, E. 1979. Hydraulic geometry, stream equilibrium, and urbanization. In D.D. Rhodes and G.P. Williams (eds.), *Adjustments of the Fluvial System*. Kendall/Hunt, Dubuque, IA.

Nilsson, C., Pizzuto, G.E., Moglen, M.A., Palmer, M.A., Stanley, N.E., Bockstael, N.E., and Thompson, L.C. 2003. Ecological forecasting and the urbanization of stream ecosystems: challenges for economists, hydrologists, geomorphologists, and ecologists. *Ecosystems* 6 (7), 243–258.

NRC (National Research Council). 2001. *Grand Challenges in Environmental Sciences*. National Academic Press, Washington, D.C.

Palmer, M.A., Covich, A.P., Finlay, B.J., Gilbert, J., Hyde, K.D., Johnson, R.K., Kairesalo, T., Lake, P.S., Lovell, C.R., Naiman, R.J., Ricci, C., Sabater, F., and Strayer, D. 1997. Biodiversity and ecosystem processes in freshwater sediments. *Ambio* 26, 571–577.

Palmer, M.A., Covich, A.P., Lake, A., Biro, P., Brooks, J.J., Cole, J., Dahm, C., Gibert, J., Goedkoop, W., Martens, K., Verhoeven, J., and Van De Bund, W. 2000. Linkages between aquatic sediment biota and life above sediments as potential drivers of biodiversity and ecological processes. *Bioscience* 50 (12), 1062–1075.

Prisloe, S., Lei, Y., and Hurd, J. 2001. Interactive GIS-based impervious surface model. In *Proceedings of the American Society of Photogrammetry and Remote Sensing*. St. Louis, MO, April 23–27.

Ridd, M.K. 1995. Exploring a V-I-S (vegetation-impervious surface-soil) model for urban ecosystems analysis through remote sensing: comparative anatomy for cities. *International Journal of Remote Sensing* 16 (12), 2165–2185.

Schueler, T. 1994. The importance of imperviousness. *Watershed Protection Techniques* 1 (3), 100–111.

Schueler, T.R. and Galli, J. 1992. Environmental impacts of stormwater ponds. In P. Kimble and T. Schueler (eds.), *Watershed Restoration Source Book*. Washington, D.C., Metropol, Wash. Counc. Gov.

Stepenuck, K.F., Crunkilton, R.L., and Wang, L. 2002. Impacts of urban land use on macroinvertebrate communities in southeastern Wisconsin streams. *Journal of American Water Resources Association* 38, 1041–1051.

Synder, M.N., Goetz, S.J., and Wright, R.K. 2005. Stream health rankings predicted by satellite derived land cover metrics. *Journal of American Water Resources Association* 41 (3), 659–677.

USEPA (U.S. Environmental Protection Agency). 2002. Biological indicators of biological health. Available at http://www.epa.gov/bioindicators/.

Walsh, C.J. 2004. Protection of in-stream biota from urban impacts: minimize catchment imperviousness or improve drainage design? *Marine and Freshwater Research* 55, 317–326.

Walsh, C.J., Papas, P.J., Crowther, D., Sim, P.T., and Yoo, J. 2004. Stormwater drainage pipes as a threat to a stream-dwelling amphipod of conservation significance, Austrogammarus australis, in south-eastern Australia. *Biodiversity and Conservation* 13, 781–793.

Wang, L., Lyons, J., and Kanehl, P. 2001. Impacts of urbanization on stream habitat and fish across multiple spatial scales. *Environmental Management* 28, 255–266.

Williams, J.D., Fuller, S.L.H., and Grace, R. 1992. Effects of impoundments on freshwater mussels (Mollusca: Bivalvia: Unionidae) in the main channel of the Black Warrior and Tombigbee Rivers in western Alabama. *Bulletin Alabama Museum of Natural History* 13, 1–10.

Yang, X. and Lo, C.P. 2002. Using a time series of satellite imagery to detect land use and land cover changes in the Atlanta, Georgia Metropolitan area. *International Journal of Remote Sensing* 23 (9), 1775–1798.

20

Using Remotely Sensed Impervious Surface Data to Estimate Population

Bingqing Liang, Qihao Weng, and Dengsheng Lu

CONTENTS

20.1 Introduction

Urbanization is continuously accelerated accompanied with increasing congregation of population in cities, with and without planned development. Reports have shown that over 45% of people worldwide live in urban areas currently [1] and this number will reach 50% by the year 2010 [2]. Accurate, up-to-date, detailed, and spatially explicit estimation of population at different scales is required to support urban land management decision making and planning. Much of traditional methods for population estimation is based on census data, which are recognized as a labor-intensive and expensive task and to have difficulty in updating database regularly [3–5]. As a cost-effective data acquisition technology, remote sensing has been increasingly used in estimating population in recent years in response to the flourishing of various remotely sensed data [3,4,6].

Research on population estimation based on remotely sensed data can be tracked back as early as in the 1950s and became more and more popular since 1970s. Many remotely sensed images collected from different sensors have been used to estimate population. With various spatial resolutions, they are especially applicable at a certain scale for the study. For instance, high spatial resolution aerial photography is useful for population estimation at microscale [6–8], whereas low spatial resolution data, such as those from Defense Meteorological Satellite Program Operational Linescan System (DMSP-OLS) are suitable for global or regional scale measurements. However, if a medium scale such as at a city level is concerned, images with medium spatial resolution, such as those obtained from Landsat TM/ETM+ and Terra's ASTER sensors, should be considered. Research has proved that such data are efficient and effective in predicting population in city or county levels [3–5,9,10]. Lo [6] summarized several approaches commonly used in population estimation with remotely sensed data: counting the dwelling units, using per-pixel spectral reflectance, measuring urban areas, and using land-use information. The application of these methods is actually in response to different analytical scales, with the first two applicable at small areas (1 km^2 or less) and the last two for larger (or regional) and medium scales, respectively [9]. It should be noted that the application of various satellite data is not limited by their coarse spatial resolution, if they are. With the advancement in image-processing techniques and the combination of suitable ancillary data, low spatial resolution data can also have the potential in population studies conducted at detailed scales [11,12].

Because remotely sensed images are scale-dependent, population models derived from such data are also subject to the impact of scale. Lo [13] used DMSP-OLS nighttime lights data to model the Chinese population and population densities at three different spatial scales: province, county, and city. When different models (allometric growth models and linear regression model) were applied, it was found that the image data showed promise in estimating population at all three levels. However, the best models were obtained at the city level. Qiu et al. [5] carried out a bi-scale study of the decennium urban population growth from 1990 to 2000 in the north Dallas–Fort Worth Metroplex using models developed with remote sensing and GIS techniques. Both models yielded comparable results with that obtained from a more complex commercial demographics model at the city as well as the census tract levels, yet the GIS model remained robust to the scale change because of its insensitiveness to the spatial scale. The remote sensing model was attenuated when moved to the census tract from the city level. Harvey [9] suggested that population models could produce reliable estimates for large areal units rather than the analytical units at the same scale. Nevertheless, the scale effect on population modeling has not been completely understood, and further efforts toward this direction are needed.

Modeling population based on remotely sensed data remains a challenge primarily due to the spatial incompatibility problem between input and output data [10]. First, population information (population counts or population density) is not directly related to spectral characteristics of surface features. Although research has proved that population density is closely correlated with spectral reflectance values of image pixels [14,15], it is argued that these pixel values are unable to differentiate areas with various population densities [3,16]. High population density may be located in high-rise residential buildings in the city downtown such as those supercities in the Eastern World or the multistory apartments in the city uptown like those cities in the Western countries, whereas low population density may be observed in commercially or agriculturally used lands [3]. Strategies to deal with this problem have been proposed to combine textural information [16] or arithmetic operation of image channels [9,10]. Secondly, it creates a new challenge to apply remotely sensed data to model a quantitative variable like population counts or population density since previous remote sensing analyses are routinely used for qualitative measurement. New approaches are called for rather than the purely statistical techniques such as maximum likelihood classification or image segmentation algorithms. Finally, ground reference data often fail to fully use the relative "detailed" information contained in image data for population modeling, in that the former usually has lower spatial resolutions than the latter. As a result, variables extracted from remotely sensed data are aggregated to the same scale as the ground reference data level when developing population models. Such models more likely suffer from the problem of inefficiently modeling extreme cases. Almost all population models based on a single sampling data have reported to overestimate low population (density) areas but underestimate high population (density) areas [3,4,10]. In order to correct this error, Li and Weng [3] stratified population density into three categories as low, medium, and high density, and developed models individually. Yet the spatial continuity of population data was no longer held if such a stratified method was applied [3]. Besides, remote sensing is mainly regarded as an efficient tool to investigate population distribution, which is commonly represented by population density. Yet statistical analyses, which are often involved with population modeling, are designed to predict population counts rather than population density [12]. Although these two variables are convertible, using area-based data to develop models by a point-based technique is error-induced. It was suggested that model performance can be improved if the model became more complex [9,10]. However, collecting multiple inputs is complicated and expensive, and often more difficulties are encountered when building such models.

Although remote sensors do not collect population data directly, their significance relies on the multifacet ancillary information provided, and thus serves as a feasible solution to all the problems mentioned here.

Like many other human phenomena, the distribution of population is closely intervened by various human activities such as lighting, dwelling, transportation networks, urban sprawl, and land uses. All this information can be obtained through the technology of remote sensing. Previous studies have demonstrated that there is a solid correlation between population data and different remote sensing variables [8–11,17,18]. Impervious surface (including roads, buildings, parking lots, etc.) is one of the variables that can be extracted from remote sensing images. Impervious surface has been emerging as a key environmental indicator for sustainable urban development and natural resource planning in recent years [19]. In addition to its applications in land-use classification [20,21], urban thermal features mapping [20], and nonpoint source pollution monitoring [22], the maps of impervious surfaces are useful in measuring socioeconomic factors such as population density and social conditions [23]. Lu et al. [4] identified several advantages of impervious surface data when applying to estimate population: stable, almost season- and atmosphere-independent, and land-use-dependent. The current research uses impervious surface data to estimate population density.

Census data have been extensively applied in population modeling either directly or indirectly [12]. However, these data actually show a de jure population that reports only usual residents of a given area [12], which often link to lands that are primarily used for residential homes. In order to model residential population successfully, applying a suitable boundary to separate residential and nonresidential areas is essential. A common method to solve this problem is to use classified images [4]. However, obtaining a high-quality classified image is not easy, especially with medium or coarse spatial resolution remotely sensed data [20]. In contrast, zoning, which primarily indicates land use, may serve as a suitable alternative in identifying residential lands. In practice, zoning is one of the several tools used by urban planners to control new development from harming existing landscapes. Many factors are considered when creating zoning polygons, for example, maximum building height and density, extent of impervious surface and open space, and land-use types and activities [24]. Zoning data are thus capable of revealing detailed land-use information for a given city. These data are often available from local government departments. The residential area identified by zoning data is rarely seen in past population studies. The current research uses these data to delineate the residential site of the study area.

Great efforts have been dedicated to modeling population with various remotely sensed images, yet rarely has research been conducted at multiple scales. Although Lu et al.'s study [4] has proved that the impervious surface data can be used to effectively estimate residential population at the block group (BG) level, the accuracy of using such models at different census levels is not certain. This study aims to estimate the residential population in Marion County, Indiana, at three census scales (block, BG, and tract) using a high-quality impervious surface derived from a Landsat ETM+ image.

20.2   Study Area

The study area is Indianapolis, located in Marion County, Indiana, in the United States (Figure 20.1). As the nation's 12th largest city, Indianapolis is the geographical center of Indiana and the capital of the state. According

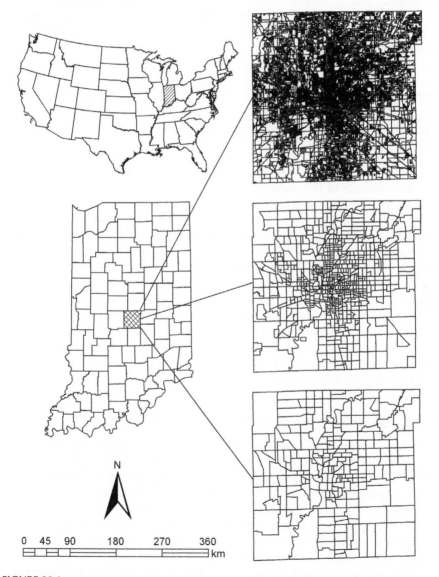

FIGURE 20.1

A map of the study area: Marion County, Indiana, United States, at three census units (block, block group, and tract).

to the census of 2000, the population of the city is 860,454—a 7.9% increase over 1990. Around 627,835 households and 232,619 families were residing in the city. The population density was 855.1 per km^2. Situated on the Tipton Till Plain, Indianapolis has the possibility to develop in all directions. Like the typical urban sprawl presented in other American cities, the city continuously enlarged at the expense of adjacent agricultural areas and forestlands during the past 187 years. Knowing how many people live within a specific geographic area or administrative unit is essential to public organizations such as city departments of transportation, tax assessor offices, department of parks and recreation, and private companies like utility companies and real-estate companies.

20.3 Data and Methodology

20.3.1 Datasets

A Landsat 7 ETM+ image dated June 22, 2000 was used for this study. It was collected about the same time when the census and zoning data were created, showing good promise in the current study. Using the 1:24,000 topographic maps, the image was first geocorrected to a common Universal Transverse Mercator (UTM) coordinate system. All reflective bands and the thermal band were then resampled using the nearest neighbor algorithm with 30 m pixel size. The resultant root-mean-square error (RMSE) was found to be less than 0.2 pixel. This image was used to develop the impervious surface. Two types of GIS data were also applied, including the Census 2000 data and the zoning data layer. The 2000 population census data have been attached to three shape files indicating three levels of census: block, BG, and tract. The zoning data are obtained from the Indianapolis Mapping and Geographic Information System (IMAGIS) (http://www6. indygov.org/imagis/index.htm).

20.3.2 Extraction of Impervious Surfaces from Landsat ETM+ Imagery

Extracting impervious surfaces from images is challenging primarily because of the heterogeneity of urban environment. Consequently, images with limited spectral and spatial resolutions often fail to map this characteristic explicitly. Many methods have been adopted to derive impervious surfaces from remotely sensed data. The most popular method currently used is linear spectral mixture analysis (LSMA), which assumes that the spectrum measured by sensors is a linear combination of spectra from pure surface feature types, called endmember [25].

All the reflective bands in the ETM+ image were used to derive impervious surfaces. Basically, several steps were involved in developing the impervious surface: first, using the minimum noise fraction transform to extract the majority of information from original images into the first three components; second, identifying endmembers (i.e., vegetation, high albedo, low albedo, and soil) by combining the conventional image-based endmember selection method with the scatter plots of the first three resultant components; third, unmixing the six ETM+ reflective bands into four fraction images based on a constrained least-square solution; fourth, removing nonimpervious surface pixels from both the high- and low-albedo fraction images by incorporating land surface temperature data; and finally combining the modified low- and high-albedo fraction images to get the final impervious surfaces—Imp (Figure 20.2). The accuracy assessment shows the overall RMSE and system error of the resultant impervious surface as 9.22% and 5.68%, respectively. Detailed descriptions for developing this impervious surface image can be found in Lu and Weng [20].

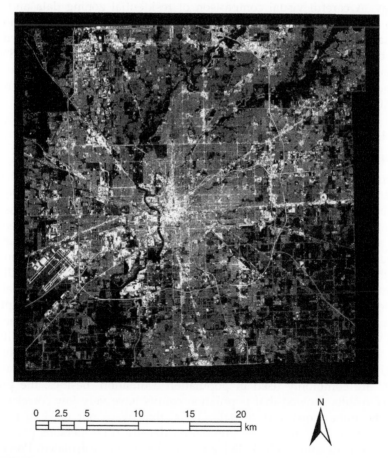

FIGURE 20.2
Distribution of impervious surface (Imp) derived from Landsat ETM+ 2000 Image in the study area.

20.3.3 Zoning Data Processing

The original zoning data layer used in this study has 6908 polygons categorized into 10 different zoning groups. A detailed thematic map illustrating 15 residential levels was first created by aggregating all polygons labeled with "D" (means "dwelling"). Since DA and DP are primarily used for agriculture and future development, respectively, they were not included in the detailed residential map. Besides, as DS mainly indicates areas having extremely low population density in suburban areas, it was also excluded. Hence, the final residential map has only 12 residential levels ranging from D1 to D12 and they were grouped into 4 categories based on their comprehensive use: very low density, low density, medium density, and high density (Figure 20.3). These polygons were united as one polygon showing all residential areas for the study area. These data were then used to delineate the residential impervious surface (RImp) through the GIS overlay. A careful visual comparison of residential zoning data and the RImp image indicates that the application of the former effectively removes the majority of the nonresidential impervious pixels in the latter. Based on three types of census units (will be discussed later), it is observed only 8% of blocks, 0.3% of BGs, and 0% of tracts that contain impervious surface but without dwelling units were unable to be deleted. These census units were identified as outliers and would be removed when building models.

20.3.4 Statistical Analysis

20.3.4.1 Model Development

The census data were used as the analytical unit in the study. At each level, the census data were grouped into two sample datasets: one is the modeling dataset for developing the population estimation models and the other is the validation dataset for accuracy assessment. Only 30% of the total census units at three census levels were randomly selected to build models, and a 2.5 standard deviation combining with scatter plots of population density and the impervious surface variable was used to identify the outliers. Correspondingly, the remaining 70% of the census data were the validation data for model assessment. Table 20.1 summarizes the statistical descriptive variables for the modeling samples. Because the sizes of polygons for census units vary with urban land-use patterns, people reside in polygons differently. In order to avoid the size effect on the actual population distribution at a given census unit, population density instead of individual population counts are commonly used in population estimation. Besides, the preliminary results showed that population counts have very low correlations with the impervious surface data. Thus, they were not used to model population in this study. The population density (PD—persons per km^2) for each census unit at block, BG, and tract levels was calculated. Previous research indicated that the application of both the square root and the logarithmic forms helped to improve population modeling [3,4,9]. Hence, the two transformations of population density—SPD and LPD—were also computed.

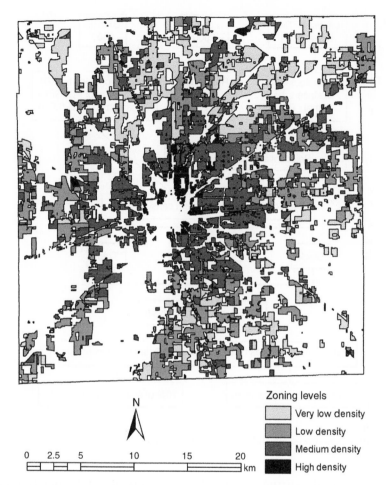

FIGURE 20.3
Map of four residential zoning categories: very low density (suburban single family), low density (low intensity single and two family), medium density (medium intensity single and multifamily), and high density (center urban high-rise apartment).

TABLE 20.1

Summary of Statistical Descriptions of Modeling Samples at Three Census Levels

Census Levels	Total Cases	Samples	Minimum	Maximum	Mean	Standard Deviation
Block	13,989	2,963[a] (4,196[b])	1.79	10,491.53	2,074.93	1,584.11
BG	658	183[a] (197[b])	12.62	4,442.26	1,541.62	976.12
Tract	212	61[a] (63[b])	104.49	3,415.85	1,303.16	785.69

[a] Samples that removed outliers and finally used for modeling.
[b] Samples selected based on random sampling techniques.

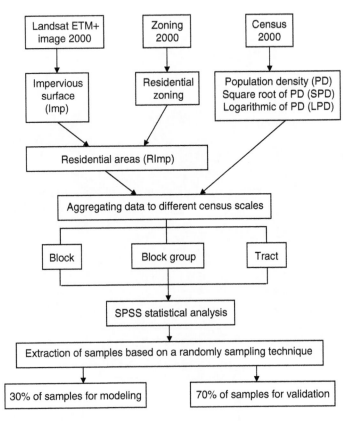

FIGURE 20.4
A flowchart of data processing and modeling development.

Data used for the current research have two formats: impervious surfaces are in raster format whereas residential zoning and census data are vector-based. The two classes of data were integrated after registering all vector data to have the same projection as the images. The statistical mean values of RImp (MRImp) were then calculated for each census zone at the levels of census block, BG, and census tract.

All variables were exported into SPSS to conduct statistical analysis. Pearson correlation analyses were first performed among all variables at each census level. Three groups of regression models were then constructed by using PD, SPD, and LPD as the dependent variable and mean RImp (MRImp) as the independent variable at different levels. A flowchart illustrating the whole process is shown in Figure 20.4.

20.3.4.2 Model Assessment

Accuracy assessment is essential for complete model development, and population modeling is not the exception. Calculating residuals derived from established models is a very common way, in addition to frequently used R^2 values, to evaluate the model performance. For any individual cases, the relative error (RE) is defined as

$$RE = (P_e - P_g)/P_g \times 100 \tag{20.1}$$

where P_g and P_e stand for reference and estimated values, respectively. The model's overall performance was then determined by mean RE (MRE) with the following formula:

$$MRE = \frac{\sum_{k=1}^{n} |RE_k|}{n} \tag{20.2}$$

Since mean values are easily affected by extreme values and may not appear representative of the central region for skewed datasets such as residuals, a median RE (MedRE) value was also used to measure the model performance. The MRE and MedRE were calculated for each validated model at three census levels. In order to further identify the bias generated from the population models, the error of total (ET) in percentage was also calculated.

Three residual maps were created based on the best estimation models at each level for geographical analysis of predicted errors. Scatter plots of predicted and reference variables and relative errors and reference variables derived from such models using validating data were also used for the analysis.

20.4   Results

20.4.1   Models of Residential Population Density

Pearson correlation analyses were first employed to examine the bivariate relationships for all seven variables at each census scale. Table 20.2 exhibits the correlation matrix of three population density parameters and the impervious surface variables generated at three levels. All these correlation coefficients were significant at 0.01 (99%) level, indicating that these relationships were significant and the variables were linearly and positively related. Clearly, the strongest correlations were 0.710, 0.595, and 0.480 between SPD and MRImp for the census block and between PD and MRImp for the BG and census tract level, respectively. It seems the SPD was a better descriptor than LPD in representing PD since it generally had a higher correlation with the independent variable—MRImp—at all levels. Besides, correlation coefficients became larger as the scales increased between any combination of the three population parameters and the MRImp. The highest correlation was at the tract level, followed by the BG and finally the block level. Higher aggregation levels such as census tract were of benefit to remove more extreme values and to narrow the variance

TABLE 20.2

Summary of All Models at Each Census Level

Census Levels	Corr.	R^2	Regression Model	MRE	MedRE	ET (%)
Block	0.428	0.183	PD = −1219.335 + 6987.572 × MRImp	396.15	37.12	−12.69
	0.480	0.230	SPD = 2.096 + 85.737 × MRImp	51.48	19.45	−3.16
	0.466	0.218	LPD = 5.410 + 0.466 × MRImp	30.71	25.52	+1.42
BG	0.595	0.354	PD = −1995.631 + 7378.838 × MRImp	275.80	32.77	−3.91
	0.592	0.351	SPD = −12.403 + 104.295 × MRImp	44.75	18.71	+0.62
	0.555	0.308	LPD = 4.354 + 5.785 × MRImp	10.41	6.16	+1.14
Tract	0.710	0.504	PD = −3021.746 + 9550.366 × MRImp	55.96	27.15	+4.01
	0.706	0.499	SPD = −29.567 + 141.658 × MRImp	22.74	14.32	+3.53
	0.672	0.452	LPD = 3.134 + 8.492 × MRImp	6.36	4.28	+1.33

Notes: Corr.—the Pearson correlation coefficient between PD (SPD, LPD) and MRImp; MRE—mean value of relative error; MedRE—median value of relative error; ET (%)—error of total, which is the total population estimation error based on the overall dataset in the study area (an addition (+) symbol means overestimated whereas a minus (−) symbol means underestimated); PD—population density; SPD—square root of population density; LPD—natural log of population density.

of datasets. With relatively stable covariance between population density and the areas of impervious surfaces, their corresponding correlation coefficients will increase with increasing aggregation levels.

Using the three population parameters as the dependent variable and the MRImp as the independent variable, a total of nine population density models were built. Table 20.2 also presents the results of all the modeling and model validation. R^2 is interpreted as the square root of the correlation between observed and predicted values. Therefore, higher correlation coefficients are promising in constructing better models by the two correlated variables. R^2 is frequently used as an important indicator of the strength of the linear regression relationships. Table 20.2 shows that the the best models developed at different levels were the SPD-derived model at the block scale and the PD-derived models at the BG and the tract scales with an R^2 value of 0.230, 0.354, and 0.504, respectively. It is evident that the poor model was developed based on smaller analytical units (like the block level), whereas better models are associated with larger analytical units (like the BG or tract levels). Hence, the SPD-derived model at the block level should create more estimation errors than by two PD-derived models at the other two levels. With increasing areal units, population density models became better and better and the best model created at the tract level was able to account for over 50% of the variance in population density.

Two of the best models at BG and census tract levels were developed using the same dependent variable (PD). However, the best model at the block level was created by SPD and PD turned out to lead to the poorest model. This is probably because the strength of PD, which is the "original" population modeling variable, and MRImp in modeling population were attenuated by smaller areal units. Once a certain size of the areal unit is reached, the correlation between these two variables can be greatly improved, leading to better population models. Hence, if the analytical units are large enough, the population density variable does not need to use other transforms, such as the square root, logarithmic, and inverse, to guarantee a better population model.

20.4.2 Validation of Residential Population Density Models

The validation results for all developed population density models at three census scales are summarized in Table 20.2. It is expected that good models should have small MRE and MedRE values. Because the square root and logarithmic transforms diminish the original population density data quantitatively, the resultant MRE and MedRE values from the former are smaller than those from the latter. Generally speaking, the difference between MRE and MedRE decreased consistently from models using LPD as the dependent variable, then SPD, and finally PD at each census scale. Although models developed by LPD are the poorest in terms of R^2 values, they surprisingly generate good validation results. Conversely, the PD-derived models could generate higher R^2 values, yet their validation results tended to be the poorest. This is probably because the logarithmic transform greatly reduces the original population density data and this change is greater than that resulted from the square root transform. Meanwhile, the difference between MRE and MedRE values indicates that extreme values of population density had a significant impact on the models. PD tends to create the most extremes at all three scales, followed by SPD, then LPD. As analytical scale changes from block to BG to tract, the scale effect was apparent with respect to both MRE and MedRE. With the same dependent variables, the model's relevant MRE and MedRE values reduced correspondingly as well as their difference. Therefore, population models were systematically improved by areal units with increasing size.

Based on the model developed from each census scale, the population of individual census units can be calculated, and the total population of the whole city can be summed up. The error of total in percentage was computed with the difference between estimated total and census reference total by the census reference total for the whole city. An addition symbol indicates the overall overestimation, whereas a minus symbol represents an overall underestimation. Clearly, almost all population models tended to overestimate the total population, no matter which census level was considered. Only three of them underestimated the total population: the PD- and SPD-derived models at the block level and the PD-derived model at the BG level. Overall, all population models developed at all levels were able to produce comparable results. The best model generated the smallest gap between the estimated and reference total, which is the SPD-derived model at the BG scale, whereas the poorest model produced the huge difference, which is the PD-derived model at the block level. With a medium analytical scale, the model developed at the BG level surprisingly produced the closest number to the total census reference by using all three population parameters. The SPD-derived model at this level far exceeded the one constructed at the census tract level. The models developed at the tract level only produced "middle" results; even they were the best in terms of R^2, MRE, and MdRE. Although it is diagnosed that the poorest models were all developed at the census block level with the values of R^2, MRE, and MedRE, the SPD model at this scale actually produced a closer result to the reference value than that from the census tract level.

Scatter plots of reference and predicted population density are illustrated in Figure 20.5. Obviously, the two variables did not present a very clear linear relation at the census block level (Figure 20.5a). When more aggregated data were used as the analytical units, the linear correlation between reference and predicted population density became apparent (Figure 20.5b and c). Models were thus improved, especially for medium population density areas. All models seem to have a common problem. Lower population density areas were often overestimated. In contrast, high population density areas tend to be underestimated. A close look at Figure 20.6a shows that the model at the block level tended to overestimate the results if population density is less than ~100 (persons per km^2). This threshold nears about 200 and 400 (persons per km^2) for models produced at the BG and tract levels, respectively (Figure 20.6b and c). Conversely, the models would underestimate the results if population density is greater than ~500 (persons per km^2) at the block level, 1500 (persons per km^2) at the BG level, and 1800 (persons per km^2) at the tract level (Figure 20.6).

Residual maps showing the distribution of residuals at the residential areas resulted from the best model at each census level were also created (Figures 20.7 through 20.9). The visual comparison between residual maps and airphoto of the study area indicates that census units that were significantly underestimated are mainly covered by multistoried apartments. Although high population density is reported by census data, the relatively much "smaller" impervious surface of multistoried buildings is insufficient to reveal this pattern. Meanwhile, areas that were greatly overestimated are dominated by high impervious surfaces such as highways, playground racetracks, and parking lots, but not related to any dwelling unit. This indicates the problems when using zoning data to delineate residential/nonresidential areas on an image data. Although zoning data are designed to indicate land-use information and they try to reveal such information as detailed as possible, it is primarily a GIS variable that is area-based. Consequently, the detailed information as presented by any single zoning is limited by the size of the polygon. Once a certain kind of land-use type locates within the zoning but is to be got rid of, there is no way to remove it. For example, it is technically impossible to delete a highway that runs across a zoning polygon with the zoning GIS data layer alone. Additionally, no matter how many types of information are endued to a zoning polygon by different attributes, it can only represent certain

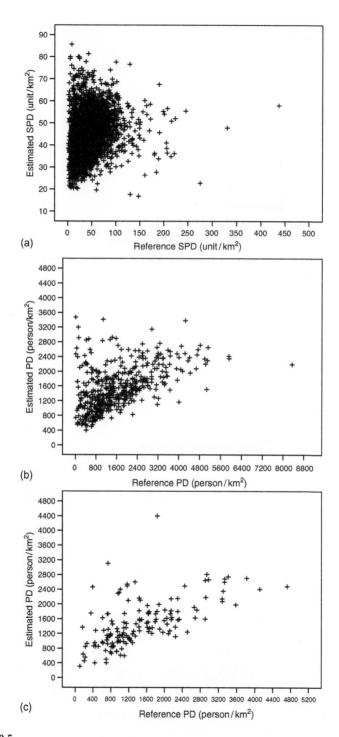

FIGURE 20.5
Scatter plots of estimated and reference SPD (or PD) from the model constructed at the census block level (a), at the block group level (b), and at the census tract level (c).

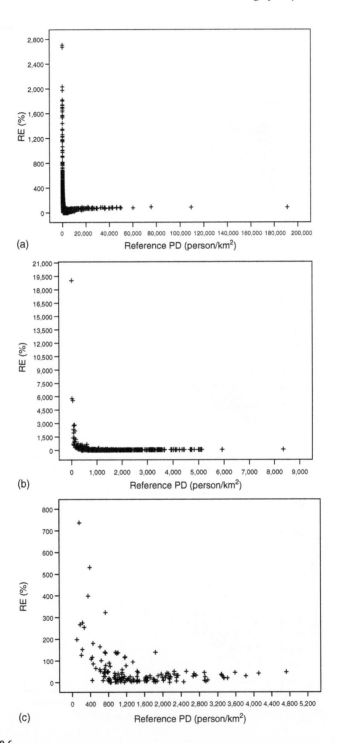

FIGURE 20.6
Scatter plots of relative error (%) and reference PD from the model constructed at the census block level (a), at the block group level (b), and at the census tract level (c).

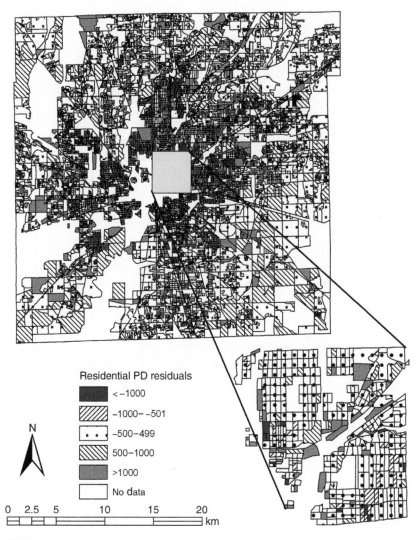

FIGURE 20.7
Distribution of residential population density residuals calculated from the model constructed at the census block level.

information with a single GIS layer. This property induces problems especially when it is to be integrated with a pixel-based data such as images. Besides, the visual comparison between a residential zoning map and three residual maps was also conducted. It showed that the highly over- or underestimated census units in the central city all fall into high- and medium-density zoning categories, which are primarily used for suburban and inner-city high-rise apartments (Figure 20.3). Conversely, in the city's surrounding rural area, census blocks, BGs, and tracts that are greatly

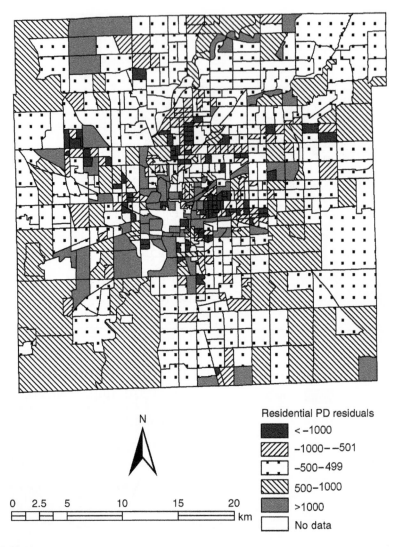

FIGURE 20.8

Distribution of residential population density residuals calculated from the model constructed at the block group level.

over- or underestimated tend to fall into zonings that are not primarily residential. For any single unit of these census data, only a small portion is identified to be covered by residential impervious surface. When using these data in modeling, significant biases would occur. The remaining census units that stand in between tend to be moderately over- and underestimated, which is consistent with the distribution of low- and medium-density residential zonings.

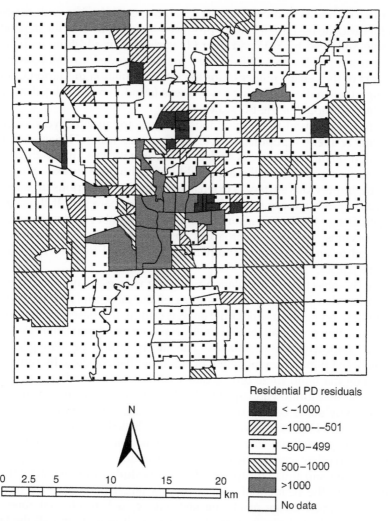

Residential PD residuals

- ■ < −1000
- ▨ −1000−−501
- ∷ −500−499
- ▧ 500−1000
- ▦ >1000
- □ No data

N

0   2.5   5   10   15   20
km

FIGURE 20.9

Distribution of residential population density residuals calculated from the model constructed at the census tract level.

20.5 Discussions and Conclusions

This study investigates the feasibility of using impervious surface data to conduct a census-based multiple scale analysis of residential population modeling at a city (or county) level. With different criteria (R^2, MRE, and MedRE), the best three population models were all developed at the census tract level on the basis of different dependent variables. According to the R^2 value, the best model was developed from SPD at the census block scale and from PD at the BG and census tract level, respectively. Yet with the MRE and MedRE, the best models were produced by using LPD as the dependent variable, followed by SPD and then PD variables. In addition, with an ET value around 1(%), the LPD-derived models tended to estimate closer numbers to the census reference for the total population in Marion County, although the smallest ET value (0.62%) was found to be associated with the SPD-derived model at the BG scale.

Overall, it is found that impervious surfaces are an effective variable in modeling residential population at all three census scales. However, acquiring a high-quality impervious surface is not an easy task due to the impervious surface per se and the availability of techniques in extracting such surfaces. Impervious surface is extensively mixed with various urban areas, especially those that are spectrally confused with bare soil. It thus creates a big challenge to identify a suitable single endmember to denote all kinds of impervious surfaces. Approaches to solve such problems have been suggested to apply multiple endmember LSMA and to use hyperspectral imagery [20,26].

Since the objective of population modeling is to estimate population (or population density) in residential areas of the city, applying an accurate threshold to remove nonresidential areas is critical. It is observed that zoning data have the potential to effectively separate residential/nonresidential areas in remotely sensed images. The incorporation of zoning data in census-based population modeling brings other benefits, too. By comparing the zoning map and three census datasets, the correlation between census population distribution and specific land-use types is directly given. It thus explains why some census units' population density is more likely to be mis-estimated. However, because of the vector-based data format, the size of zoning polygons limits the degree of detailed information that can be extracted. In other words, any detailed information within a single polygon is aggregated and thus "disappeared." In this study, if the residential zone identified by the residential zoning map is largely occupied by residentially used lands, image pixels masked out by such residential zones can represent residential pixels correctly. However, if more than half the lands of the residential zone is not used for dwelling, many image pixels can be incorrectly masked out as residential pixels and thus attenuate the performance of using impervious surfaces to model residential population. A more efficient threshold is needed to isolate true residential pixels so that the results of remote sensing-based population model can be greatly improved.

The difference between MRE and MedRE shows that the performance of population density models was strongly affected by extreme values (extreme low and high population density). A common problem found in all three models is that they significantly underestimated the high-density areas but overestimated the low-density areas. This is because the population model was developed in such a way that it assumes a general uniform population density per census unit at the same level. Hence, none of the models performed well at extreme values.

With different criteria (R^2, MER, MedRE, and ET), the best model is identified to be with different population parameters. It is difficult to say which population parameter is superior to others in producing better population density models. In modeling development, the original form of population density seems to hold promise in building acceptable models with large R^2 values, once the analytical units are large enough such as the BG and census tract units. In model validating, however, the transforms of population density like the square root or the logarithm are required to obtain an ideal MER, MedRE, and ET values.

The significance of the census-based scale effect described by various sizes of areal units on modeling population is obvious. Generally speaking, the performance of population modeling increased with increasing analytical scales from census block to BG, and it reached the best at census tract. Data predicted by microscale-based models as such from the block level tend to produce wide biases and thus it is less possible for the predicted and reference values to hold a linear correlation at this scale. Nevertheless, it should be noted that different model assessment criteria tend to indicate different information. For example, the ET criteria reveals that all population density models developed at the BG level could produce the most comparable estimates as the census reference values, even better than those models derived from the census tract scale. Clearly, the census block level is the most inappropriate scale to estimate population for the whole city. Yet with the transformation of the square root for the dependent variable, the performance of the model developed at the block level could be greatly improved and exceeded the model constructed at the tract level. Thus, applying multiple assessment standards is necessary to make a general judgment for research regarding different models with different variables and at different scales.

References

1. United Nations, Urban and rural areas 1996, ST/ESA/SER.A/166, Sales No. E. 97 XIII.3, 1997.
2. Christopherson, R.W., *Elemental Geosystems*, 5th ed., Pearson Education, Inc., Upper Saddle River, 2007, p. 104.
3. Li, G. and Weng, Q., Using Landsat ETM+ imagery to measure population density in Indianapolis, Indiana, USA, *Photogrammetric Engineering and Remote Sensing*, 71, 947, 2005.
4. Lu, D., Weng, Q., and Li, G., Residential population estimation using a remote sensing derived impervious surface approach, *International Journal of Remote Sensing*, 27, 3553, 2006.
5. Qiu, F., Woller, K.L., and Briggs, R., Modeling urban population growth from remotely sensed imagery and TIGER GIS road data, *Photogrammetric Engineering and Remote Sensing*, 69, 1031, 2003.
6. Lo, C.P., *Applied Remote Sensing*, Longman, New York, 1986.

7. Cowen, D.J. et al., The design and implementation of an integrated GIS for environmental applications, *Photogrammetric Engineering and Remote Sensing*, 61, 1393, 1995.

8. Lo, C.P. and Welch, R., Chinese urban population estimates, *Annals of the Association of American Geographers*, 67, 246, 1977.

9. Harvey, J.T., Estimating census district populations from satellite imagery: some approaches and limitations, *International Journal of Remote Sensing*, 23, 2071, 2002.

10. Harvey, J.T., Population estimation models based on individual TM pixels, *Photogrammetric Engineering and Remote Sensing*, 68, 1181, 2002.

11. Dobson, J.E. et al., LandScan: a global population database for estimating populations at risk, *Photogrammetric Engineering and Remote Sensing*, 66, 849, 2000.

12. Wu, S., Qiu, X., and Wang, L., Population estimation methods in GIS and remote sensing: a review, *GIScience and Remote Sensing*, 42, 80, 2005.

13. Lo, C.P., Modeling the population of China using DMSP operational linescan system nighttime data, *Photogrammetric Engineering and Remote Sensing*, 67, 1037, 2001.

14. Iisaka, J. and Hegedus, E., Population estimation from Landsat imagery, *Remote Sensing of Environment*, 12, 259, 1982.

15. Lo, C.P., Automated population and dwelling unit estimation from high resolution satellite images: a GIS approach, *International Journal of Remote Sensing*, 16, 17, 1995.

16. Webster, C.J., Population and dwelling unit estimates from space, *Third World Planning Review*, 18, 155, 1996.

17. Ogrosky, C.E., Population estimation from satellite imagery, *Photogrammetric Engineering and Remote Sensing*, 41, 707, 1975.

18. Sutton, P. et al., Census from heaven: an estimate of the global human population using night-time satellite imagery, *International Journal of Remote Sensing*, 22, 3061, 2001.

19. Ji, M. and Jensen, J.R., Effectiveness of subpixel analysis in detecting and quantifying urban imperviousness from Landsat Thematic Mapper imagery, *Geocarto International*, 14, 33, 1999.

20. Lu, D. and Weng, Q., Use of impervious surface in urban land-use classification, *Remote Sensing of Environment*, 102, 146, 2006.

21. Phinn, S. et al., Monitoring the composition of urban environments based on the vegetation-impervious surface-soil (VIS) model by subpixel analysis techniques, *International Journal of Remote Sensing*, 23, 4131, 2002.

22. Flanagan, M. and Civco, D.L., Subpixel impervious surface mapping, in *Proceedings of the 2001 ASPRS Annual Convention* (CD-ROM), American Society for Photogrammetry and Remote Sensing, Bethesda, MD, 2001.

23. Wu, C. and Murray, A.T., Estimating impervious surface distribution by spectral mixture analysis, *Remote Sensing of Environment*, 84, 493, 2003.

24. Wilson, J.S. et al., Evaluating environmental influences of zoning in urban ecosystems with remote sensing, *Remote Sensing of Environment*, 86, 303, 2003.

25. Roberts, D.A. et al., Mapping chaparral in the Santa Monica mountains using multiple endmember spectral mixture models, *Remote Sensing of Environment*, 65, 267, 1998.

26. Weng, Q., Lu, D., and Liang, B., Urban surface biophysical descriptors and land surface temperature variations, *Photogrammetric Engineering and Remote Sensing*, 72, 1275, 2006.

Authors

Marvin E. Bauer is a professor of remote sensing in the Department of Forest Resources, College of Agricultural, Food and Natural Resource Sciences at the University of Minnesota. His current research is focused on development and applications of digital satellite remote sensing for mapping and monitoring land and water resources in Minnesota. He also teaches classes on remote sensing of natural resources and environment and digital remote sensing, and serves as editor of the journal, *Remote Sensing of Environment*. He is a fellow of the American Society of Photogrammetry and has received the NASA Distinguished Service Award and the Minnesota GIS/LIS Lifetime Achievement Award.

Toby N. Carlson, PhD, Imperial College, University of London, is a retired (emeritus) professor of meteorology at the Pennsylvania State University. He had been an adjunct member of the Environmental Institute, College of Earth and Mineral Sciences. Prof Carlson's scientific contributions, approximately 90 papers published in refereed journals, reflect a wide range of interests: synoptic and dynamic meteorology, radiative transfer, severe local storms, plant–atmosphere interactions, aerosol transport and chemistry, remote sensing of land surface properties and surface energy processes, and, most recently, applications of remote sensing to the study of urban sprawl and small watershed runoff. He has developed methods for assessing surface soil water content and surface energy fluxes using remotely sensed image data. In 1991 Prof Carlson published a widely used book on meteorology (*Mid Latitude Weather Systems*). Recently, he has created two new web products related to his current interest in land surface processes: a course in Land Surface Processes (Simsphere) and a database of impervious surface area and fractional vegetation cover determined from Landsat 5 digital imagery at 25 m resolution for all of Pennsylvania, 1986 and 2000 (to be found at the PASDA Web site); he has created a web-based tool that allows users to assess the health (nutrient load) and surface runoff for a user-defined stream basin in the Chesapeake Bay Basin.

Fabio Dell'Acqua obtained a first-class honors degree in electronics engineering at the University of Pavia in 1996. In 1999 he received his PhD in remote sensing at the University of Pavia, investigating shape analysis techniques to process meteorological images. In the first half of the year 2000, he worked as a research associate at the Vision Laboratory of the University of Edinburgh in Edinburgh, Scotland, on analysis and reconstruction of range data (EU TMR–CAMERA). From July 2000 to November 2001 he was a postdoctoral fellow at the University of Pavia; in December 2001 he obtained a position as an assistant professor.

He is currently working on processing and interpretation of satellite images. His fields of interest include earth observation, radar data processing, shape analysis, and retrieval of images from archives.

Giorgio Franceschetti has been full professor of electromagnetic wave theory at the University of Naples, Italy, since 1968. He was a visiting professor at the University of Illinois, Urbana-Champaign, in 1976 and 1977, and at the University of California at Los Angeles in 1980 and 1982. He was a research associate at Caltech, Pasadena, in 1981 and 1983, a visiting professor at National Somali University in 1984, Somalia, and a visiting professor at the University of Santiago de Compostela, Spain, in 1995. He has published several books and more than 150 refereed papers in the field of applied electromagnetics (reflector antennas, transient phenomena, shielding, non-linear propagation, and scattering) and, more recently, in the field of SAR data processing and simulation. He has lectured in several summer schools in China, Great Britain, Holland, Italy, Spain, Sweden, and the United States. Prof Franceschetti was a Fulbright Scholar at Caltech, Pasadena, in 1973. He was the recipient of several national and international awards. He was director of IRECE, a Research Institute of the Italian National Council of Research and member of the board of the Italian Space Agency. He is currently also an adjunct professor at UCLA, and a distinguished visiting scientist at the Jet Propulsion Laboratory, Pasadena, United States. He is a *life fellow* of the IEEE.

Paolo Gamba is currently associate professor of telecommunications at the University of Pavia, Italy. He received the Laurea degree in electronic engineering "cum laude" from the University of Pavia, Italy, in 1989. He also received from the same university the PhD degree in electronic engineering in 1993.

He is the organizer and technical chair of the biennial GRSS/ISPRS Joint Workshops on "Remote Sensing and Data Fusion over Urban Areas" from 2001 to 2007. He is and has been the guest editor of special issues of *ISPRS Journal of Photogrammetry and Remote Sensing*, *IEEE Transactions on Geoscience and Remote Sensing*, and *International Journal of Information Fusion and Pattern Recognition Letters*.

He has been the chair of Technical Committee 7 "Pattern Recognition in Remote Sensing" of the International Association for Pattern Recognition from October 2002 to October 2004. He is a senior member of IEEE, associate editor of IEEE Geoscience and Remote Sensing Letters, and currently the chair of the Data Fusion Committee of the IEEE Geoscience and Remote Sensing Society.

He has published more than 40 papers in peer-reviewed journals on urban remote sensing and presented more than 100 papers at workshops and conferences.

James Gerjevic completed his BA and MA in geography at the University of Northern Iowa in 2004. He did a master thesis on "Extraction of Transportation Infrastructure from Hyperspectral Remote Sensing Data" and currently he is working as a GIS specialist in Union Pacific Railway at Omaha, Nebraska.

Robert R. Gillies is the director of the Utah Climate Center at Utah State University (USU) and state climatologist for the State of Utah. He is an associate professor in meteorology and geography in the Department of Plants, Soils and Climate at USU. His research focuses on land surface processes and remote sensing integrating the fields to study various aspects of the earth's environments. Recent research projects have been concerned with, impervious surface area analysis in Logan, Utah, the spatial analysis of vector-borne disease outbreaks like West Nile virus, the linked micromaps of chronic wasting disease in mule deer, and plant phenology in the timing of tree budburst within urban environments. As director of the climate center, Dr Gillies has been instrumental in using data management and visualization technologies (e.g., linked micromaps) to provide straightforward data access and interpretation of climate data resources. He has published widely in book chapters and refereed research papers. He has twice been the recipient of best scientific paper in remote sensing from the American Society of Photogrammetry and Remote Sensing.

Armin Gruen is professor and head of the chair of photogrammetry at the Institute of Geodesy and Photogrammetry, Federal Institute of Technology (ETH) Zurich, Switzerland since 1984.

He graduated 1968 as Dipl.-Ing. in geodetic science and obtained his doctorate degree 1974 in photogrammetry, both from the Technical University Munich, Germany. From 1969 to 1975 he worked as assistant professor, and until 1981 as associate professor, at the Institute of Photogrammetry and Cartography, Technical University Munich. From 1981 to 1984 he acted as associate professor at the department of geodetic science and surveying, The Ohio State University, Columbus, Ohio, USA.

Prof Gruen has held lecturing and research assignments at the University of Armed Forces, Munich, Germany, Helsinki University of Technology, Finland, Universita degli Studi di Firenze, Italy, Stanford Research Institute, Menlo Park, USA, Department of Geodesy, Technical University Delft, Netherlands, Asian Institute of Technology (AIT), Bangkok, Thailand, Department of Geomatics, University of Melbourne, Australia, and at the Center for Space and Remote Sensing Research, National Central University, Jhongli City, Taiwan. He has lectured at the university level since 1969, with photogrammetry and remote sensing as major subjects, and surveying, cartography, and adjustment calculus as minor subjects.

He served as the head of the department of geodetic sciences 1996–1997 and as the dean of faculty "Rural Engineering and Surveying" of ETH Zurich (1996–98). He is promoter and director of the postdiploma course "Spatial Information Systems" at ETH Zurich. Through the Commission for Remote Sensing he is a member of the Swiss National Academy of Natural Sciences. He is member of the editorial boards of several scientific journals. He has published more than 250 articles and papers and is editor and coeditor of 15 books and conference proceedings. He has organized and co-organized/cochaired over 30 international conferences and he has served as a consultant to various government agencies, system manufacturers, and engineering firms in Germany, Japan, Switzerland, USA, and other countries. He is cofounder of CyberCity AG, Zurich, Switzerland.

He served as the president of ISPRS Commission V (ISPRS—International Society of Photogrammetry and Remote Sensing), as ISPRS council member (Second Vice President), council member of IUSM (International Union of Surveys and Mapping), and as chairman of the ISPRS Financial Commission. He is currently chairman of the ISPRS International Scientific Advisory Committee (ISAC), international member of the Fourth Academic Committee of the State Key Laboratory of Information Engineering in Surveying, Mapping and Remote Sensing (LIESMARS), Wuhan University, China, and member of the Executive Board of the Digital Earth Society.

He is member of the calibration/validation team and principal investigator for the PRISM sensor on JAXA's ALOS satellite.

His major international awards and honors include the Otto von Gruber Gold Medal (ISPRS, 1980), Talbert Abrams Award Grand Trophy (ASPRS, 1985 and 1995), with Honorable Mention 1989, Fairchild Award (ASPRS, 1994), Miegunyah Distinguished Fellowship Award of the University of Melbourne, 1999, ISPRS U.V. Helava

Award 2000, E.H. Thompson Award 2005, Corresponding Membership of the German Geodetic Commission, Munich, Honorary Professorships of the Wuhan Technical University of Surveying and Mapping, Wuhan, China and Yunnan Normal University, Kunming, China, Honorary Member of the Japan Society of Photogrammetry and Remote Sensing, and Fellow Professor, Center for Space and Remote Sensing Research, National Central University, Jhongli City, Taiwan.

His main current research interests include: Automated object reconstruction with digital photogrammetric techniques, building and line feature extraction, 3D city modeling, image matching for DTM generation, and object extraction, three-line linear array sensor modeling, industrial quality control using vision techniques, motion capture, body and face reconstruction for animation, imaging techniques for generation and control of VRs/VEs, especially for cultural heritage recording and modeling, processing of very high resolution satellite images, photogrammetric UAV research.

Karin Hedman received an MSc in space engineering from the Lulea University of Technology, Sweden, in 2004. From 2004 until 2006, she has worked as a scientific collaborator at the Institute of Photogrammetry and Cartography at Munich University of Technology. She is currently with the Institute of Astronomical and Physical Geodesy as a course coordinator of the international Master's Program in "Earth Oriented Space Science and Technology." Since 2004, Karin Hedman has been working on her PhD. Her research focuses on urban SAR remote sensing and data fusion.

Martin Herold received his first graduate degree (diploma in geography) in 2000 from the Friedrich Schiller University of Jena, and the Bauhaus University of Weimar, Germany, and his PhD at the Department of Geography, University of California–Santa Barbara in 2004. Dr Herold is currently coordinating the ESA GOFC-GOLD Land Cover project office at the Friedrich Schiller University of Jena, Germany. In his earlier career, his interests were in multifrequency, polarimetric, and interferometric SAR-data analysis for land surface parameter derivation and modeling. He joined the Remote Sensing Research Unit, University of California, Santa Barbara, in 2000 where his research focused on remote sensing of urban areas and the analysis and modeling of urban growth and land-use change processes.

Dr Herold's most recent interests are in international coordination and cooperation toward operational terrestrial earth observations with specific emphasis on the harmonization and validation of land-cover datasets. Martin Herold is an expert in the field of remote sensing, and digital processing and modeling of geographic data as well as local and regional planning. His publications include more than 20 peer-reviewed journal articles and book chapters. During his university career, he taught several classes in processing and analysis of remote sensing and spatial data, imaging spectrometry, GIS, and modeling of spatial processes.

Stefan Hinz graduated in 1998 from the Munich University of Technology with a Dipl-Ing degree in geodesy and geoinformation and, in 2003, he received the Dr-Ing degree with "summa cum laude" for his PhD work on automatic extraction of urban road networks from aerial images. From 1998 to 2003, he was a scientific collaborator at the chair for photogrammetry and remote sensing of the Munich University of

Technology for Photogrammetry and Cartography. Since then, he holds the position of a research assistant at the remote sensing technology unit of the same institute. He is steering committee member of the institute's Joint Research Lab "Image Understanding for High Resolution Remote Sensing" and serves as head of the Helmholtz Young Investigators Group embedded in Joint Research Lab. From March to June 1999, he was with the Institute of Robotics and Intelligent Systems of the University of Southern California at Los Angeles. The research conducted at the Institute of Robotics and Intelligent Systems was funded by the German Academic Exchange Service by a grant. His research interests include various aspects of computer vision, image understanding, and remote sensing—especially the automatic extraction of objects from optical, infrared, and radar remote sensing data—with major emphasis on the characterization of dynamic processes like the estimation of traffic flow parameters from airborne and spaceborne images as well as long-term surface subsidence monitoring with persistent scatterer radar-interferometry. Since 1998, he has published more than 60 contributions, of which more than 20 are peer-reviewed articles.

Xuefei Hu is currently a PhD student in the Department of Geography, Geology and Anthropology, Indiana State University, under the guidance of Dr Qihao Weng. He received his bachelor and master degrees from the China University of Geosciences. His research interests focus on estimation of impervious surface from remote sensing imageries, use of artificial neural networks, and GIS.

Antonio Iodice received the Laurea degree "cum laude" in electronic engineering and the PhD degree in electronic engineering and computer science, both from the University of Naples "Federico II," Naples, Italy, in 1993 and 1999, respectively.

In 1995, he received a grant from the Italian National Council of Research to be spent at the Istituto di Ricerca per l'Elettromagnetismo e i Componenti Elettronici, Naples, Italy, for research in the field of remote sensing. He was with Telespazio S.p.A., Rome, Italy, from 1999 to 2000. Since 2000, he has been with the Department of Electronic and Telecommunication Engineering of the University of Naples "Federico II," where he is currently professor of electromagnetics.

His main research interests are in the field of microwave remote sensing and electromagnetics: modeling of electromagnetic scattering from natural surfaces and urban areas, simulation and processing of synthetic aperture radar signals, and SAR interferometry.

Prof Iodice is author or coauthor of more than 100 papers published in refereed journals or in proceedings of international and national conferences. He is a senior member of the IEEE.

Bingqing Liang is currently a PhD student at Indiana State University under the guidance of Dr Qihao Weng. She obtained her BS in geography education from South China Normal University, China in 2002 and MA in geography from Indiana State University in 2005. Her research interest is the application of remote sensing and GIS to urban analysis at multiple scales, including urban temperature modeling, urban quality assessment, and residential population estimation.

Brian C. Loffelholz was a research fellow with the Department of Forest Resources and Remote Sensing and Geospatial Analysis Laboratory, University of Minnesota, where his work focused on analysis and classification of Landsat, IKONOS, and

QuickBird imagery for mapping land cover and forests. He is now a GIS analyst with the Wisconsin Department of Agriculture, Trade and Consumer Protection.

Dengsheng Lu received a BA in forestry from the Zhejiang Forestry University, China in 1986; MA in forestry from the Beijing Forestry University, China in 1989; and PhD in physical geography from Indiana State University, United States in 2001. He worked at the Center for the Study of Institutions, Population, and Environmental Change and the Anthropological Center for Training and Research on Global Environmental Change, Indiana University, Indiana, as a postdoctoral fellow and research scientist during 2001 and 2006. He is currently working in the School of Forestry and Wildlife Sciences, Auburn University, Alabama. His research focuses on land-use/cover change, biomass/carbon estimation, human–environment interaction, and urban impervious surface. He is the author of more than 40 peer-reviewed journal articles and book chapters.

Travis Maxwell completed his BSc degree at the Royal Military College of Canada in 1997. He completed his MSc in engineering at the University of New Brunswick, Canada in 2005 with his thesis focused on object-oriented classification of very high resolution satellite imagery. He is currently working as a geomatics specialist for the Canadian Department of National Defence.

Assefa M. Melesse earned his BSc (1986) and MEngSc (1991) degrees in agricultural engineering from the Alemaya University (Ethiopia) and the National University of Ireland, respectively. Dr Melesse also obtained an ME and PhD in agricultural and biological engineering with two concentrations (hydrological sciences and GIS) and three minors (civil engineering, environmental engineering, and remote sensing) in 2000 and 2002, respectively. He served at the Alemaya University for more than 10 years in various capacities including as assistant professor from 1995 to 1998. He also served as faculty at Earth Systems Science Institute, University of North Dakota, from 2002 to 2004 and is currently an assistant professor at the Department of Environmental Studies, Florida International University. Dr Assefa M. Melesse is a water resources engineer with background in remote sensing and geospatial applications to hydrologic modeling. His areas of research and experience include spatially distributed hydrologic modeling, ecohydrology, spatial surface water balance modeling, surface and groundwater interactions modeling, spatial evapotranspiration mapping using remote sensing, water–energy–carbon fluxes coupling and modeling, remote sensing hydrology, and land-cover change detection and scaling. Dr Melesse is a registered professional engineer in the State of Florida and has published over 30 peer-reviewed journal articles and is the editor-in-chief of the *Journal of Spatial Hydrology.*

Rama Prasada Mohapatra was born and brought up in a region flanked by the Eastern Ghats (Hill Ranges) of India, where the hills, streams, jungles, the underprivileged people, and the associated spatiotemporal diversity motivated him to pursue a carrier in geography. He received his BA, MA, and MPhil degrees in geography from the Utkal University, Orissa in 1995, 1997, and 1998, respectively. After completing his MPhil, he was trained in GIS and remote sensing from India's premier organizations dealing with space research and thematic mapping. Later on, he associated himself with institutions such as the Indian Institute of Technology, Kharagpur, India; the

University of Western Ontario, Canada; the Utkal University, Orissa, India; and so on to advance his research and teaching career. Currently, he is a doctoral student in the Department of Geography, University of Wisconsin–Milwaukee. His research interest includes remote sensing, GIS, artificial intelligence, and their applications in modeling land-use land-cover changes. He has presented several research articles in various conferences including the American Association of Geographers and also has been the recipient of several awards.

Renaud Péteri received an engineering degree in physics and digital images from the Ecole Nationale Supérieure de Physique de Strasbourg, France and an MS in photonic and image processing from the University of Strasbourg in 2000. In 2003, he received a PhD degree from the "Ecole des Mines de Paris" in image and signal processing.

He was awarded an ERCIM postdoctoral research grant for spending 1.5 years at the Hungarian Academy of Sciences, Hungary (2004) and at the Mathematics and Computer Science Institute in Amsterdam, the Netherlands (2005). Since September 2005, he is an associate professor in computer science at the University of La Rochelle, France. Dr Péteri is a regular reviewer for several international journals and conferences in the field of remote sensing and image processing.

Lindi J. Quackenbush received BSurv and BSc degrees from the University of Melbourne, Australia in 1994, and MS and PhD degrees in remote sensing and photogrammetry from the State University of New York College of Environmental Science and Forestry (SUNY-ESF) in 1998 and 2004, respectively. Her research interests are in the fields of remote sensing and image processing, particularly focused on spatial techniques for both urban and forest classification. She is currently an assistant professor in environmental resources and forest engineering at the State University of New York College of Environmental Science and Forestry. Dr Quackenbush was the recipient of the Robert E. Altenhofen Memorial Scholarship Award by the American Society for Photogrammetry and Remote Sensing in 1997.

Thierry Ranchin received a PhD degree from the University of Nice Sophia Antipolis, France, in the field of applied mathematics, in 1993 and his "Habilitation à diriger les recherches" in 2005.

After a postdoctoral fellowship in a company in Tromso, Norway, he joined the remote sensing group of Ecole des Mines de Paris in the fall of 1994. He was an invited scientist from the University of Jena, Germany in 1998. He has a patent for sensor fusion and more than 100 publications, communications in international symposia, and chapters of books or articles in journals with peer-review committees in the field of remote sensing of the Earth system and in the field of image processing. He was the coeditor of the series of conferences "Fusion of Earth Data."

Dr Ranchin received the Autometrics Award in 1998 and the Erdas Award in 2001 from the American Society for Photogrammetry and Remote Sensing for articles on data fusion.

He is the cochair of the Energy Community of Practices of the Global Earth Observation System of Systems initiative. Since January 2007, he is the head of the Observation, Modelling, Decision Team of the Centre for Energy and Processes at Ecole des Mines de Paris.

Daniele Riccio received "cum laude" the Laurea degree in electronic engineering at the University of Naples, "Federico II", in 1989.

He is a professor of electromagnetics and remote sensing at the University of Naples, "Federico II," Italy. He was a research scientist at the Istituto di Ricerca sull'Elettromagnetismo e i Componenti Elettronici of the Italian National Council of Research and at the Department of Electronic and Telecommunication Engineering of the University of Naples, "Federico II," Italy. In 1994 and 1995, he was also guest scientist at the DLR High-Frequency Institute, Munich, Germany.

His research activity is witnessed by 150 papers published in the fields of microwave remote sensing, synthetic aperture radar simulation, modeling and information retrieval for land, oceanic and urban scenes, as well as in the application of fractal geometry to electromagnetic scattering from natural surfaces and to remote sensing.

Prof Riccio has won several fellowships from private and public companies (SIP, Selenia, CNR, CORISTA, CRATI) for research in the remote sensing field. He is a senior member of the IEEE.

Uwe Stilla received a diploma (Dipl-Ing) in electrical engineering from Gesamthochschule Paderborn, Germany, in 1980 and in 1987 he received an additional diploma (Dipl-Ing) in biomedical engineering from the University of Karlsruhe, Germany. From 1990 until 2004, he was with the Institute of Optronics and Pattern Recognition (FGAN-FOM), a German research establishment for defense-related studies. In 1993, he received his PhD (doctor of engineering) from the University of Karlsruhe with work in the field of pattern recognition. Since 2004, Prof Uwe Stilla is the head of the Department of Photogrammetry and Remote Sensing, and currently director of the Institute of Photogrammetry and Cartography at Munich University of Technology. He is currently the dean of Student Affairs of the Diploma, Bachelor's, and Master Program "Geodesy and Geoinformation" in the Faculty of Civil Engineering and Geodesy. Additionally, he is involved in the master's course "Earth Oriented Space Science and Technology" and "Land Management and Land Tenure." He has the chair of the ISPRS working group "Road Extraction and Traffic Monitoring" and is steering committee member of the institute's joint research lab "Image Understanding for High Resolution Remote Sensing." His research focuses on image analysis in the field of photogrammetry and remote sensing. He published more than 130 contributions, of which more than 40 are peer-reviewed articles.

Ramanathan Sugumaran is an associate professor of geography and GeoTREE director at the University of Northern Iowa. He has over 15 years of experience in remote sensing, GIS, GPS spatial decision support systems (SDSS) applications for natural resources and environmental planning and management. He is and has been working with federal, state, local, and tribal government agencies for the past 10 years and developed several SDSS tools and techniques. He has undertaken several research projects funded mostly by NASA, USDA, Missouri DNR, U.S. Fish and Wildlife, Iowa DOT/Midwest Transportation Consortium, Raytheon Corporation, and Iowa Space Consortium. Since 1999, he has directly participated in over $5M of funded research. He has published more than 15 journal articles and presented more than 70 papers in national and international conferences.

Matthew Voss received his BS in journalism from Iowa State University. He has also earned a BA in geography from the University of Northern Iowa, and is currently completing his MA in geography at the University of Northern Iowa. Matthew is currently using the data fusion approach with hyperspectral imagery and LIDAR data to identify tree species on the University of Northern Iowa campus.

Xixi Wang received his BS and MS in civil engineering (water resources and hydrology) from the Tsinghua University, Beijing, China, in 1989 and 1993, respectively, and PhD in agricultural engineering (soil and water) from Iowa State University, Ames, Iowa, United States, in 2001. Currently, Dr Wang is a research scientist in the Energy & Environmental Research Center at the University of North Dakota. His research focuses on hydrology and hydraulics, watershed analysis and modeling, precision resource management, GIS-based hydromodeling, multivariate spatial statistics, and decision support system. He has published a number of research papers in peer-reviewed journals. In addition, Dr Wang received awards from the American Society of Agricultural and Biological Engineers as an outstanding reviewer and Tsinghua University as an excellent educator. Further, because of his innovative research, Dr Wang was presented with awards by the Chinese Science Foundation and several other funding agencies.

Yeqiao Wang is a professor of terrestrial remote sensing in the Department of Natural Resources Science at the University of Rhode Island. He received his PhD in natural resources management and engineering from the University of Connecticut in 1995. He was on the faculty of the University of Illinois at Chicago between 1995 and 1999 prior to his employment at the University of Rhode Island. His research interests and teaching responsibilities are in remote sensing of natural resources, analysis, and mapping. Particular areas of research interests include remote sensing of dynamics of landscape and impacts of land-cover/land-use change on the environment. He has been leading various research projects and his major study areas include the Northeast and Midwest United States, East Africa, and Northeast China.

Birgit Wessel received a diploma degree (Dipl-Ing) in geodesy from the University of Hanover in 2000 and a Dr-Ing degree from Munich University of Technology, Munich, in 2006 for her PhD work on "Automatic Extraction of Roads from SAR images." From 2000 to 2005, she was scientific collaborator at the Institute for Photogrammetry and Cartography at the Munich University of Technology. Since 2005, she has been with the German Aerospace Center and is currently subsystem engineer for the German interferometric TanDEM-X satellite mission. Her research focuses on SAR remote sensing including image analysis, characterization of objects, and SAR geocoding.

Bruce Wilson is a limnologist and research scientist with the Minnesota Pollution Control Agency. He has led several water quality assessment efforts of major lake and river systems in Minnesota and has provided technical assistance to over 30 locally led watershed management projects. His current work includes satellite remote sensing of lake and river water quality, impervious mapping, and monitoring the effectiveness of stormwater Best Management Practices. He is a past president of the North American Lake Management Society.

Changshan Wu is an assistant professor in the Department of Geography, University of Wisconsin–Milwaukee. Dr Wu's current research area involves GIS and remote sensing applications in population estimation, urban development, housing, and transportation analysis.

George Xian is currently working for Science Applications International Corporation at the U.S. Geological Survey Center for Earth Resources Observation and Science. He received a bachelor of science from the Yunnan University in Kunming, China, a master of science from the Colorado State University, and a PhD from the University of Nevada, Reno. He has worked for several projects related to satellite remote sensing information system, urban environment, and urban dynamic research conducted by USGS, NASA, and US EPA since 1997 at EROS. He has conducted urban growth and environmental influence research using remote sensing information for several metropolitan areas including Chicago, Illinois; Tampa, Florida; Las Vegas, Nevada; Seattle, Washington; Atlanta/Columbus, Georgia; and Detroit, Michigan. His research interests include urban land-use and land-cover change detection and impacts on regional climate change and air quality. He has published several articles on these topics.

Xinsheng Zhang is a senior scientist at the EarthData International, Inc. He is an expert in GIS-based urban development simulation modeling as well as data system development.

Yun Zhang is an associate professor in the Department of Geodesy and Geomatics Engineering at the University of New Brunswick, Canada. He holds a BSc degree from Wuhan Technical University of Surveying and Mapping (1982), completed his master's studies at the East China Normal University, Shanghai (1989), and received his PhD degree from the Free University of Berlin, Germany (1997). Dr Zhang is an expert in remote sensing and photogrammetry, with substantial research experience in remote sensing, image processing, and digital photogrammetry. He is the sole inventor of three patent-pending technologies, coinventor (with his student) of one patent-pending technology. In the past seven years, he has developed many innovative technologies; three of them have been used by internationally leading imaging and mapping companies. The breakthrough image fusion algorithm—Pansharp—developed by Dr Zhang, has been integrated into PCI Geomatica and DigitalGlobe's production line, resulting in worldwide application of the algorithm. Dr Zhang is the sole or first author of 22 peer-reviewed journal papers and coauthor of a dozen of other peer-reviewed journal papers. He is the recipient of the Talbert Abrams Grand Award of the American Society for Photogrammetry and Remote Sensing (2005) and several other awards.

Guoqing Zhou earned his PhD in 1995 from the Department of Remote Sensing and Information Engineering of Wuhan University, and is now working at Old Dominion University as associate professor. Dr Zhou has published 3 books and 150 publications. Dr Zhou has worked on 48 projects as PI or Co-PI, for which the total funding has exceeded several million dollars. Dr Zhou serves as panel reviewer of NASA Office of Space Science, U.S. National Science Foundation, chair, editorial board, editorial advisor, editor-in-chief of several journals, and serves as chair and cochair of working group of many international societies and organizations. Dr Zhou has won five prestigious international awards.

Yuyu Zhou is a PhD student at the Department of Natural Resources Science, University of Rhode Island. He received bachelor and master of science degrees in geography from the Beijing Normal University, China. His current research focuses on the impacts of increasing impervious surface areas on watershed and coastal ecosystems through hydrological modeling.

Index

Printed and bound by CPI Group (UK) Ltd, Croydon, CR0 4YY

24/10/2024

01778302-0015